Molecular Research in Rice: Agronomically Important Traits 2.0

Molecular Research in Rice: Agronomically Important Traits 2.0

Editors

Kiyosumi Hori
Matthew Shenton

MDPI • Basel • Beijing • Wuhan • Barcelona • Belgrade • Manchester • Tokyo • Cluj • Tianjin

Editors
Kiyosumi Hori
National Agriculture and
Food Research Organization
(NARO)
Japan

Matthew Shenton
National Agriculture and
Food Research Organization
(NARO)
Japan

Editorial Office
MDPI
St. Alban-Anlage 66
4052 Basel, Switzerland

This is a reprint of articles from the Special Issue published online in the open access journal *International Journal of Molecular Sciences* (ISSN 1422-0067) (available at: https://www.mdpi.com/journal/ijms/special_issues/Rice_Traits_Edition2).

For citation purposes, cite each article independently as indicated on the article page online and as indicated below:

LastName, A.A.; LastName, B.B.; LastName, C.C. Article Title. *Journal Name* **Year**, *Volume Number*, Page Range.

ISBN 978-3-0365-4943-9 (Hbk)
ISBN 978-3-0365-4944-6 (PDF)

Cover image courtesy of Kiyosumi Hori

Contents

About the Editors

Kiyosumi Hori

Crop Gene Function Group, Institute of Crop Science, National Agriculture and Food Research Organization (NARO) from 2021–present. Department of Integrated Biosciences, Graduate School of Frontier Science, The University of Tokyo from 2019–present. Rice Applied Genomics Unit, Institute of Crop Science, NARO from 2016–2021. Rice Applied Genomics Research Unit, National Institute of Agrobiological Sciences from 2006–2016.

Matthew Shenton

Crop Gene Function Group, Institute of Crop Science, National Agriculture and Food Research Organization (NARO) from 2021–present. Breeding Materials Development Unit, Institute of Crop Science, NARO from 2018–2021. Meiji University School of Agriculture, Bioinformatics Laboratory, 2017–2018. Plant Genetics Laboratory, National Institute of Genetics, Japan, 2009–2017. Iwate Biotechnology Research Center, 2006–2009

Preface to "Molecular Research in Rice: Agronomically Important Traits 2.0"

This volume presents recent significant research achievements concerning the molecular genetic basis of agronomic traits in rice. Continuing research efforts are necessary to elucidate the genetic networks and molecular mechanisms controlling agronomically important traits. We hope that contents in this volume will help to expand the knowledge of recent advances in the genetics of and breeding research in rice.

Kiyosumi Hori and Matthew Shenton
Editors

International Journal of *Molecular Sciences*

MDPI

Editorial

Current Advances and Future Prospects for Molecular Research for Agronomically Important Traits in Rice

Kiyosumi Hori * and Matthew Shenton *

National Agriculture and Food Research Organization, Institute of Crop Science, Tsukuba 305-8518, Japan
* Correspondence: horikiyo@affrc.go.jp (K.H.); matt.shenton@affrc.go.jp (M.S.)

Citation: Hori, K.; Shenton, M. Current Advances and Future Prospects for Molecular Research for Agronomically Important Traits in Rice. *Int. J. Mol. Sci.* **2022**, *23*, 7531. https://doi.org/10.3390/ijms23147531

Received: 22 June 2022
Accepted: 5 July 2022
Published: 7 July 2022

Rice (*Oryza sativa* L.) is the most important food crop in the world, being a staple food for more than half of the world's population. The significance of rice as a crop and as a model species promoted overseas collaborations leading to the formation of the International Rice Genome Sequencing Project, and rice became the first crop genome to be sequenced in 2004 [1]. After the release of the first complete rice genome of 'Nipponbare', researchers have constructed several reference genomes for other rice cultivars including '93-11', 'IR64', 'Zhenshan 97', and 'Minghui 63', and collected large sets of whole-genome sequence data comprising more than 3000 rice cultivars distributed worldwide [2,3]. These fundamental genomic data resources are easily accessed via public databases such as the MSU Rice Annotation Project (http://rice.uga.edu/, accessed on 4 July 2022), the RAP-DB (https://rapdb.dna.affrc.go.jp/, accessed on 4 July 2022), and the TASUKE+ (https://tasuke-plus.dna.affrc.go.jp/, accessed on 4 July 2022) [4–6]. Recent advances in molecular genomics and genetics tools are represented in other databases including the RiceXpro transcriptome atlas (https://ricexpro.dna.affrc.go.jp/, accessed on 4 July 2022) and the Rice Expression Database (RED) (http://expression.ic4r.org/, accessed on 4 July 2022) [7]. Mutant populations can be found at the RMD (http://rmd.ncpgr.cn, accessed on 4 July 2022) and the KitBase (https://kitbase.ucdavis.edu/, accessed on 4 July 2022) [8]. Molecular research tools are well-established in rice, and are increasing our knowledge about genetic factors at QTLs and responsible genes controlling agronomic traits. Analysis of rice genome sequences has shown that there are more than 37,000 genes in the ~400 Mbp rice genome [1,4,5]. However, as of 2022, there are only about 3000 genes whose molecular functions are characterized in detail, e.g., the developmental stages and tissues where they are expressed from germination to harvest, and how they are co-expressed with other genes and proteins [9]. The individual molecular functions of most of the remaining genes are still unknown so far. Therefore, continuing many research efforts in gene functional analysis is necessary to elucidate the genetic networks and molecular mechanisms controlling agronomically important traits.

Recent improvements in living standards have increased the worldwide demand for high-yielding and high-quality rice cultivars. To develop novel cultivars with superior agronomic performance, we need to understand the molecular basis of agronomically important traits related to grain yield, grain quality, disease resistance, and abiotic-stress tolerance. Molecular biology techniques can reveal the complex mechanisms involved in the control of these agronomic traits [10,11]. After editing the previous Special Issue in the International Journal of Molecular Science [9], we have continued to collect recent significant studies that identified genetic factors and revealed their molecular contributions to rice agronomic traits in the Special Issue "Molecular Research in Rice: Agronomically Important Traits 2.0".

Rice grain yield consists of four main components: the number of panicles per plant, the number of grains per panicle, the percentage of ripening grains, and 1000-grain weight. Changes in panicle architecture have been associated with improved grain yield. In rice, panicle architecture is mainly determined by the spikelet and branch arrangement. Spikelets and branches are initiated and developed from inflorescence meristems. Li et al. showed that the *Verticillate Primary Branch 1* (*VPB1*) gene is preferentially expressed in the inflorescence meristem and is associated with control of primary branch arrangement [12]. The *VPB1* gene encodes a BELL-like homeodomain (BLH) protein, which alters the expression of other genes involved in inflorescence meristem identity and hormone-signaling pathways by binding to the promoter region of the *OsBOP1* gene. Yin et al. summarized recent advances concerning genes involved in controlling the grain number per panicle (GNPP) in rice [13]. They collected a total of 36 genes affecting GNPP, such as *DEP1*, *IPA1*, *APO1*, *TAW1*, and *SP1*, and classified these GNPP-related genes based on their molecular functions such as positive or negative regulation of GNPP by rachis branch development, phase transition from rachis branch meristem to spikelet meristem, and spikelet specialization. Park et al. focused on grain shape and grain weight. They detected a large-genetic-effect QTL on rice chromosome 1, and estimated that the candidate gene of this QTL is the *OsBRKq1* gene [14]. The *OsBRKq1* gene encodes brassinosteroid leucine-rich repeat-receptor kinase protein, is expressed at the spikelet differentiation stage, and positively affects grain size. Usman et al. created long-grain mutant lines of the *GS3* gene by using CRISPR/Cas9-mediated genome editing [15]. The genome-edited lines showed a grain length and 1000-grain weight increased by about 25%. These novel alleles of the *GS3* gene altered the expression of other genes including those annotated with cysteine synthase, cysteine proteinase inhibitor, vacuolar protein sorting-associated, ubiquitin, and DNA ligase GO (gene ontology) terms.

The understanding of the genetic basis for various abiotic stress resistance and tolerance is important for rice cultivation, because these stress conditions cause significant decreases in yield and grain quality. Bang et al. characterized the *Rice Chloroplast RNA-binding Protein 1* (*OsCRP1*) gene, which is essential for the stabilization of RNAs from the NAD(P)H dehydrogenase complex in rice chloroplast [16]. They clearly indicated that the *OsCRP1*-overexpressing lines showed higher cyclic electron transport activity and elevated ATP levels for photosynthesis, compared with wild-type plants. Additionally, the *OsCRP1*-overexpressing lines significantly enhanced drought- and cold-stress tolerance. Huang et al. reported that the *Rice WRKY Transcription Factor 55* (*OsWRKY55*) gene is involved both in drought responses and in plant growth regulation [17]. They created *OsWRKY55*-overexpression lines, and revealed that the overexpression lines showed faster water loss and a greater accumulation of hydrogen peroxide (H_2O_2) and superoxide radicals (O_2^-) in leaves under drought-stress conditions, and consequently decreased drought resistance compared with wild-type plants. They revealed protein–protein interactions of the OsWRKY55 protein with mitogen-activated protein kinases (MAPKs) OsMPK7, OsMPK9, OsMPK20-1, and OsMPK20-4 that could be induced by drought conditions, and showed binding activity to the promoter region of the *OsAP2-39* gene that controls cell size and plant height. Wu et al. used marker-assisted selection to introduce the *Sub1A* gene and develop breeding lines with strong submergence tolerance [18]. The developed lines exhibited desirable agronomic traits including high grain yield and quality, and showed strong drought and submergence tolerance. Abiotic stresses induced by high or low temperatures are also an important issue for rice cultivation because they cause significant decreases in grain yield and quality. Hori et al. detected two quantitative trait loci (QTLs) for the determination of flowering time according to seasonal temperature conditions [19]. The *Hd16* gene is one of the genetic factors associated with robustness of flowering time to environmental fluctuations. These research efforts are helpful in dissecting the genetic basis of stress resistance and tolerance in future breeding programs.

Nitrogen is an essential macronutrient that plays a critical role in the growth and development of rice plants. Improvement in nitrogen use efficiency is required to achieve the Sustainable Development Goals in agriculture and crop production. Kabange et al. showed that the nitrate reductase (NR) gene of *OsNR2* and nitrate transporter (NRT) genes of *OsNIA1* and *OsNIA2* were differentially expressed between the root, leaf, and stem in rice [20]. The results in their study suggested that the *NR* and *NIR* genes had tissue-specific molecular functions in response to potassium chlorate ($KClO_3$) and the reduction activities of the ammonium transporters and glutamate synthase. Sato et al. developed a novel three-dimensional (3D)-imaging method with single-cell resolution [21]. The authors successfully captured shoot, floret, and root apical meristems at cellular resolution, and revealed the 3D distribution of auxin-signaling pathway proteins in the columella, quiescent center, and multiple cell rows of the root apical meristem.

The significant studies mentioned here demonstrate the importance of the research community in understanding and explaining the molecular genetic basis of agronomically important traits in rice. To develop novel rice cultivars showing good agronomic performance and strong climate resilience in the future, we have to identify further important genes, elucidate their molecular functions, and design desirable genotypes based on individual and interaction effects of those genes. In addition to the traits described in the previous and present Special Issues, there are many other agronomically important traits that should be explored using molecular genetics and biological research. In addition, the wide genetic diversity that exists in rice genetic resources including wild rice species (e.g., *Oryza rufipogon* Griff., *Oryza longistaminata* Chev. et Roehr.) has not been fully exploited in genetic analysis and breeding programs. Although several disease resistance genes have been introduced from wild species into rice cultivars [22], wild rice species also possess strong abiotic-stress resistance such as drought resistance, submergence tolerance, and anaerobic resistance. Some varieties exhibit early-morning flowering, thus avoiding anthesis at high temperature conditions closer to midday [23]. Further exploitation of this wide genetic diversity should be a priority. The sharing of research results among researchers is necessary to address problems of food security against a background of increasing demand around the world.

Following the construction of the first rice genome reference sequence, whole-genome reference sequences have been published in other crop species such as wheat, barley, maize, and soybean, as well as in vegetables, fruit trees, and ornamental plants [24]. Comparative genome sequence and homology analysis can identify highly conserved gene families among various crop species such as those involved in plant hormone synthesis and those controlling tissue development and stress responses. The gene-regulatory network for flowering time determination by florigens is also well-conserved among plant species [25], as are the polyphenol biosynthesis pathways associated with the coloring of edible parts. These pathways are targets for increasing functional food components with antioxidative effects, and providing responsiveness to ambient temperatures in almost all cereal, vegetable, and fruit tree crops. Abundant genome sequence information and annotated gene lists in rice will facilitate the improvement in various agronomic traits, not only in rice, but also in other monocot cereal crops, as well as in other dicotyledonous crop species.

Funding: This research received no external funding.

Conflicts of Interest: The authors declare no conflict of interest.

References

1. International Rice Genome Sequencing Project. The map-based sequence of the rice genome. *Nature* **2005**, *436*, 793–800. [CrossRef] [PubMed]
2. Wang, S.; Gao, S.; Nie, J.; Tan, X.; Xie, J.; Bi, X.; Sun, Y.; Luo, S.; Zhu, Q.; Geng, J.; et al. Improved 93-11 genome and time-course transcriptome expand resources for rice genomics. *Front. Plant Sci.* **2022**, *12*, 769700. [CrossRef] [PubMed]
3. Wang, W.; Mauleon, R.; Hu, Z.; Chebotarov, D.; Tai, S.; Wu, Z.; Li, M.; Zheng, T.; Fuentes, R.R.; Zhang, F.; et al. Genomic variation in 3010 diverse accessions of Asian cultivated rice. *Nature* **2018**, *557*, 43–49. [CrossRef] [PubMed]
4. Ouyang, S.; Zhu, W.; Hamilton, J.; Lin, H.; Campbell, M.; Childs, K.; Thibaud-Nissen, F.; Malek, R.L.; Lee, Y.; Zheng, L.; et al. The TIGR Rice Genome Annotation Resource: Improvements and new features. *Nucleic Acids Res.* **2007**, *35*, D883–D887. [CrossRef] [PubMed]
5. Sakai, H.; Lee, S.S.; Tanaka, T.; Numa, H.; Kim, J.; Kawahara, Y.; Wakimoto, H.; Yang, C.C.; Iwamoto, M.; Abe, T.; et al. Rice Annotation Project DataBase (RAP-DB): An integrative and interactive database for rice genomics. *Plant. Cell Physiol.* **2013**, *54*, e6. [CrossRef] [PubMed]
6. Kumagai, M.; Nishikawa, D.; Kawahara, Y.; Wakimoto, H.; Itoh, R.; Tabei, N.; Tanaka, T.; Itoh, T. TASUKE+: A web-based platform for exploring GWAS results and large-scale resequencing data. *DNA Res.* **2019**, *26*, 445–452. [CrossRef] [PubMed]

7. Sato, Y.; Takehisa, H.; Kamatsuki, K.; Minami, H.; Namiki, N.; Ikawa, H.; Ohyanagi, H.; Sugimoto, K.; Antonio, B.A.; Nagamura, Y. RiceXPro Version 3.0: Expanding the informatics resource for rice transcriptome. *Nucleic Acids Res.* **2013**, *41*, D1206–D1213. [CrossRef]
8. Jain, R.; Jenkins, J.; Shu, S.; Chern, M.; Martin, J.A.; Copetti, D.; Duong, P.Q.; Pham, N.T.; Kudrna, D.A.; Talag, J.; et al. Genome sequence of the model rice variety KitaakeX. *BMC Genomics* **2019**, *20*, 905. [CrossRef]
9. Hori, K.; Shenton, M. Recent advances in molecular research in rice: Agronomically important traits. *Int. J. Mol. Sci.* **2020**, *21*, 5495. [CrossRef]
10. Song, S.; Tian, D.; Zhang, Z.; Hu, S.; Yu, J. Rice genomics: Over the past two decades and into the future. *Genomics Proteomics Bioinform.* **2018**, *16*, 397–404. [CrossRef]
11. Biswal, A.K.; Mangrauthia, S.K.; Reddy, M.R.; Yugandhar, P. CRISPR mediated genome engineering to develop climate smart rice: Challenges and opportunities. *Semin. Cell Dev. Biol.* **2019**, *96*, 100–106. [CrossRef] [PubMed]
12. Li, M.; Fu, D.; Xu, T.; Wu, C. VPB1 encoding BELL-like homeodomain protein is involved in rice panicle architecture. *Int. J. Mol. Sci.* **2021**, *22*, 7949. [CrossRef]
13. Yin, C.; Zhu, Y.; Li, X.; Lin, Y. Molecular and genetic aspects of grain number determination in rice (*Oryza sativa* L.). *Int. J. Mol. Sci.* **2021**, *22*, 728. [CrossRef] [PubMed]
14. Park, J.R.; Resolus, D.; Kim, K.M. *OsBRKq1*, related grain size mapping, and identification of grain shape based on QTL mapping in rice. *Int. J. Mol. Sci.* **2021**, *22*, 2289. [CrossRef] [PubMed]
15. Usman, B.; Zhao, N.; Nawaz, G.; Qin, B.; Liu, F.; Liu, Y.; Li, R. CRISPR/Cas9 guided mutagenesis of *Grain Size 3* confers increased rice (*Oryza sativa* L.) grain length by regulating cysteine proteinase inhibitor and ubiquitin-related proteins. *Int. J. Mol. Sci.* **2021**, *22*, 3225. [CrossRef] [PubMed]
16. Bang, S.W.; Lee, H.S.; Park, S.H.; Lee, D.K.; Seo, J.S.; Kim, Y.S.; Park, S.C.; Kim, J.K. *OsCRP1*, a ribonucleoprotein gene, regulates chloroplast mRNA stability that confers drought and cold tolerance. *Int. J. Mol. Sci.* **2021**, *22*, 1673. [CrossRef]
17. Huang, K.; Wu, T.; Ma, Z.; Li, Z.; Chen, H.; Zhang, M.; Bian, M.; Bai, H.; Jiang, W.; Du, X. Rice transcription factor OsWRKY55 is involved in the drought response and regulation of plant growth. *Int. J. Mol. Sci.* **2021**, *22*, 4337. [CrossRef]
18. Wu, Y.P.; Wang, S.M.; Chang, Y.C.; Ho, C.; Hsu, Y.C. Submergence gene *Sub1A* transfer into drought-tolerant *japonica* rice DT3 using marker-assisted selection. *Int. J. Mol. Sci.* **2021**, *22*, 13365. [CrossRef]
19. Hori, K.; Saisho, D.; Nagata, K.; Nonoue, Y.; Uehara-Yamaguchi, Y.; Kanatani, A.; Shu, K.; Hirayama, T.; Hirayama, T.; Yonemaru, J.-I.; et al. Genetic elucidation for response of flowering time to ambient temperatures in Asian rice cultivars. *Int. J. Mol. Sci.* **2021**, *22*, 1024. [CrossRef]
20. Kabange, N.R.; Park, S.Y.; Lee, J.Y.; Shin, D.; Lee, S.M.; Kwon, Y.; Cha, J.K.; Cho, J.H.; Duyen, D.V.; Ko, J.M.; et al. New insights into the transcriptional regulation of genes involved in the nitrogen use efficiency under potassium chlorate in rice (*Oryza sativa* L.). *Int. J. Mol. Sci.* **2021**, *22*, 2192. [CrossRef]
21. Sato, M.; Akashi, H.; Sakamoto, Y.; Matsunaga, S.; Tsuji, H. Whole-tissue three-dimensional imaging of rice at single-cell resolution. *Int. J. Mol. Sci.* **2022**, *23*, 40. [CrossRef] [PubMed]
22. Ronald, P.C.; Albano, B.; Tabien, R.; Abenes, L.; Wu, K.S.; McCouch, S.; Tanksley, S.D. Genetic and physical analysis of the rice bacterial blight disease resistance locus, *Xa21*. *Mol. Gen. Genet.* **1992**, *236*, 113–120. [CrossRef] [PubMed]
23. Hirabayashi, H.; Sasaki, K.; Kambe, T.; Gannaban, R.B.; Miras, M.A.; Mendioro, M.S.; Simon, E.V.; Lumanglas, P.D.; Fujita, D.; Takemoto-Kuno, Y.; et al. *qEMF3*, a novel QTL for the early-morning flowering trait from wild rice, *Oryza officinalis*, to mitigate heat stress damage at flowering in rice, *O. sativa*. *J. Exp. Bot.* **2015**, *66*, 1227–1236. [CrossRef] [PubMed]
24. Chen, F.; Song, Y.; Li, X.; Chen, J.; Mo, L.; Zhang, X.; Lin, Z.; Zhang, L. Genome sequences of horticultural plants: Past, present, and future. *Hortic. Res.* **2019**, *6*, 112. [CrossRef] [PubMed]
25. Eshed, Y.; Lippman Zachary, B. Revolutions in agriculture chart a course for targeted breeding of old and new crops. *Science* **2019**, *366*, eaax0025. [CrossRef]

International Journal of
Molecular Sciences

MDPI

Article

VPB1 Encoding BELL-like Homeodomain Protein Is Involved in Rice Panicle Architecture

Mu Li *, Debao Fu, Tingting Xu and Changyin Wu *

National Key Laboratory of Crop Genetic Improvement, Huazhong Agricultural University, Wuhan 430070, China; debaofu@webmail.hzau.edu.cn (D.F.); xtt8831@webmail.hzau.edu.cn (T.X.)
* Correspondence: limu@webmail.hzau.edu.cn (M.L.); cywu@mail.hzau.edu.cn (C.W.); Tel.: +86-136-3868-9237 (M.L.); +86-027-8728-1887 (C.W.)

Abstract: Inflorescence architecture in rice (*Oryza sativa*) is mainly determined by spikelets and the branch arrangement. Primary branches initiate from inflorescence meristem in a spiral phyllotaxic manner, and further develop into the panicle branches. The branching patterns contribute largely to rice production. In this study, we characterized a rice *verticillate primary branch 1*(*vpb1*) mutant, which exhibited a clustered primary branches phenotype. Gene isolation revealed that *VPB1* was a allele of *RI*, that it encoded a BELL-like homeodomain (BLH) protein. *VPB1* gene preferentially expressed in the inflorescence and branch meristems. The arrangement of primary branch meristems was disturbed in the *vpb1* mutant. Transcriptome analysis further revealed that *VPB1* affected the expression of some genes involved in inflorescence meristem identity and hormone signaling pathways. In addition, the differentially expressed gene (DEG) promoter analysis showed that *OsBOPs* involved in boundary organ initiation were potential target genes of VPB1 protein. Electrophoretic mobility shift assay (EMSA) and dual-luciferase reporter system further verified that VPB1 protein bound to the promoter of *OsBOP1* gene. Overall, our findings demonstrate that *VPB1* controls inflorescence architecture by regulating the expression of genes involved in meristem maintenance and hormone pathways and by interacting with *OsBOP* genes.

Keywords: inflorescence architecture; BLH homedomain protein; branching pattern; verticillate primary branch; transcriptome analysis; hormone pathways

Citation: Li, M.; Fu, D.; Xu, T.; Wu, C. *VPB1* Encoding BELL-like Homeodomain Protein Is Involved in Rice Panicle Architecture. *Int. J. Mol. Sci.* **2021**, *22*, 7909. https://doi.org/10.3390/ijms22157909

Academic Editors: Kiyosumi Hori, Matthew Shenton and Marcello Iriti

Received: 15 June 2021
Accepted: 20 July 2021
Published: 24 July 2021

1. Introduction

Inflorescence is the clusters of flowers arranged on a stem, and it comprises a main branch and lateral branches with a complicated arrangement [1–3]. The inflorescence architecture of higher plants contributes not only to plant morphology but also to plant reproduction, and further affecting the final grain yield in crops [4]. The panicle-type inflorescences are characteristics of grasses such as maize (*Zea mays*) and rice (*Oryza sativa*) [5]. Maize has two types of inflorescences, male tassel and female ear, which are different in morphology and branching pattern [6]. Rice inflorescence, also known as 'panicle', during panicle development, and shoot apical meristem (SAM) is transformed into the inflorescence meristem (IM) after transition from vegetative phase to reproductive phase, IM successively generates the primary and secondary branch meristem (PBM and SBM), floret meristem (FM), and spikelet meristem [7]. The main stem of rice panicle has primary and secondary branches, which are arranged in a spiral phyllotaxy [8]. Thus, the panicle branching patterns determine rice panicle architecture and eventually affect grain yield in rice [9].

So far, a large number of genes involved in regulating inflorescence architecture in rice have been identified, such as *LAX PANICLE1* (*LAX1*) and *LAX2* participating in the formation of axillary meristem (AM) in rice [10,11] and *ABERRANT PANICLE ORGANIZATION 1* (*APO1*) positively regulating the number of spikelets and primary branches and affecting the attributes of floral organs and the identity of flowers [12]. *APO2* has been reported to regulate the transition from rice vegetative growth to reproductive growth and to control the development of panicle branches, and it can directly interact with *APO1* to control the inflorescence and flower development [13]. The functional loss of either *FLORAL ORGAN NUMBER1* (*FON1*) or *FON2* causes the enlargement of the floral meristem, thus resulting in the increased floral organs [14,15]. *ABERRANT SPIKELET AND PANICLE1* (*ASP1*; also known as *OsREL2*) regulates different aspects of rice development and physiological responses, such as the development of panicles, branches, and spikelets [16,17]. *FON2* and *ASP1* are involved in the negative regulation of stem cell proliferation in both inflorescence meristems and flowers [18]. *TILLERS ABSENT1* (*TAB1*) plays an important role in initiating the rice axillary meristems, but this gene is not involved in maintaining the established meristem [19]. *TAW1* regulates inflorescence development by enhancing the activity of inflorescence meristems to inhibit the transformation from inflorescence meristems to spikelet meristems [20]. Those above-mentioned genes mainly control the length and the number of branches and meristem maintenance. However, our knowledge of the genetic mechanisms underlying branching patterns including branch phyllotaxy and internode elongation in rice remains limited.

Interestingly, the three-amino-acid-loop-extension (TALE) class of homeoproteins falls into two subfamilies, *KNOTTED1-like* homeobox (*KNOX*) and *BELL1-like* homeobox (*BLH*), which have been reported to control meristem formation and maintenance, organ position in plant, and organ morphogenesis [21]. For example, in *Arabidopsis thaliana*, two paralogous *BLH* genes, *PENNYWISE* (*PNY*) (also known as *BELLRINGER* (*BLR*), *REPLUMLESS* (*RPL*), or *VAAMANA* (*VAN*)) and POUND-FOOLISH (*PNF*), play significant roles in maintaining the SAM and the development of the inflorescence architecture [22–29]. Loss-of-function *PNY* gene causes the altered phyllotaxy, including irregular internode elongation, clusters of branches and flowers on the stem, and eventually reducing apical dominance [30]. Furthermore, *PNY* is involved in the establishment of normal phyllotaxis by repressing the expression of *PME5* (pectin methylesterase) in the meristem and the maintenance of phyllotaxis by activating *PME5* in the internode [31]. BLH proteins can interact with KNOX proteins to form heterodimer. For example, PNY interacts with the SHOOTMERISTEMLESS (STM) and BREVIPEDICELLUS (BP). The double mutant *bp/pny* exhibits synergistic phenotype of the short internodes interspersed with the long internodes and the increased branches [30]. The interaction between PNY and STM maintains the boundary between floral primordia and inflorescence meristem, and the SAM function in *Arabidopsis* requires both PNY and STM [32,33]. In addition, ChIP-seq results reveal that *PNY* interacts with many of the key genes regulating stem morphogenesis and controling the oriented growth by directly repressing organ boundary genes [34]. In maize, the two *BLH* genes, *BLH12* and *BLH14*, are close homologs of *PNY* and *PNF*, and double mutant *blh12/blh14* causes abnormality in internode pattern and vascular bundles anastomosis as well as indeterminate branch formation in the tassel [35].

In rice, one *BLH* gene *qSH1* is a main quantitative trait locus of seed shattering [36]. In addition, another *BLH* gene *SH5* induces seed shattering by facilitating abscission-zone development and inhibiting lignin biosynthesis, and SH5 can interact with KNOX protein OSH15 to induce grain shattering by repressing lignin biosynthesis-related genes [37,38]. One recent study has reported that gene *RI* encoding a BLH transcription factor affects primary branch pattern mainly by regulating the arrangement and initiation time of the primary branch meristems, the *BLH* gene family is essential for regulating inflorescence structure in plant [39]. However, the molecular mechanism by which these genes regulate the branch arrangement pattern remain largely unknown in rice.

In this study, we characterized the rice *verticillate primary branch 1* (*vpb1*) mutant, which displayed a clustered primary branch phenotype. Gene isolation experiment revealed that *VPB1* was a allele of *RI*, and it encoded a BLH transcription factor. Further experiments demonstrated that *VPB1* negatively regulated the expression of *OsBOP1* gene to construct panicle architecture in rice. Transcriptome analysis indicated that *VPB1* was likely to negatively regulate the expression of genes involved in auxin hormonal pathways to form the normal inflorescence architecture. Our results provide new insights into the branching patterns in rice.

2. Results

2.1. Inflorescence Phenotypes in Vpb1 Mutant

To identify the key regulators that control panicle architecture formation in rice, we screened two recessive and allelic mutants which exhibited abnormal panicles from rice T-DNA insertion mutant library. We designated them as *verticillate primary branch 1-1 (vpb1-1)* and *vpb1-2* (Figure S1). Compared with wild-type inflorescence, the *vpb1* mutant inflorescence exhibited the clustered primary branch phenotype, indicating the primary branches are initiated in a verticillate manner (Figure 1A–D). Our findings are consistent with a previous report that mutant phenotype of *RI* [39]. To investigate *vpb1* inflorescence quantitatively, we counted the number of inflorescence branches in the wild type and mutant. The primary branches number of *vpb1* mutant panicle was increased by 26.8%, and the secondary branches number was decreased by 32.8%, compared to the wild-type inflorescence (Figure 1E,F). Quantitative analysis of *vpb1* mutant panicle indicated that the length of rachis and the number grains of panicle were respectively reduced by 56.5% and 27% compared with wild types (Figure 1G,H). The clustered panicle appearance and the reduction in spikelet number in the *vpb1* mutant might be attributable to the reduced rachis length and the decreased number of secondary branches. Moreover, the *vpb1* mutants exhibited a defect in producing the inflorescence meristem.

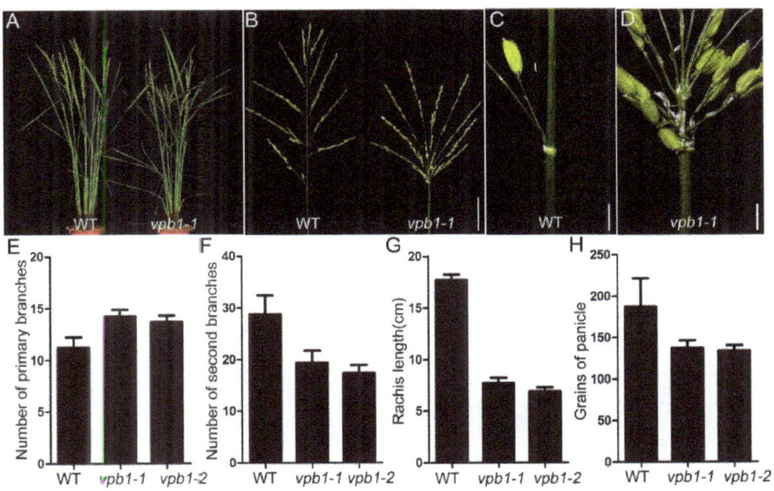

Figure 1. Phenotypic characterization of *vpb1-1* mutant. (**A**) Mature wild-type plants (left) and the *vpb1-1* mutant (right). (**B**) Mature panicles of wild-type (left) and *vpb1-1* mutant (right). (**C**,**D**) Close-up view of the branch site of the primary branches in wild-type (**C**) and *vpb1-1* mutant (**D**). (**E**–**H**) Quantitative traits of wild-type and *vpb1* mutant panicles. Vertical bars indicate standard deviations, $n = 15$. (**E**) The numbers of primary branches in wild type and *vpb1* mutant. (**F**) The numbers of secondary branches in wild type and *vpb1* mutant. (**G**) Rachis length of wild type and *vpb1* mutant. (**H**) The numbers of grains of panicle in wild type and *vpb1* mutant. Scale bars, 4 cm in (**B**); 2 cm in (**C**,**D**).

To further examine the defects of *vpb1* panicles, we used scanning electron microscope (SEM) to determine the time when the panicle development of *vpb1-1* plants first differed from that wild type plants. SEM results indicated no significant morphological difference between *vpb1* and the wild-type SAMs in the vegetative stage and reproductive stage except the primary branch meristem (PBM) formation stage (Figure 2 and Figure S2). The wild type PBMs were initiated in a regular spiral pattern (Figure 2A). By contrast, *vpb1* mutant PBMs were initiated in an irregular pattern and they might be simultaneously initiated from the inflorescence meristems (Figure 2D). The lateral view of PBMs showed that the height of the PBM cluster of *vpb1* was lower than that of wild type (Figure 2B,E). These results confirmed that the primary branch meristems of *vpb1* mutant displayed an abnormal arrangement on inflorescence meristem. We hypothesized that the disordered primary branch meristems might be caused by the abnormal development of inflorescence. To test this hypothesis, we especially used the paraffin section method to examine morphological characteristics of panicles, we found that the inflorescence meristem of *vpb1* mutant was extremely defective (Figure 2C,F). Therefore, we considered that the disordered phyllotactic pattern of *vpb1* inflorescence might be due to the disturbed arrangement of the primary branch meristems. *VPB1* functioned as a determinant factor to regulate inflorescence meristem activity during panicle morphogenesis.

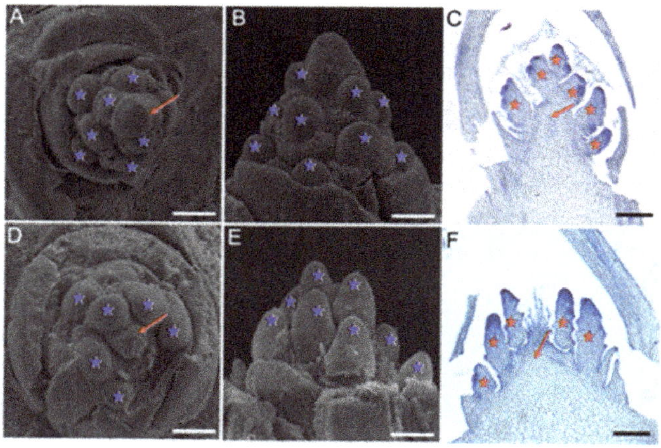

Figure 2. Morphological analysis of wild-type and *vpb1* inflorescence. (**A**,**B**) Scanning electron microscope (SEM) images of PBMs at their initiation stage in wild-type. (**D**,**E**) Scanning electron microscope (SEM) images of PBMs at their initiation stage in *vpb1* mutant. (**C**,**F**) Paraffin section images showing the inflorescence of the wild type (**C**) and *vpb1* (**F**). The arrow and asterisks indicate inflorescence and primary branch meristems, respectively. Scale bar, 100 μm.

2.2. Map-Based Cloning of VPB1

We constructed a mapping population by crossing the original *vpb1* mutant with *indica* variety Dular. Of 1200 F_2 plants, 288 exhibited a *vpb1*-like phenotype, and chi-square test results indicated that a segregation ratio of the *vpb1* mutant plants and wild-type plants was 1:3. These results demonstrated that the phenotype of *vpb1* mutant was controlled by a recessive single gene. To clone gene *VPB1* through a map-based approach, Primary gene mapping showed that *VPB1* locus was located between the molecular markers RM3575 and RM7448 on chromosome 5, and we then fine-mapped the locus to a 38.5-kb region between markers RM3295 and IN22.30 (Figure 3A). Within this region, five genes were predicted in the Nipponbare genome (TIGR Rice Genome Annotation Database) (Table S1). PCR-based sequencing and bioinformatics analyses of this 38.5-kb region fragment revealed that a 433-bp DNA fragment was inserted into the second exon of the candidate gene *LOC_Os05g38120* in *vpb1-1* mutant to generate a premature stop codon, and that a 7-bp nucleotide deletion in the second exon in *vpb1-2* led to amino acid frameshift (Figure S3). *LOC_Os05g38120* composed of four exons and five introns encoded a homeodomain protein (Figure 3A and Table S1). To verify whether the clustered primary branch phenotype was caused by the DNA insertion and deletion in *LOC_Os05g38120*, a pair of gene-specific primers P1 and P2 were used to detect the genotype of the F_2 population derived from the cross of *vpb1* with WT. Cosegregation analysis of an F_2 population indicated that all the *vpb1-1* plants with homozygous DNA insertion showed the phenotype of the clustered primary branch, and the other plants without DNA insertion or with heterozygous DNA insertion showed normal panicle morphology (Figure 3B), and all the *vpb1-2* plants with homozygous DNA deletion showed the phenotype of the clustered primary branch, and the other plants without DNA deletion or with heterozygous DNA deletion showed normal panicle morphology (Figure S3). Therefore, these results suggested that *LOC_Os05g38120* was determined as the candidate gene of *VPB1*, which was a new allele of *SH5/RI* [37,39].

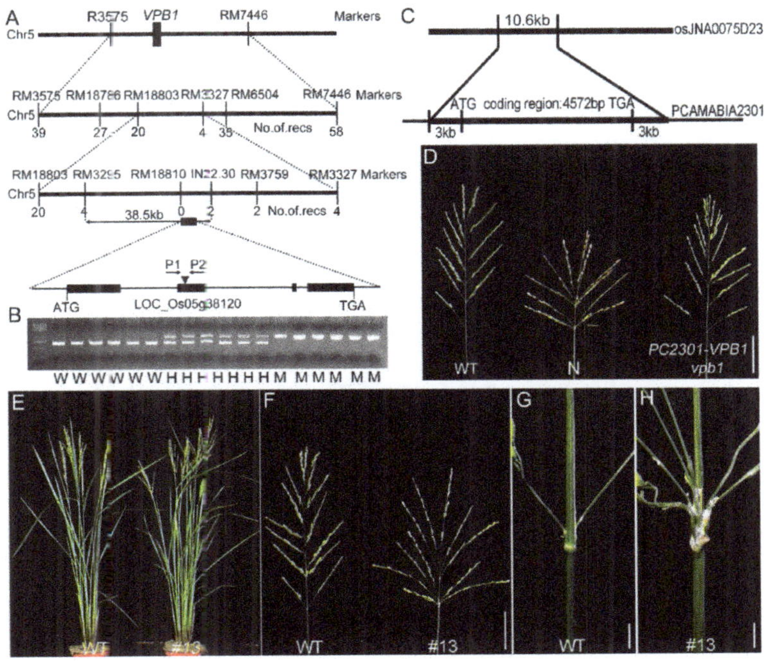

Figure 3. Positional cloning of the gene responsible for the *vpb1* mutation. (**A**) Fine mapping of the *VPB1* on chromosome 5. The *VPB1* locus was narrowed to a ~38.5-kb genomic DNA region between markers RM3295 and IN22.30. recs is the number of recombinants. The structure of *VPB1*, showing the mutation site of *vpb1*. Closed boxes indicate the coding sequence, and lines between boxes represent introns. (**B**) Cosegregation analysis of a F_2 population derived from a cross of *vpb1* x WT (ZH11) via PCR using the primers (P1, P2) shown in (**A**). M: mutant; H: hetero; W: wild type. (**C**) Schematic diagram of the pC2301-*VPB1* construct. (**D**) Genetic complementation of *vpb1*. N indicates negative control. Scale bar, 4 cm. (E-H) Performance of *VPB1* positive and negative transgenic plants generated using the CRISPR/Cas9 strategy. (**E**) Mature wild-type plants (left) and the #13 mutant (right). (**F**) Mature panicles of wild-type (left) and #13 mutant (right). Scale bar, 4 cm. (**G**,**H**) Close-up view of the branch site of the primary branches in wild-type (**G**) and #13 mutant (**H**). Scale bar, 2 cm.

To test *VPB1* whether could complement the mutant phenotype, we constructed a vector. This vector fragment containing the coding sequence of *VPB1* flanked by a 3000-bp upstream fragment of the start codon and a 3000-bp downstream fragment of the stop codon was cloned into pCAMBIA2301 (Figure 3C). This vector was transformed into *vpb1* mutant callus, and 31 independent transgenic plants were obtained. The abnormal inflorescence phenotype of *vpb1* of these 31 transgenic plants was fully rescued by this constructed pC2301-*VPB1*, whereas that of 12 plants transformed with empty vector (negative control) remained unrescued (Figure 3D). Additonally, we generated function-deficient mutants in the ZH11 background using the CRISPR system (Figure S4) [40], and these mutants displayed reduced rachis length and verticillate primary branches (Figure 3E–H). Afterwards, we transformed vector pC1301S-VPB1-GFP with green fluorescent protein (GFP) fused to the C terminus of VPB1 into rice ZH11 (WT) callus, and obtained multiple independent lines overexpressing *VPB1*, their phenotypes were similar to those of wild-type (Figure S5). Moreover, in the young panicle, the expression of *VPB1* was relatively lowly expressed in mutant, compared to that in wild-type plants (Figure S6A). The immunoblot assay with an anti-VPB1 antibody revealed that the accumulation of VPB1 protein in the young panicle (2–3mm) was greatly reduced in *vpb1-1* and *vpb1-2* (Figure S6B). These results suggested that the mutation of *VPB1* was responsible for abnormal panicle morphology of *vpb1*.

2.3. *VPB1 Encodes a BELL1-Type Transcription Factor*

Bioinformatic analysis revealed that the amino acid sequence of VPB1 contains a conserved BELL domain, indicating that VPB1 is one member of the BLH family. Members of BLH family regulate many key developmental processes in plants [21,27,35,41,42]. Thirteen members of the BLH family have been identified in *Arabidopsis* and 17 members in rice [38]. These BLH proteins domain had three extra amino acids (Proline[P], tyrosine[Y], Proline [P]) between the first and the second helix (Figure S7A). To examine the relationship between VPB1 and other BLH proteins, we used amino acid sequences of VPB1 and other BLH proteins in rice and *Arabidopsis* to construct a phylogenetic tree (Figure S7B). The result revealed that the VPB1 protein was highly homologous to *Arabidopsis* PNY and PNF. Gene *LOC_Os05g38120* has been reported to be *SH5*, phylogenetic analysis also revealed that the VPB1 was highly homologous to qSH1, and that both SH5 and qSH1 were responsible for the formation of seed abscission layer in rice [36,37]. Moreover, the alignment and motif analysis of VPB1 homologue in rice and *Arabidopsis* showed that VPB1 contained the intermediate BLH domain composed of SKY and BELL regions and the C-terminal homeobox domain, and it was relatively conservative in various plant species (Figure S7C).

2.4. *Expression Pattern of VPB1*

To reveal the role of *VPB1* in inflorescence development, we explored its expression pattern. The qRT-PCR analysis indicated that *VPB1* was expressed in all tested tissues, including young leaf, mature leaf, leaf sheath, panicle, root, and stem; especially, it was expressed more highly in panicle than in other tissues (Figure 4A). RNA in situ hybridization further showed that *VPB1* transcripts were detectable at different inflorescence development stages in wild-type, and that *VPB1* was highly expressed in shoot apical meristem, primary and secondary branch meristem (Figure 4B,C,E,F). This agrees with the results by Ikeda et al. (2019) [39]. As expected, *VPB1* expression was hardly detectable when sense probe was used as a negative control (Figure 4D,G). The expression pattern analysis of both *VPB1* suggested that the *VPB1* gene played a critical role in establishing and maintaining meristem in rice.

2.5. *Subcellular Localization and VPB1 Transcriptional Activity*

Consistent with the function of VPB1 as a transcription factor, through the subcellular localization prediction tool Plant-mPLoc [43], VPB1 was predicated to be located in the nucleus. To test this prediction, VPB1 was fused with YFP, and Ghd7 (a nuclear protein) was fused with cyan fluorescent protein (CFP). The obtained two fusion plasmids were transiently expressed in rice protoplasts, and the fluorescence signal assay indicated that VPB1 and Ghd7 were co-localized to the nucleus (Figure 5A), suggesting that VPB1 was a nuclear protein.

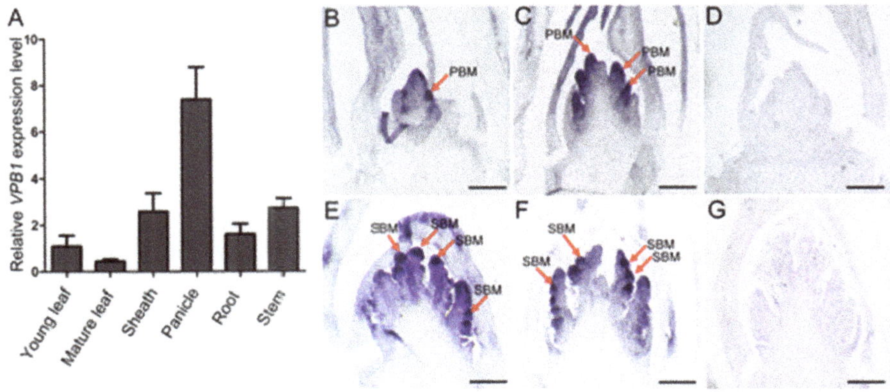

Figure 4. Expression pattern of *VPB1*. (**A**) RT–qPCR of organ-specific *VPB1* expression in WT plants. Including young leaf, mature leaf, sheath, panicle (1–2 mm), root, stem. Data are mean ± SD (n=3 biological replicates). (**B–G**) In situ hybridization of *VPB1*. (**B**) Whole a developing inflorescence at the stage of SAM; (**C**) Whole a developing inflorescence at the stage of primary branch meristem (PBM) differentiation; (**E**,**F**) Whole a developing inflorescence at the stage of secondary branch meristem (SBM) differentiation. (**D**,**G**) Sense probe as control. The red arrow points to the branch meristem. Scale bars, 100 μm.

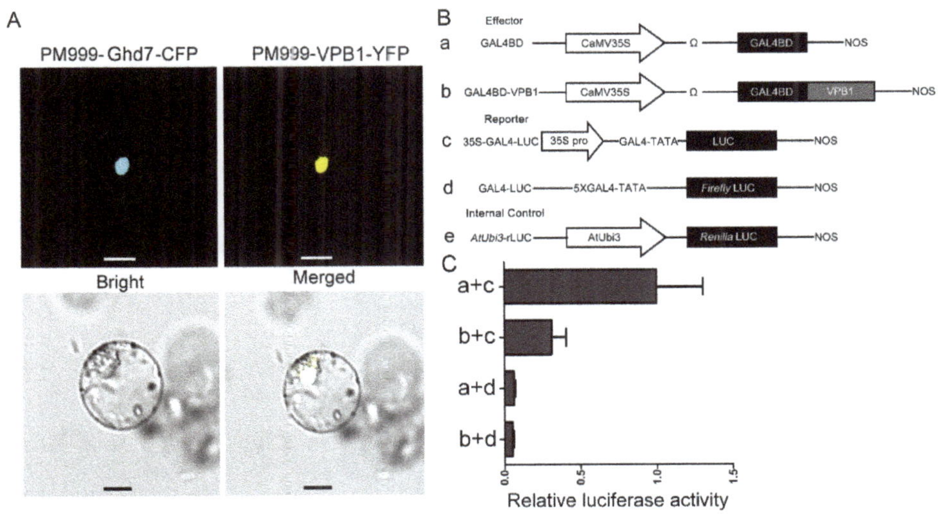

Figure 5. Subcellular localization and transcription activity of VPB1. (**A**) Subcellular localization of VPB1 protein. The VPB1-YFP fusion protein co-localized with the Ghd7 nucleus marker in rice protoplasts. Scale bars, 10 μm. (**B**) Scheme of the constructs used in the protoplast co-transfection assay. (**C**) Transcriptional activity assay of *VPB1*. The activity of 35S-GAL4-LUC and GAL4-LUC was used as the reporter, and rLUC activity was used as an internal control. The fLUC/rLUC ratio represents the relative luciferase activity. Data are mean ± SD (n = 3 independent replicates).

We next investigated the transcriptional activity of VPB1 using a dual-luciferase reporter system. We constructed an effector GAL4BD-VPB1, and the firefly luciferase gene driven by CaMV35S enhancer contained five copies of the GAL4 binding element, and it was used as a reporter, and the renilla luciferase gene driven by a *AtUbi3* promoter was used as the internal control (Figure 5B). The results showed that the effector GAL4BD-VPB1 had significantly lower relative luciferase activity than the GAL4BD, but no significant difference in relative luciferase activity was observed between the reporter GAL4-fLUC and the GAL4BD (Figure 5C). Based on this result, we concluded that VPB1 could actively mediate transcriptional repression.

2.6. *VPB1 Affects the Expression of Genes Involved in Inflorescence Development and Hormone Pathways*

To reveal the molecular mechanism of inflorescence development in *vpb1* mutant, we analyzed gene expression levels in the young panicle (1–2 mm) of *vpb1-1* mutant and wild type plants at the stage of PBM initiation by RNA-Seq with Q value ≤ 0.05 and fold change ≥ 1.5 as the cutoff criteria. We identified differentially expressed genes (DEGs) between wild type and mutants in three biological replicates. A total of 2028 genes were upregulated, and 2418 genes were downregulated in *vpb1-1* mutant, compared with wild type (Table S2 and Figure 6A,B). Further gene ontology (GO) analyses revealed that these DEGs were enriched in multiple biological processes, including transcription regulation, plant hormone signal transduction, flower development, shoot system development regulation, meristem maintenance, internode patterning, organ growth, and metabolism processes (Figure 6C), suggesting that *VPB1* participated in a complex regulation network of rice inflorescence development.

Auxin signaling and transport have been reported to be important determinants of inflorescence development in *Arabidopsis* [34]. Our DEG analysis revealed that *VPB1* mainly participated in the auxin pathway and affected the genes related to meristem activity and inflorescence development. For example, genes *OsMADS1*, *OsMADS3*, *OsMADS6*, *OsMADS8*, and *OsMADS58* have been reported to interact with each other to promote flower development, which is very important for the maintenance of flower meristem identity and the formation of flower organ [44], and genes *GNP1*, *OsNPY2*, *SHAT1*, *FON1*, *ASP1*, *SHO1*, *OsSNB*, and *OsPIL1* are associated with the abscission zone development, meristem activity and fate, internode patterning, and inflorescence morphology [18,45–48]. To verify RNA-seq results, qPCR was used to analyze auxin pathway-related 7 genes and the above-mentioned 15 genes in the young panicle (2mm) of WT and *vpb1* plants. Our data indicated that the results of RNA-seq and qPCR were consistent, seven *ARFs* genes in the auxin pathway were strongly upregulated in *vpb1* mutant at young panicle stage (Figure 7A), and *MADS-box* genes and eight genes mainly involved in the maintenance of meristem activity were significantly different between wild type and *vpb1* mutant (Figure 7B). Taken together, RNA-seq results indicated that *VPB1* ensured the formation of normal panicle architecture by regulating the expression of the genes related to auxin pathways and inflorescence meristem development.

Results indicated that *VPB1* suppressed the expression of *OsBOP1*.

2.7. *VPB1 Negatively Regulates OsBOP1 Expression*

Evidence suggests that the BEL-type proteins regulate downstream target gene transcription by recognizing a core motif of these genes' promoters in *Arabidopsis* [49]. In our study, VPB1 encoded BLH proteins belonging to TALE family. Thus, to identify potential target genes of VPB1 protein, we downloaded TALE family binding core motifs (TFmatrixID_0278, Figure 8A) from PlantPAN 3.0 website [50], and we screened RNA-seq-obtained DEGs containing the core motifs from the upstream 2 kb promoter regions of DEGs with MEME FIMO [51]. The results revealed that a total of 682 DEGs with core motifs were screened, including 309 upregulated and 373 downregulated genes (Table S3). Since *VPB1* was transcriptional repressor, we further analyzed these 309 upregulated genes, and we found that genes *OsBOP* genes were related to meristem development. Therefore, we speculated that *OsBOPs* might be a potential target gene of *VPB1*.

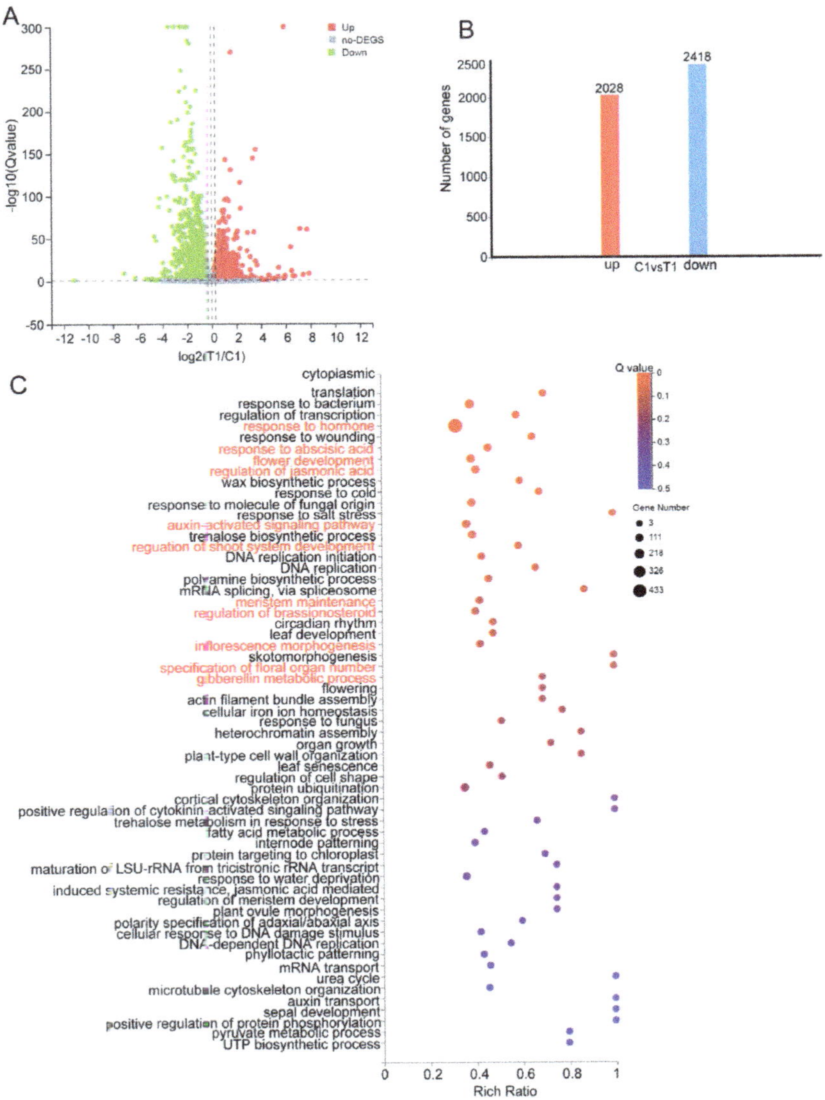

Figure 6. Differentially expressed genes statistics and GO analysis. (**A**) Volcano plots were used to visualize RNA sequencing (RNA—Seq) data. Each point corresponds to a reference sequence (Ref Seq) Gene. Red and green represent upregulated and downregulated genes in *vpb1* lines compared with WT. T1: Mutant treatment group, C1: Wild type control group. (**B**) Statistics of the number of differentially expressed genes. Red represents upregulated DEGs, blue represents downregulated DEGs. (**C**) Gene ontology (GO) analysis functional categories of genes that differed in abundance between *vpb1* and WT. Biological pathways related to hormone signaling and inflorescence architecture are indicated on the left and marked in red. Points of different color and size represent gene numbers.

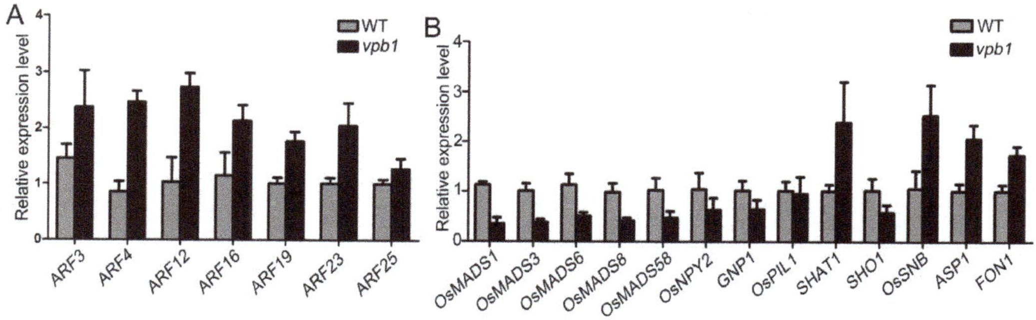

Figure 7. *VPB1* regulates genes involved in hormone signaling and meristem maintenance. (**A**) Expression analysis of the auxin responsive factor genes in WT and *vpb1* young panicles. Data are mean ± SD. (n = 3 biological replicates). (**B**) Expression analysis of the related to meristem maintenance genes in WT and *vpb1* young panicles. Data are mean ± SD. (n = 3 biological replicates).

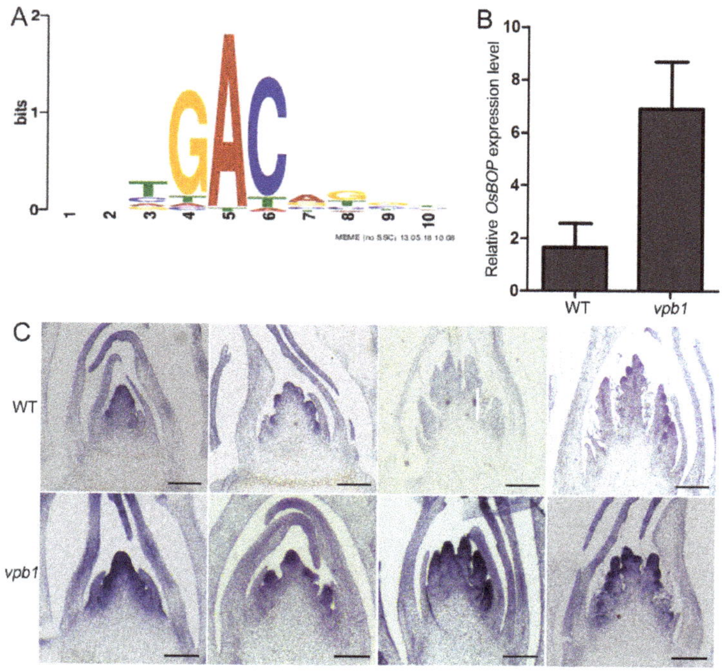

Figure 8. *VPB1* negatively regulates *OsBOP1* expression. (**A**) Putative TFBS for VPB1. (PlantPan3.0: http://plantpan.itps.ncku.edu.tw/index.html (accessed on 21 November 2020). (**B**) RT–qPCR analysis of *OsBOP1* expression in WT and *vpb1* young panicles (2 mm). Data are mean ± SD. (n = 3 biologically independent replicates). (**C**) In situ hybridization of *OsBOP1* mRNA in WT and *vpb1* in different stages of inflorescence development. Scale bars, 100 μm.

To test whether the *OsBOP* expression was directly regulated by VPB1 protein, we first compared the expression patterns of *OsBOP1* in WT and *vpb1*. The qRT-PCR analysis revealed that in young panicle (1–2 mm), the *OsBOP1* expression level was higher in *vpb1* than in WT (Figure 8B). Consistently, in situ hybridization experiments detected a broader expression of *OsBOP1* in the SAM, PBMs and SBMs in *vpb1* mutant plants than in WT plants with its expression expanded throughout the PBMs and SBMs (Figure 8C). These

Then, we examined the ability of VPB1 protein to bind to the promoter region of *OsBOPs* using electrophoretic mobility shift assay (EMSA). Promoter analysis revealed that *OsBOPs* contained TALE family core motif (Table S3). Thus, *OsBOP1* promoter fragment (50 bp) containing the 10-bp sequence CATGACAGAT and *OsBOP2* promoter fragment (50 bp) containing the 10-bp sequence TATGACAGAT were selected for EMSA. We constructed MBP protein and MBP-VPB1 fusion protein, and by using them, we detected the shifted bands which combined MBP-VPB1 fusion protein and the probes with CATGACAGAT and TATGACAGAT in the *OsBOPs* promoter region, but not the shifted bands of MBP protein (Figure 9A).

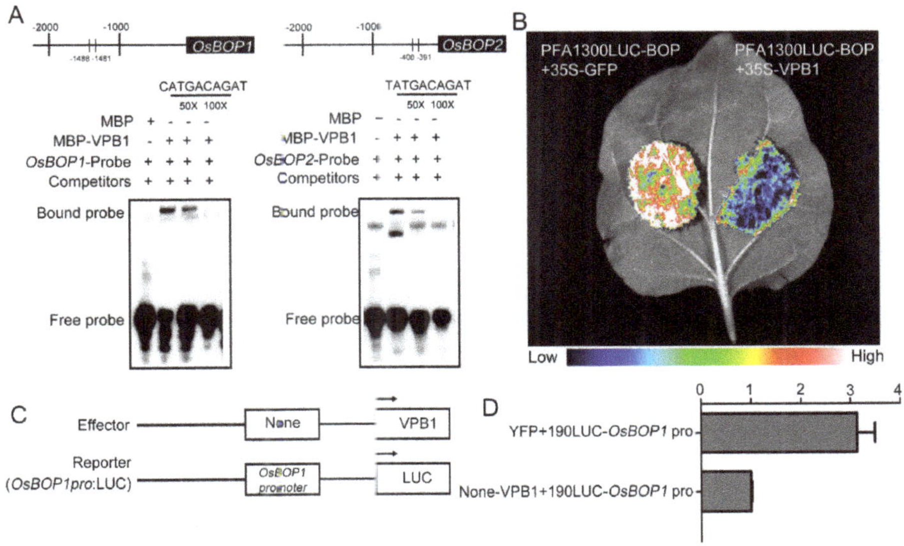

Figure 9. VPB1 is the transcriptional repressor of *OsBOP1*. (**A**) Schematic diagram of the OsBOP1/2 promoter showing the potential VPB1 binding sites and EMSA of MBP and MBP—VPB1 recombinant proteins incubated with biotin—labeled probes of *OsBOP1* and *OsBOP2*. Numbers above the diagram indicate the distance away from ATG. Competition for binding was performed using 50× and 250× competitive probes; MBP was used as a negative control. (**B**) Analysis of the binding ability of VPB1 with the *OsBOP1* promoter transiently expressed in tobacco leaves by transient expression regulation assays, showing that VPB1 protein suppresses the expression of *OsBOP1*. (**C**) Scheme of the constructs used in the protoplast dual luciferase reporter assays. (**D**) Dual luciferase reporter assays in rice protoplasts shows that the VPB1 protein suppresses the expression of LUC gene through binding to the *OsBOP1* promoter. Data are mean ± SD (n = 3 independent replicates).

Additionally, we attempted to confirm VPB1 binding ability in *Nicotiana benthamiana* leaves using transient expression assays. Strong signals were detected in tobacco leaves when pro*OsBOP1*: LUC was transformed, but only weak signals were detected when VPB1 protein was coexpressed with pro*OsBOP1*: LUC (Figure 9B). This result indicated that VPB1 could directly bind to the *OsBOP1* promoter to repress its expression. Finally, dual luciferase reporter assays in rice protoplasts showed that VPB1 could suppress the expression of LUC gene by binding to the *OsBOP1* promoter (Figure 9C,D). In addition, we created a double mutant *vpb1/osbop1*, and found that the morphology of *osbop1* single mutant plants was normal, but the *vpb1/osbop1* double mutant plants exhibited similar phenotype with the *vpb1* mutant plant, indicating inflorescence architecture defects caused by *vpb1* mutation were not rescued (Figure S8). Importantly, our data demonstrated that *VPB1* controled the inflorescence development by directly negatively regulating the expression of *OsBOP* genes.

3. Discussion

3.1. *VPB1* Regulates the Initiation and Arrangement of Primary Branch Meristems

The normal development of the primary branch meristems is important for the inflorescence architecture of rice [8]. Morphological analysis at the stage of primary branch development indicated that in *vpb1* mutant plants, the initiation timing and arrangement of the primary branch meristems were abnormal, that inflorescence meristem was damaged, and that the activity of the inflorescence meristem was reduced, resulting in the clustered primary branch meristems, but the secondary branch meristems and spikelets were less affected, suggesting that *VPB1* mainly maintained the activity of inflorescence meristem and regulated the phyllotactic pattern of the primary branches. Similarly, we found that *VPB1* was expressed in shoot apical meristem in the early stage of panicle development, and specifically expressed in the primary and secondary branch meristems. Moreover, *PNY* gene is essential for the formation of meristems and the maintenance of activity in *Arabidopsis* [23]. Collectively, these observations indicated that *VPB1* gene ensured the initiation and fine arrangement of the branch meristem by maintaining the activity of the inflorescence meristem.

3.2. *VPB1* Belongs to a Functionally Conserved Gene Family

Numerous previous studies have reported that *BLH* genes influence many aspects of plant morphology and typically maintain the meristem activity essential for organ formation [21,35,39,42], but little is known about their involvement in the regulation of panicle morphology development in rice. In this study, we isolated the key regulator *VPB1* encoding a BLH protein, and we found that the functional loss of *VPB1* resulted in clustered primary branches and short rachis. Thus, it could be concluded that *VPB1* played an important role in inflorescence formation. Positional cloning revealed that *VPB1* was identical to the previously reported *RI* gene which was identified as the ortholog of *Arabidopsis PNY* and maize *BLH12/14*, indicating that *VPB1* was involved in forming normal inflorescence architecture by regulating the phyllotactic pattern [30,35,39]. These findings supported that the BLH transcription factors had partially conserved functions in regulating the inflorescences in dicots and monocots.

3.3. The *VPB1* Gene Participates in a Complex Molecular Pathway to Regulate Panicle Development

TALE genes are well-known to play critical roles in regulating inflorescence architecture by affecting plant hormones [21]. For example, in *Arabidopsis*, *PNY* has been reported to directly target the auxin transport- and signaling pathway- related genes [34]. The mutually combined transcriptional regulators ETT, IND, BP, RPL, and SEU regulate the transcription of genes responsible for inflorescence development and auxin polar transport to facilitate proper auxin distribution in inflorescence in brassicaceae [52]. Our RNA-seq results showed that VPB1 was a powerful regulatory protein, and it significantly affected the genes related to the auxin, brassinosteroid (BR), abscisic acid, and gibberellin pathways (Figure 6C). Interestingly, *CPB1* (a new allele of *D11*) has been reported to encode a cytochrome protein P450 which is involved in BR biosynthesis pathway, and *cpb1* mutant plants also exhibit a clustered primary branch phenotype, compared to wild type plants [53]. Therefore, we guessed that *VPB1* might regulate the expression of *CPB1* gene during inflorescence development. We further analyzed the expression levels of auxin-related genes (*ARFs)* in WT and *vpb1* young panicles by qRT-PCR (Figure 7A). Our qRT-PCR results were consistent with RNA-seq data. Based these results, we speculated that the distribution or content of auxin in the *vpb1* mutant has changed, reducing the activity of the inflorescence meristem, and ultimately leading to the disorder of the initiation and arrangement of the branch meristem, the mechanism underlying *VPB1* regulation of branch arrangement in relation to auxin action is important issues to be resolved in our future studies.

Our data indicated the phenotype of the *vpb1* mutant plant might be caused by the reduced inflorescence meristem activity. Notably, our DEG analysis revealed that *VPB1* regulated multiple genes involved in the meristem identity maintenance and inflorescence development. The expressions of these genes exhibited significant difference between wild type and *VPB1* mutant (Figure 7B). The possible reason for such difference might lie in that the *VPB1* made these genes unable to be normally expressed in meristems, thus causing the failure in maintaining inflorescence meristem growth. Alternatively, the inhibition of inflorescence meristem activity might be associated with a change in cell wall components, as reported in *Arabidopsis* [31]. The regulation mechanism by which the change in cell wall components affects meristem activity remains to be further investigated in future studies.

3.4. VPB1 Regulates Inflorescence Development by Directly Binding to OsBOP1

This study indicated that *VPB1* was a transcriptional repressor. Our RNA-seq data of *vpb1* young panicle revealed that a total of 2028 genes were upregulated (Table S2). Of these upregulated genes, some genes were found to contain the conserved TALE core motifs, such as *OsBOP* genes. Previous studies have shown that *BOP1* and its highly homologous gene, *BOP2*, are involved in floral patterning, abscission zone formation, and bract suppression, and control of axillary bud growth and inflorescence development in plants [54–56]. Three *BOP* genes (*OsBOP1*, *OsBOP2*, and *OsBOP3*) in rice determine the leaf sheath: blade ratio by activating proximal sheath differentiation and suppressing distal blade differentiation, and these three genes are related to the microRNA156/SPL pathway [57]. Pioneering work in *Arabidopsis* has shown that *PNY* directly binds to *BOP1*, *BOP2*, and *KNAT6* to inhibit their expressions, eventually to regulate inflorescence development [49,58]. Our dual-luciferase reporter system and EMSA confirmed that the expressions of these genes were repressed by VPB1, and that the expression level of *OsBOP1* involved in the boundary organ initiation pathway was significantly upregulated in *vpb1* mutant young panicle. Consistently, RNA in situ hybridization assay indicated that *VPB1* suppressed expression of *OsBOPs*. However, inflorescence architecture defects caused by *vpb1* mutation were not rescued in the *vpb1/osbop1* double mutant plants. Previous research has shown that *OsBOP* genes in rice redundantly control leaf development [57]. Considering this, we speculated that three *OsBOP* genes might also redundantly control inflorescence architecture, in addition to regulating *OsBOP1*, *VPB1* might also regulate other downstream target genes to control panicle development. Based on these findings, it could be concluded that VPB1 protein could directly interact with the promoter of these *OsBOP* genes and suppress their transcriptions to maintain the normal development of inflorescence meristem. The genetic relationship between *VPB1* and *OsBOP* genes will be the focus of our future research.

3.5. The Role of BLH-KNOX Dimer Functions in Inflorescence Development

The interaction between BLH and KNOX homeobox proteins to form heterodimers has been widely reported [32,59], and these two proteins can form complexes and participate in meristem maintenance and the plant growth and development regulation. For example, PNY physically interacts with BP to form BP-PNY complex required by normal inflorescence architecture development [30]. Our study found VPB1 interacted with OSH1 and OSH15 (Figure S9), which are consistent with previously reported that SH5 can interact with OSH15 protein [38]. These findings indicated that the mechanism by which *BLH* and *KNOX* transcription factors regulated inflorescence architecture in rice was similar to that in *Arabidopsis*. Overall, VPB1 interacted with the typical genes of the KNOX family OSH1/OSH15 to form a protein complex, thus regulating panicle architecture development in rice.

Notably, the *vpb1* mutant identified in this study represents a new allele of the rice gene *SH5* regulating seed shattering [37,38]. Based on these results, we speculated that the *BLH* genes may play different roles and participate in different biological processes across rice varieties. Thus, identifying favorable alleles of *VPB1* will enrich our knowledge of panicle architecture in rice.

4. Materials and Methods

4.1. Plant Materials and Growth Conditions

This study used *Oryza sativa* subspecies *Japonica* "Zhonghua11" (ZH11) as rice materials. The mutant *vpb1* was derived from our T-DNA insertional mutant library (http://rmd.ncpgr.cn/ (accessed on 13 July 2016)). Plants were cultivated under natural long day (LD) conditions during the rice growing season in the experimental field of Huazhong Agriculture University, Wuhan, China, and they were moved to a greenhouse during the winter. All transgenic plants were grown under similar growth conditions.

4.2. Scanning Electron Microscopy

In scanning electron microscopy assay, young panicles from WT and *vpb1-1* mutants at different developmental stages were dissected, and immediately fixed in solution containing 70% ethanol, 3.7% formaldehyde, and 5% acetic acid for 24 h at 4 °C overnight. Tissues were dehydrated with a concentration series of ethanol from 25% to 100% and air-dried. After ethanol dehydration and drying, the samples were coated with gold by using an E-100 ion sputter, and then observed under a scanning electron microscope (S570, Hitachi, Tokyo, Japan).

4.3. Histological Sectioning

For paraffin sectioning, young panicles from wild type plant and *vpb1-1* mutant plant at different developmental stages were dissected. The samples were fixed in FAA solution at ratio of formaldehyde: glacial acetic acid: ethanol = 1:1:18, v/v/v at 4 °C for 24 h. Subsequently, the samples were dehydrated and cleared in a graded series of ethanol and xylene. The samples were microtome sectioned at the thickness of 5 μm. Afterwards, the sections were stained with 0.5% toluidine blue at room temperature for 30 min, and they were observed with a light microscope.

4.4. Map-Based Cloning of VPB1

To determine the *vpb1* locus, we crossed the *vpb1* mutant with *indica* variety Dular to obtain F_1 plants, and generated an F_2 mapping population through F_1 self-crossing. For rough mapping, 15 F_2 *vpb1* plants and 15 WT plants were used to establish two DNA pools. A total of 1200 independent individuals from the F_2 population were adopted for fine mapping. The five genes were screened from 38.5 kb regions between two genetic markers on the physical map. Genotyping analysis of the *vpb1* co-segregating population was performed by PCR with the primers *VPB1*-CS-P1 and *VPB1*-CS-P2. PCR was conducted as follows: pre-denaturation at 95 °C for 5 min, followed by 32 cycles of denaturation 95 °C for 45 s, annealing at 58 °C for 45 s, and extension at 72 °C for 1 min. Subsequently, PCR products were verified by sequencing.

4.5. Plasmid Construction and Rice Transformation

To prepare the complementation vector, we extracted ZH11 BAC clone OSJNA0075D23, and used PCR to amplify this clone into three fragments and obtained a about 10.6 kb foreign fragment consisting of the entire *VPB1* gene coding region, one 3 kb fragment in front of the ATG, and another 3 kb fragment behind the stop code. We connected this foreign fragment to the PCAMBIA2301 vector by the Gibson Assembly Master Mix (NEB, catalog, E2611L). For overexpression of *VPB1*, the full-length cDNA sequence of *VPB1* was amplified with primer pair *VPB1*-OX-F/*VPB1*-OX-R, and then cloned into pCAMBIA1301S by KpnI-XbaI digestion. For overexpression of *OsBOP1*, the full-length cDNA sequence of *OsBOP1* was amplified with primer pair *OsBOP1*-OX-F/*OsBOP1*-OX-R, and then cloned into pCAMBIA1301S by KpnI-BamHI digestion. Two 20-bp fragments targeting *LOC_Os05g38120* were designed to generate *VPB1* knockout mutants by using CRISPR/Cas9 vector system [40]. The target fragment was inserted into the binary vector pYLCRISPR/Cas9-MH. The above constructs were introduced into

Agrobacterium tumefaciens EHA105 and homozygous callus from *vpb1* mutuant plant and wild type plant (ZH11), as previously reported [60]. All the primers were listed in Table S4.

4.6. Total RNA Isolation and qRT-PCR Analyses

Total RNA was extracted with TRIzol reagent (Invitrogen, Shanghai, China). The 3 μg of RNA was treated with RNase-free DNaseI (Invitrogen). Subsequently, we synthesized first-strand cDNA with oligo (dT)18 primer (TaKaRa, Kyoto, Japan) and M-MLV reverse transcriptase (Invitrogen, Shanghai, China). The qRT-PCR was performed with SYBR Green Master MIX (Roche) in a total 10 μL reaction system on the Applied Biosystems ViiA 7 Real-Time PCR system according to the manufacturer's instructions. Data were normalized into the internal rice *ubiquitin* (*UBQ*) gene. The relative quantification method (2(-Delta Delta CT)) was used for data analysis. All primers were listed in Table S4.

4.7. In Situ Hybridization

Sample fixation and sectioning were performed as described above, followed by hybridization and immunological detection in the previously reported method [61]. The gene-specific primers were used to amplify the probes of *VPB1*, *OSH1*, and *OsBOP1* by PCR. The forward and reverse primers were fused with T7 and SP6 promoters, respectively. SP6 and T7 RNA polymerases were used to transcribe the antisense and sense probes in vitro, respectively, using the digoxigenin-labeled nucleotide mixture (Sigma-Aldrich, St. Louis, MO, USA).

4.8. Subcellular Localization

To construct the subcellular localization plasmids, primers *VPB1*-pM999-F and *VPB1*-pM999-R with KpnI-XbaI digestion sites were used to amplify the full-length cDNA of *VPB1*, and then amplified product was inserted into pM999-YFP vector. The obtained constructs were transformed into rice protoplasts isolated from two weeks etiolated seedlings and incubated at 23 °C for 12 ± 16 h. After incubation, the fluorescence of transformed protoplasts was observed with a confocal laser scanning microscope (TCS SP2; Leica, Weztlar, Germany).

4.9. Transcriptional Activity Analysis

Dual-Luciferase Reporter assay system (Promega, Madison, WI, USA) was used to analyze the transcriptional activity of VPB1 in rice protoplasts prepared from etiolated seedlings [62]. We used the GAL4-responsive vector as a reporter, which was produced by fusing the firefly *LUC* gene driven by the CaMV 35S promoter, five copies of the GAL4 binding site in tandem, and a minimal TATA box, and used the *Renilla luciferase* gene driven by *Arabidopsis thaliana UBIQUITIN3* promoter as internal control. The full-length coding sequence of *VPB1* was amplified using the primers GAL4BD-VPB1-F and GAL4BD-VPB1-R (Table S4) with EcoRI-SalI sites, and the amplified product was inserted into the vector that contained GAL4BD where it acted as an effector. In each transcriptional activity assay, we co-transformed the reporter, effector, and internal control into rice protoplasts in a ratio of 5:5:1 and incubated them at 23 °C for 12 ± 16 h. After incubation, the relative luciferase activity was measured in the DLR assay system with the TECAN Infinite M200 microplate reader.

To assess the specific binding ability of *OsBOP1* promoter, we prepared rice protoplasts from two-week-old fully green plant of ZH11 variety [63]. We inserted the coding sequence of *VPB1* into the NONE vector with the EcoRI-SalI sites to obtain an effector plasmid. Then, we amplified a 2000-bp upstream fragment of the *OsBOP1* promoter, and inserted the amplified product into 190-LUC vector with the HindIII sites to construct the *OsBOP1*:LUC reporter vector. The *Renilla luciferase* gene driven by CaMV 35S was used as internal control. In each transcriptional activity assay, we co-transformed 5 μg of effector plasmid DNA and 5 μg of reporter plasmid DNA into rice protoplasts. All primers were presented in Table S4.

4.10. RNA-Seq Analysis

We isolated total RNA from 2 mm young panicles of WT plants and vpb1 mutant plants. The experiment had three biological replicates. RNA-seq library was constructed and sequenced using DNBSeq at the Wuhan Genome Institute (BGI) (China). The clean reads were mapped to the rice reference genome (Os-Nipponbare-Refrence-IRGSP-1.0, MSU7) using Hisat2 (http://ccb.jhu.edu/software/hisat2/index.shtml (accessed on 27 October 2020). Q value $\leq$ 0.05 and fold-change (|Log2 ratio|) >1.5 were considered as statistically significantly different. The GO analysis of DEGs was performed using agriGO [64].

4.11. EMSA

Promoter *OsBOP1* with core motif CATGACAGAT and promoter *OsBOP2* with core motif TATGACAGAT were produced by annealing oligonucleotides with biotin 5′-end labeled *OsBOP1*-EMSA-F/R, *OsBOP2*-EMSA-MF/MR, respectively. In each reaction, we incubated 50 fmol biotin-labeled probes with the MBP-VPB1 protein in the binding buffer containing 10 μM$ZnCl_2$, 10 mM Tris, 50 mM KCl, 1 μg/μL poly (dI-dC), 1 mM DTT, 0.05% NP-40, and 0.1% BSA, 2.5% glycerol on ice for 30 min by using the LightShift Chemiluminescent EMSA kit. EMSA was performed as previously reported [61].

4.12. The Transient Expression System in Tobacco

To construct the tobacco transformation plasmids, primers PFA1300-BOP1-F/R with KpnI-SalI digestion sites were used to amplify a 2000-bp upstream fragment of the *OsBOP1* promoter, and then amplified product was inserted into PFA1300-LUC vector, primers 35S-GFP-VPB1-F/R with KpnI-BamHI digestion sites were used to amplify the full-length cDNA of *VPB1*, and then amplified product was inserted into 35S-CGFP vector. The vector combination 35S-CGFP-VPB1/PFA1300-LUC-BOP1, 35S-CGFP/PFA1300-LUC-BOP1 were transformed into *Nicotiana benthamiana* leaves by *Agrobacterium*, repeated three times, cultured at room temperature for two days, and the whole leaves were injected with 1 mM luciferase substrate. Images were visualized on Tanon-5200 Chemiluminescent Imaging System (Tanon Science and Technology).

4.13. Accession Numbers

Sequence data used in this study were downloaded from the Rice Genome Annotation Project website (MSU) and TAIR library. The accession numbers of genes were as follows: *VPB1* (*LOC_Os05g38120*), *qSH1* (*LOC_Os01g62920*), *OSH1* (*LOC_Os03g51690*), *OSH15* (*LOC_Os07g03770*), *OSH71* (*LOC_Os05g03884*), *OsBOP1* (*LOC_Os01g72020*), *OsBOP2* (*LOC_Os11g04600*), *OsBOP3* (*LOC_Os12g04410*), *PNY* (*AT5G02030*), *BOP1* (*AT3G57130*), and *BOP2* (*AT2G41370*).

5. Conclusions and Future Prospects

In conclusion, we show that a BELL-like homeodomain protein, VPB1, is involved in the regulation of panicle architecture in rice. *VPB1* loss-of-function mutants exhibited the clustered primary branch phenotype and the length of rachis was reduced. Map-based cloning revealed that *VPB1* is identical to previously reported *SH5/RI* gene [38,39]. While the *SH5/RI* gene and its protein as an interactor with KNOX protein with key roles in rice panicle development has been reported in gene expression studies, few studies provided a molecular mechanism of the panicle branching patterns in rice based on mutant analysis. This study fills that gap in knowledge and provides evidence that *VPB1* regulates the expression of related genes involved in inflorescence meristem development and auxin pathways, and directly inhibits the expression of lateral organ gene *OsBOP1*, maintaining the balance of inflorescence meristem and lateral meristem development, thereby ensuring the fine arrangement of panicle branches. Therefore, these results indicate that *VPB1* is a key gene for the normal arrangement of panicle branches in rice.

Indeed, VPB1 can interact with OSH1 and OSH15 to form heterodimers, indicating that VPB1 may regulate the panicle branching patterns in rice by functioning as heterodimers with KNOX proteins, but few of their functions have been identified. Is it possible to recruit more proteins after the formation of heterodimers to participate in the regulation of panicle development? We hope that their functions can be revealed in our future studies. In addition, VPB1, as a transcription factor with DNA binding ability, obtained 682 genes containing core sequences by analyzing the promoter sequences of the differential genes. What are its downstream target genes besides *OsBOP1* gene? It will be interesting to determine how *VPB1* genetically interacts with these genes to regulate rice panicle morphogenesis in future research.

Supplementary Materials: The following are available online at https://www.mdpi.com/article/10.3390/ijms22157909/s1.

Author Contributions: Conceptualization, C.W. and M.L.; Methodology, M.L., D.F. and T.X.; Validation, C.W. and M.L.; Formal analysis, C.W., M.L. and D.F.; Investigation, M.L.; Resources, D.F. and T.X.; Data curation, C.W. and M.L.; Writing—original draft preparation, M.L.; Writing—review and editing, C.W. and M.L.; Visualization, M.L.; Supervision, C.W.; Project administration, C.W.; Funding acquisition, C.W. All authors have read and agreed to the published version of the manuscript.

Funding: This work was supported by the National Natural Science Foundation of China (31630054, 31425018, 31821005), and Program for Chinese Outstanding Talents in Agricultural Scientific Research.

Institutional Review Board Statement: Not applicable.

Informed Consent Statement: Not applicable.

Data Availability Statement: The data presented in this study are available on request from the corresponding author.

Acknowledgments: Great gratitude goes to Ping Liu from Foreign Language College, Huazhong Agriculture University, Wuhan, China for her work at English editing and language polishing.

Conflicts of Interest: The authors declare no conflict of interest.

References

1. Benlloch, R.; Berbel, A.; Serrano-Mislata, A.; Madueno, F. Floral initiation and inflorescence architecture: A comparative view. *Ann. Bot.* **2007**, *100*, 659–676. [CrossRef]
2. Gao, X.-Q.; Wang, N.; Wang, X.-L.; Zhang, X.S. Architecture of wheat inflorescence: Insights from rice. *Trends Plant Sci.* **2019**, *24*, 802–809. [CrossRef]
3. McSteen, P.; Leyser, O. Shoot branching. *Annu. Rev. Plant Biol.* **2005**, *56*, 353–374. [CrossRef] [PubMed]
4. Wang, Y.; Li, J. Molecular basis of plant architecture. *Annu. Rev. Plant Biol.* **2008**, *59*, 253–279. [CrossRef]
5. Teo, Z.W.N.; Song, S.; Wang, Y.-Q.; Liu, J.; Yu, H. New insights into the regulation of inflorescence architecture. *Trends Plant Sci.* **2014**, *19*, 158–165. [CrossRef] [PubMed]
6. Bortiri, E.; Hake, S. Flowering and determinacy in maize. *J. Exp. Bot.* **2007**, *58*, 909–916. [CrossRef]
7. Ikeda, K.; Sunohara, H.; Nagato, Y. Developmental course of inflorescence and spikelet in rice. *Breed. Sci.* **2004**, *54*, 147–156. [CrossRef]
8. Itoh, J.-I.; Nonomura, K.-I.; Ikeda, K.; Yamaki, S.; Inukai, Y.; Yamagishi, H.; Kitano, H.; Nagato, Y. Rice plant development: From zygote to spikelet. *Plant Cell Physiol.* **2005**, *46*, 23–47. [CrossRef]
9. Xing, Y.; Zhang, Q. Genetic and molecular bases of rice yield. *Annu. Rev. Plant Biol.* **2010**, *61*, 421–442. [CrossRef] [PubMed]
10. Oikawa, T.; Kyozuka, J. Two-step regulation of LAX PANICLE1 protein accumulation in axillary meristem formation in rice. *Plant Cell* **2009**, *21*, 1095–1108. [CrossRef]
11. Tabuchi, H.; Zhang, Y.; Hattori, S.; Omae, M.; Shimizu-Sato, S.; Oikawa, T.; Qian, Q.; Nishimura, M.; Kitano, H.; Xie, H.; et al. LAX PANICLE2 of rice encodes a novel nuclear protein and regulates the formation of axillary meristems. *Plant Cell* **2011**, *23*, 3276–3287. [CrossRef]
12. Ikeda, K.; Ito, M.; Nagasawa, N.; Kyozuka, J.; Nagato, Y. Rice *ABERRANT PANICLE ORGANIZATION* 1, encoding an F-box protein, regulates meristem fate. *Plant J.* **2007**, *51*, 1030–1040. [CrossRef]
13. Ikeda-Kawakatsu, K.; Maekawa, M.; Izawa, T.; Itoh, J.-I.; Nagato, Y. *ABERRANT PANICLE ORGANIZATION 2/RFL*, the rice ortholog of Arabidopsis *LEAFY*, suppresses the transition from inflorescence meristem to floral meristem through interaction with *APO1*. *Plant J.* **2012**, *69*, 168–180. [CrossRef]

14. Suzaki, T.; Sato, M.; Ashikari, M.; Miyoshi, M.; Nagato, Y.; Hirano, H.-Y. The gene *FLORAL ORGAN NUMBER1* regulates floral meristem size in rice and encodes a leucine-rich repeat receptor kinase orthologous to *Arabidopsis* CLAVATA1. *Development* **2004**, *131*, 5649–5657. [CrossRef] [PubMed]
15. Suzaki, T.; Toriba, T.; Fujimoto, M.; Tsutsumi, N.; Kitano, H.; Hirano, H.-Y. Conservation and diversification of meristem maintenance mechanism in *oryza sativa*: Function of the *FLORAL ORGAN NUMBER2* gene. *Plant Cell Physiol.* **2006**, *47*, 1591–1602. [CrossRef]
16. Kwon, Y.; Yu, S.-I.; Park, J.-H.; Li, Y.; Han, J.-H.; Alavilli, H.; Cho, J.-I.; Kim, T.-H.; Jeon, J.-S.; Lee, B.-H. OsREL2, a rice TOPLESS homolog functions in axillary meristem development in rice inflorescence. *Plant Biotechnol. Rep.* **2012**, *6*, 213–224. [CrossRef]
17. Yoshida, A.; Ohmori, Y.; Kitano, H.; Taguchi-Shiobara, F.; Hirano, H.-Y. *ABERRANT SPIKELET AND PANICLE1*, encoding a TOPLESS-related transcriptional co-repressor, is involved in the regulation of meristem fate in rice. *Plant J.* **2012**, *70*, 327–339. [CrossRef]
18. Suzuki, C.; Tanaka, W.; Hirano, H.-Y. Transcriptional corepressor ASP1 and CLV-like signaling regulate meristem maintenance in Rice. *Plant Physiol.* **2019**, *180*, 1520–1534. [CrossRef] [PubMed]
19. Tanaka, W.; Ohmori, Y.; Ushijima, T.; Matsusaka, H.; Matsushita, T.; Kumamaru, T.; Kawano, S.; Hirano, H.-Y. Axillary meristem formation in rice requires the *WUSCHEL* ortholog *TILLERS ABSENT1*. *Plant Cell* **2015**, *27*, 1173–1184. [CrossRef] [PubMed]
20. Yoshida, A.; Sasao, M.; Yasuno, N.; Takagi, K.; Daimon, Y.; Chen, R.; Yamazaki, R.; Tokunaga, H.; Kitaguchi, Y.; Sato, Y.; et al. *TAWAWA1*, a regulator of rice inflorescence architecture, functions through the suppression of meristem phase transition. *Proc. Natl. Acad. Sci. USA* **2013**, *110*, 767–772. [CrossRef] [PubMed]
21. Hamant, O.; Pautot, V. Plant development: A TALE story. *C.R. Biol.* **2010**, *333*, 371–381. [CrossRef] [PubMed]
22. Bhatt, A.M.; Etchells, J.P.; Canales, C.; Lagodienko, A.; Dickinson, H. VAAMANA—a BEL1-like homeodomain protein, interacts with KNOX proteins BP and STM and regulates inflorescence stem growth in Arabidopsis. *Gene* **2004**, *328*, 103–111. [CrossRef] [PubMed]
23. Byrne, M.E.; Groover, A.T.; Fontana, J.R.; Martienssen, R.A. Phyllotactic pattern and stem cell fate are determined by the *Arabidopsis* homeobox gene *BELLRINGER*. *Development* **2003**, *130*, 3941–3950. [CrossRef]
24. Kanrar, S.; Bhattacharya, M.; Arthur, B.; Courtier, J.; Smith, H.M.S. Regulatory networks that function to specify flower meristems require the function of homeobox genes PENNYWISE and POUND-FOOLISH in Arabidopsis. *Plant J.* **2008**, *54*, 924–937. [CrossRef]
25. Khan, M.; Ragni, L.; Tabb, P.; Salasini, B.C.; Chatfield, S.; Datla, R.; Lock, J.; Kuai, X.; Despres, C.; Proveniers, M.; et al. Repression of lateral organ boundary genes by PENNYWISE and POUND-FOOLISH is essential for meristem maintenance and flowering in Arabidopsis. *Plant Physiol.* **2015**, *169*, 2166–2186. [CrossRef]
26. Roeder, A.H.K.; Ferrándiz, C.; Yanofsky, M.F. The role of the REPLUMLESS homeodomain protein in patterning the *Arabidopsis* fruit. *Curr. Biol.* **2003**, *13*, 1630–1635. [CrossRef] [PubMed]
27. Smith, H.M.S.; Campbell, B.C.; Hake, S. Competence to respond to floral inductive signals requires the homeobox genes *PENNYWISE* and *POUND-FOOLISH*. *Curr. Biol.* **2004**, *14*, 812–817. [CrossRef]
28. Ung, N.; Lal, S.; Smith, H.M.S. The role of PENNYWISE and POUND-FOOLISH in the maintenance of the shoot apical meristem in Arabidopsis. *Plant Physiol.* **2011**, *156*, 605–614. [CrossRef]
29. Yu, L.; Patibanda, V.; Smith, H.M.S. A novel role of BELL1-like homeobox genes, *PENNYWISE* and *POUND-FOOLISH*, in floral patterning. *Planta* **2008**, *229*, 693–707. [CrossRef]
30. Smith, H.M.S.; Hake, S. The interaction of two homeobox genes, *BREVIPEDICELLUS* and *PENNYWISE*, regulates internode patterning in the Arabidopsis inflorescence. *Plant Cell* **2003**, *15*, 1717–1727. [CrossRef]
31. Peaucelle, A.; Louvet, R.; Johansen, J.N.; Salsac, F.; Morin, H.; Fournet, F.; Belcram, K.; Gillet, F.; Hofte, H.; Laufs, P.; et al. The transcription factor BELLRINGER modulates phyllotaxis by regulating the expression of a pectin methylesterase in *Arabidopsis*. *Development* **2011**, *138*, 4733–4741. [CrossRef]
32. Kanrar, S.; Onguka, O.; Smith, H.M.S. *Arabidopsis* inflorescence architecture requires the activities of KNOX-BELL homeodomain heterodimers. *Planta* **2006**, *224*, 1163–1173. [CrossRef]
33. Rutjens, B.; Bao, D.; van Eck-Stouten, E.; Brand, M.; Smeekens, S.; Proveniers, M. Shoot apical meristem function in Arabidopsis requires the combined activities of three BEL1-like homeodomain proteins. *Plant J.* **2009**, *58*, 641–654. [CrossRef] [PubMed]
34. Bencivenga, S.; Serrano-Mislata, A.; Bush, M.; Fox, S.; Sablowski, R. Control of oriented tissue growth through repression of organ boundary genes promotes stem morphogenesis. *Dev Cell.* **2016**, *39*, 198–208. [CrossRef]
35. Tsuda, K.; Abraham-Juarez, M.-J.; Maeno, A.; Dong, Z.; Aromdee, D.; Meeley, R.; Shiroishi, T.; Nonomura, K.-i.; Hake, S. KNOTTED1 cofactors, BLH12 and BLH14, regulate internode patterning and vein anastomosis in maize. *Plant Cell* **2017**, *29*, 1105–1118. [CrossRef] [PubMed]
36. Konishi, S.; Izawa, T.; Lin, S.Y.; Ebana, K.; Fukuta, Y.; Sasaki, T.; Yano, M. An SNP caused loss of seed shattering during rice domestication. *Science* **2006**, *312*, 1392–1396. [CrossRef] [PubMed]
37. Yoon, J.; Cho, L.-H.; Kim, S.L.; Choi, H.; Koh, H.-J.; An, G. The BEL1-type homeobox gene *SH5* induces seed shattering by enhancing abscission-zone development and inhibiting lignin biosynthesis. *Plant J.* **2014**, *79*, 717–728. [CrossRef] [PubMed]
38. Yoon, J.; Cho, L.-H.; Antt, H.W.; Koh, H.-J.; An, G. KNOX protein OSH15 induces grain shattering by repressing lignin biosynthesis genes. *Plant Physiol.* **2017**, *174*, 312–325. [CrossRef] [PubMed]

39. Ikeda, T.; Tanaka, W.; Toriba, T.; Suzuki, C.; Maeno, A.; Tsuda, K.; Shiroishi, T.; Kurata, T.; Sakamoto, T.; Murai, M.; et al. BELL1-like homeobox genes regulate inflorescence architecture and meristem maintenance in rice. *Plant J.* **2019**, *98*, 465–478. [CrossRef]
40. Ma, X.; Liu, Y.G. CRISPR/Cas9-based multiplex genome editing in monocot and dicot plants. *Curr. Protoc. Mol. Biol* **2016**, *115*, 31–36. [CrossRef]
41. Gómez-Mena, C.; Sablowski, R. *ARABIDOPSIS THALIANA HOMEOBOX GENE1* establishes the basal boundaries of shoot organs and controls stem growth. *Plant Cell* **2008**, *20*, 2059–2072. [CrossRef] [PubMed]
42. He, L.; Liu, Y.; He, H.; Liu, Y.; Qi, J.; Zhang, X.; Li, Y.; Mao, Y.; Zhou, S.; Zheng, X.; et al. A molecular framework underlying the compound leaf pattern of *Medicago truncatula*. *Nat. Plants.* **2020**, *6*, 511–521. [CrossRef] [PubMed]
43. Chou, K.-C.; Shen, H.-B. Cell-PLoc: A package of Web servers for predicting subcellular localization of proteins in various organisms. *Nat. Protoc.* **2008**, *3*, 153–162. [CrossRef]
44. Hu, Y.; Liang, W.; Yin, C.; Yang, X.; Ping, B.; Li, A.; Jia, R.; Chen, M.; Luo, Z.; Cai, Q.; et al. Interactions of *OsMADS1* with floral homeotic genes in rice flower development. *Mol. Plant* **2015**, *8*, 1366–1384. [CrossRef]
45. Itoh, J.-I.; Kitano, H.; Matsuoka, M.; Nagato, Y. *SHOOT ORGANIZATION* genes regulate shoot apical meristem organization and the pattern of leaf primordium initiation in rice. *Plant Cell* **2000**, *12*, 2161–2174. [CrossRef]
46. Wu, Y.; Wang, Y.; Mi, X.-F.; Shan, J.-X.; Li, X.-M.; Xu, J.-L.; Lin, H.-X. The QTL *GNP1* encodes GA20ox1, which increases grain number and yield by increasing cytokinin activity in rice panicle meristems. *PLoS Genet.* **2016**, *12*, e1006386. [CrossRef] [PubMed]
47. He, Y.; Yan, L.; Ge, C.; Yao, X.-F.; Han, X.; Wang, R.; Xiong, L.; Jiang, L.; Liu, C.-M.; Zhao, Y. *PINOID* is required for formation of the stigma and style in rice. *Plant Physiol.* **2019**, *180*, 926–936. [CrossRef] [PubMed]
48. Jiang, L.; Ma, X.; Zhao, S.; Tang, Y.; Liu, F.; Gu, P.; Fu, Y.; Zhu, Z.; Cai, H.; Sun, C.; et al. The APETALA2-like transcription factor SUPERNUMERARY BRACT controls rice seed shattering and seed size. *Plant Cell* **2019**, *31*, 17–36. [CrossRef] [PubMed]
49. Andrés, F.; Romera-Branchat, M.; Martínez-Gallegos, R.; Patel, V.; Schneeberger, K.; Jang, S.; Altmüller, J.; Nürnberg, P.; Coupland, G. Floral induction in Arabidopsis thaliana by FLOWERING LOCUS T requires direct repression of *BLADE-ON-PETIOLE* genes by homeodomain protein PENNYWISE. *Plant Physiol.* **2015**, *169*, 2187–2199. [CrossRef] [PubMed]
50. Chow, C.-N.; Lee, T.-Y.; Hung, Y.-C.; Li, G.-Z.; Tseng, K.-C.; Liu, Y.-H.; Kuo, P.-L.; Zheng, H.-Q.; Chang, W.-C. PlantPAN3.0: A new and updated resource for reconstructing transcriptional regulatory networks from ChIP-seq experiments in plants. *Nucleic Acids Res.* **2019**, *47*, 1155–1163. [CrossRef]
51. Grant, C.E.; Bailey, T.L.; Noble, W.S. FIMO: Scanning for occurrences of a given motif. *Bioinformatics* **2011**, *27*, 1017–1018. [CrossRef]
52. Simonini, S.; Stephenson, P.; Østergaard, L. A molecular framework controlling style morphology in *Brassicaceae*. *Development* **2018**, *145*, dev158105. [CrossRef]
53. Wu, Y.; Fu, Y.; Zhao, S.; Gu, P.; Zhu, Z.; Sun, C.; Tan, L. *CLUSTERED PRIMARY BRANCH 1*, a new allele of *DWARF11*, controls panicle architecture and seed size in rice. *Plant Biotechnol. J.* **2015**, *14*, 377–386. [CrossRef] [PubMed]
54. Khan, M.; Xu, H.; Hepworth, S.R. BLADE-ON-PETIOLE genes: Setting boundaries in development and defense. *Plant Sci.* **2014**, *215*, 157–171. [CrossRef]
55. Jost, M.; Taketa, S.; Mascher, M.; Himmelbach, A.; Yuo, T.; Shahinnia, F.; Rutten, T.; Druka, A.; Schmutzer, T.; Steuernagel, B.; et al. A homolog of *Blade-On-Petiole 1* and *2* (*BOP1/2*) controls internode length and homeotic changes of the barley inflorescence. *Plant Physiol.* **2016**, *171*, 1113–1127. [CrossRef] [PubMed]
56. Xu, C.; Park, S.J.; van Eck, J.; Lippman, Z.B. Control of inflorescence architecture in tomato by BTB/POZ transcriptional regulators. *Genes Dev.* **2016**, *30*, 2048–2061. [CrossRef]
57. Toriba, T.; Tokunaga, H.; Shiga, T.; Nie, F.; Naramoto, S.; Honda, E.; Tanaka, K.; Taji, T.; Itoh, J.-I.; Kyozuka, J. *BLADE-ON-PETIOLE* genes temporally and developmentally regulate the sheath to blade ratio of rice leaves. *Nat. Commun.* **2019**, *10*, 619–632. [CrossRef] [PubMed]
58. Ragni, L.; Belles-Boix, E.; Günl, M.; Pautot, V. Interaction of *KNAT6* and *KNAT2* with *BREVIPEDICELLUS* and *PENNYWISE* in *Arabidopsis* inflorescences. *Plant Cell* **2008**, *20*, 888–900. [CrossRef]
59. Liu, Y.; You, S.; Taylor-Teeples, M.; Li, W.L.; Schuetz, M.; Brady, S.M.; Douglas, C.J. BEL1-LIKE HOMEODOMAIN6 and KNOTTED ARABIDOPSIS THALIANA7 interact and regulate secondary cell wall formation via repression of *REVOLUTA*. *Plant Cell* **2014**, *26*, 4843–4861. [CrossRef] [PubMed]
60. Wu, C.; Li, X.; Yuan, W.; Chen, G.; Kilian, A.; Li, J.; Xu, C.; Li, X.; Zhou, D.-X.; Wang, S.; et al. Development of enhancer trap lines for functional analysis of the rice genome. *Plant J.* **2003**, *35*, 418–427. [CrossRef] [PubMed]
61. Wang, L.; Sun, S.; Jin, J.; Fu, D.; Yang, X.; Weng, X.; Xu, C.; Li, X.; Xiao, J.; Zhang, Q. Coordinated regulation of vegetative and reproductive branching in rice. *Proc. Natl. Acad. Sci. USA* **2015**, *112*, 15504–15509. [CrossRef] [PubMed]
62. Hao, Y.-J.; Song, Q.-X.; Chen, H.-W.; Zou, H.-F.; Wei, W.; Kang, X.-S.; Ma, B.; Zhang, W.-K.; Zhang, J.-S.; Chen, S.-Y. Plant NAC-type transcription factor proteins contain a NARD domain for repression of transcriptional activation. *Planta* **2010**, *232*, 1033–1043. [CrossRef] [PubMed]
63. Bart, R.; Chern, M.; Park, C.J.; Bartley, L.; Ronald, P.C. A novel system for gene silencing using siRNAs in rice leaf and stem-derived protoplasts. *Plant Methods* **2006**, *2*, 13. [CrossRef] [PubMed]
64. Tian, T.; Liu, Y.; Yan, H.; You, Q.; Yi, X.; Du, Z.; Xu, W.; Su, Z. agriGO v2.0: A GO analysis toolkit for the agricultural community, 2017 update. *Nucleic Acids Res.* **2017**, *45*, 122–129. [CrossRef] [PubMed]

International Journal of
Molecular Sciences

MDPI

Review

Molecular and Genetic Aspects of Grain Number Determination in Rice (*Oryza sativa* L.)

Changxi Yin, Yanchun Zhu, Xuefei Li and Yongjun Lin *

National Key Laboratory of Crop Genetic Improvement, Huazhong Agricultural University, Wuhan 430070, China; yinchangxi@mail.hzau.edu.cn (C.Y.); zhuyanchun@webmail.hzau.edu.cn (Y.Z.); lixuefei@webmail.hzau.edu.cn (X.L.)

* Correspondence: yongjunlin@mail.hzau.edu.cn; Tel.: +86-27-87281719

Abstract: Rice grain yield is a complex trait determined by three components: panicle number, grain number per panicle (GNPP) and grain weight. GNPP is the major contributor to grain yield and is crucial for its improvement. GNPP is determined by a series of physiological and biochemical steps, including inflorescence development, formation of rachis branches such as primary rachis branches and secondary rachis branches, and spikelet specialisation (lateral and terminal spikelets). The molecular genetic basis of GNPP determination is complex, and it is regulated by numerous interlinked genes. In this review, panicle development and the determination of GNPP is described briefly, and GNPP-related genes that influence its determination are categorised according to their regulatory mechanisms. We introduce genes related to rachis branch development and their regulation of GNPP, genes related to phase transition (from rachis branch meristem to spikelet meristem) and their regulation of GNPP, and genes related to spikelet specialisation and their regulation of GNPP. In addition, we describe other GNPP-related genes and their regulation of GNPP. Research on GNPP determination suggests that it is possible to cultivate rice varieties with higher grain yield by modifying GNPP-related genes.

Keywords: grain number per panicle; grain yield; phase transition; rachis branch; rice panicle; spikelet specialisation

Citation: Yin, C.; Zhu, Y.; Li, X.; Lin, Y. Molecular and Genetic Aspects of Grain Number Determination in Rice (*Oryza sativa* L.). *Int. J. Mol. Sci.* **2021**, 22, 728. https://doi.org/10.3390/ijms22020728

Received: 24 December 2020
Accepted: 11 January 2021
Published: 13 January 2021

1. Introduction

Rice is one of the most important food crops and feeds half of the world's population [1,2]. Because of reductions in available land and the increasing global population, increasing the rice grain yield per unit area is crucial for food security, particularly in developing countries in Asia, such as in China and India [3].

Rice grain yield is primarily determined by three traits—grain number per panicle (GNPP), grain weight and number of panicles [4]. Because the rice grain yield per unit area is high, increasing the GNPP could further improve the grain yield [4–6]. Much research has been focused on GNPP determination, and considerable progress has been made in understanding the underlying regulatory mechanism. Here, we review progress in the molecular and genetic aspects of GNPP determination in rice.

2. Panicle Development and GNPP Determination in Rice

2.1. Panical Development

The rice panicle is composed of the main rachis, rachis branches (primary rachis branches (PRBs) and secondary rachis branches (SRBs)) and spikelets (lateral and terminal spikelets) (Figure 1) [3,7]. In a few rice varieties, there are tertiary rachis branches in the panicle. GNPP determination involves the development of the inflorescence, formation of rachis branches and spikelet specialisation. During rice panicle development, the inflorescence meristem (IM) is an important regulator of GNPP formation [8]. In rice, transition to the reproductive phase involves the transformation of the shoot apical meristem (SAM) into the IM, initiating the growth of several lateral meristems as PRBs. Next, the IM loses its activity, leaving a vestige at the base of the uppermost PRB. The PRB meristem produces next-order branches as lateral meristems. The few initially formed lateral meristems grow as SRBs and later meristems directly from spikelet meristems. Lateral spikelets differentiate directly from newly formed lateral meristems, and terminal spikelets are converted from rachis branch meristems. Therefore, three factors—rachis branch formation, the transition from rachis branch meristem to spikelet meristem and spikelet specialisation—determine the overall architecture of the panicle and the GNPP in rice [9].

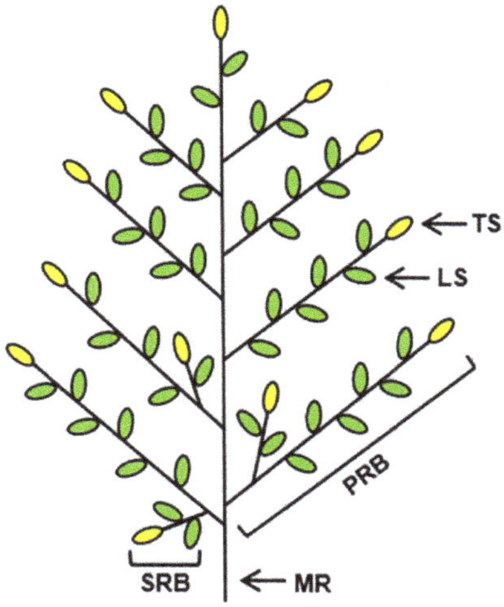

Figure 1. Panicle architecture of rice. The green ellipses show the lateral spikelets and the yellow ellipses show the terminal spikelets. LS, lateral spikelet; MR, main rachis; PRB, primary rachis branch; SRB, secondary rachis branch; TS, terminal spikelet.

2.2. GNPP Determination in Rice

Plant hormones, such as auxin, gibberellin (GA), cytokinin (CK), abscisic acid (ABA) and ethylene, are involved in regulating panicle development and GNPP in rice [10,11]. Auxin has a pivotal role in panicle development, as it is required for the initiation and maintenance of axillary meristems. Auxin is produced mainly in growing shoot apices and is transported basipetally down the site along specific transport routes through polar transport machinery. Consequently, disruption in auxin synthesis or auxin transport results in fewer rachis branches and reduced GNPP in rice [12–14]. GA can affect panicle-associated traits including panicle length, rachis branch number and GNPP in rice [15]. A previous study demonstrated that *OsCYP71D8L* controls panicle-related traits by regulating GA homeostasis. Gain-of-function of *OsCYP71D8L* leads to shorter panicles, fewer rachis branches, and reduced GNPP in rice [16]. It has been reported that the fine-tuning of bioactive CK level in the IM is a critical trait for controlling the number of rachis branches and GNPP in rice. The decreased level of bioactive CK in rice IM is usually accompanied by fewer rachis branches and reduced GNPP [17,18], and the weakened CK signalling in rice IM also results in fewer rachis branches and reduced GNPP [19,20]. This evidence suggests that CK positively regulates GNPP in rice. On the other hand, the stress hormones such as ABA and ethylene negatively regulate the GNPP in rice [11,21]. In addition, signalling cascades and responses of several plant hormones overlap, and the molecular components are often shared among them. A complex network of effectors of multiple hormonal pathways collide and communicate to regulate critical agronomic traits including GNPP in rice [10].

Increasing evidence indicates that plant hormones mediate GNPP determination mainly through the transcriptional or post-transcriptional regulation of GNPP-related genes in rice [10,22,23]. Additionally, GNPP-related genes control panicle development mainly by regulating three factors, including rachis branch formation, the transition from rachis branch meristem to spikelet meristem and spikelet specialisation [9,17,18].

3. Functional Classification of GNPP-Related Genes and Their Regulation of GNPP

GNPP-related genes control the GNPP mainly by regulating rachis branch formation, the transition from the rachis branch meristem to the spikelet meristem and spikelet specialisation. To date, numerous genes that control GNPP by regulating rachis branch formation [24–26], and several that regulate the transition from the rachis branch meristem to the spikelet meristem [9,27–30], have been investigated. However, few genes involved in regulating spikelet specialisation have been reported [31–33]. Typically, inhibition of rachis branch meristem formation and acceleration of the conversion from the rachis branch meristem to the spikelet meristem reduce the GNPP by decreasing the number of rachis branches and vice versa [30,32,34]. In addition, the inhibition of spikelet specialisation may reduce the GNPP by inhibiting spikelet formation and vice versa [32,35].

3.1. Rachis Branch Development-Related Genes and Their Regulation of GNPP

3.1.1. Positive Regulation of GNPP by Rachis Branch Development-Related Genes

LAX PANICLE (*LAX*) positively regulates the number of rachis branches and GNPP (Table 1). *LAX* encodes a plant-specific bHLH transcription factor. During rice panicle development, *LAX* is mainly expressed at the boundary region between the apical meristem and the newly formed lateral meristem, and it plays an important role in lateral meristem formation. Loss of function of *LAX* leads to a decreased number of rachis branches and a reduced GNPP. *LAX* cDNA is 1080 bp in length and contains one exon, encoding a protein of 215 amino acids. In the *lax-1* mutant, retrotransposon insertion causes a premature termination codon, and translation is terminated prematurely. In the *lax-2* mutant, there is a 36-kb deletion encompassing *LAX*. In the *lax-3* mutant, there is a 59-bp deletion in the bHLH domain of LAX. In the *lax-4* mutant, alanine 49 in the bHLH domain of LAX is changed to threonine. In the *lax-5* mutant, there is an arginine at position 50 of the bHLH domain of *LAX* [32]. Additionally, *lax-1*, *lax-4* and *lax-5* are *lax* mutants with mild changes in phenotype, with a reduced number of rachis branches, inhibited lateral spikelet development, normal terminal spikelet development and a reduced GNPP. By contrast, the panicles of severe *lax* mutants (*lax-2* and *lax-3*) have the main rachis but no rachis branches, and lateral spikelet development is completely blocked, whereas terminal spikelet development is inhibited [32]. Thus, the GNPP is reduced significantly in severe *lax* mutants. In a stronger allelic variant of the *lax* mutant, *lax1-2*, the initiation and maintenance of the rachis branch meristem, lateral spikelet meristem and terminal spikelet meristem are severely impaired. The transition from the partial rachis branch meristem to the spikelet meristem is delayed in the *lax1-2* mutant. The mild mutant *lax1-1* has no lateral spikelets but terminal spikelets and a small GNPP instead, whereas the severe mutant *lax1-2* has no spikelets and a few PRBs and SRBs [26].

Ideal Plant Architecture 1 (*IPA1*)/*WEALTHY FARMER'S PANICLE* (*WFP*)/*SOUAMOSA PROMOTER BINDING PROTEIN-LIKE 14* (*OsSPL14*) positively regulates the number of rachis branches and the GNPP (Table 1). *IPA1*/*WFP*/*OsSPL14* encodes a SOUAMOSA PROMOTER BINDING PROTEIN-LIKE protein. Rice varieties with *IPA1*/*WFP*/*OsSPL14* have an ideal plant architecture, including more rachis branches and a greater GNPP. Overexpression of *OsSPL14* increases the number of rachis branches and the GNPP [36]. Moreover, OsmiR156 interacts with and lyses OsSPL14. As a result of a point mutation, $OsSPL14^{ipa1}$ cannot be lysed by OsmiR156, resulting in a significant increase in the number of PRBs and SRBs and in the GNPP in panicles of $OsSPL14^{ipa1}$ plants [37]. Therefore, *IPA1*/*WFP*/*OsSPL14* positively regulates the number of rachis branches and the GNPP, and upregulating *IPA1*/*WFP*/*OsSPL14* or enhancing the IPA1/WFP/OsSPL14 protein level increases the number of rachis branches and the GNPP, resulting in an increased grain yield [36,37].

MONOCULM1(*MOC1*) positively regulates tiller number, rachis branch number and GNPP in rice (Table 1). The *MOC1* cDNA is 1666 bp in length and contains four exons. *MOC1* encodes a GRAS-family nuclear protein consisting of 441 amino acids with a VHIID motif and an SH2-like domain. *MOC1* is expressed mainly in the axillary buds, initiating the buds and promoting their outgrowth. In *moc1*, a 1.9-kb retrotransposon is inserted at position 948, causing a premature translation stop, resulting in a truncated 338-amino-acid fusion protein with the last 22 residues encoded by the retrotransposon sequence. The *moc1* mutant has only a main culm without tillers as a result of a defect in tiller bud formation [24]. Moreover, *MOC1* regulates the GNPP mainly by controlling IM activity and bract primordium initiation in rice. Therefore, in *moc1* mutant plants, the number of PRBs and SRBs and the GNPP are decreased by the reduced IM activity and inhibited rachis primordium initiation [24,38]. These findings indicate that *MOC1* positively regulates the tiller number, number of rachis branches and GNPP in rice.

Oryza sativa Homeobox1 (*OSH1*) positively regulates the number of rachis branches and the GNPP in rice (Table 1). *OSH1* cDNA is 1086 bp in length and contains five exons, encoding a protein product of 364 amino acids. *OSH1*, a rice homolog of *Knotted1-like Homeobox* (*KNOX*), a *KNOX* family class 1 homeobox gene, participates in SAM initiation and maintenance [39,40]. *OSH1* is self-induced and binds to five *KNOX* loci (*OSH1*, *OSH6*, *OSH15*, *OSH43* and *OSH71*), inducing their expression [22]. KNOX inhibits the expression of the GA biosynthesis gene *OsGA20ox* to reduce the GA content and promotes the expression of CK biosynthesis genes (including *OsIPT2* and *OsIPT3*) to increase the CK content [23]. KNOX induces SAM formation and maintains SAM activity by maintaining a high CK content and low GA content [22]. The self-induced expression of *OSH1* is necessary for SAM self-maintenance. Compared with the wild type, *osh1* mutant plants cannot maintain SAM activity, resulting in reduced numbers of PRBs and SRBs and a reduced GNPP [41].

Small Panicle (*SPA*) positively regulates the number of rachis branches and the GNPP (Table 1). *SPA* is functionally redundant with *LAX1* and regulates the rice axillary meristem. In *spa*, a *SPA* loss-of-function mutant, the rachis branches are short and abnormal, most base PRBs are missing and the numbers of SRBs and lateral spikelets are decreased significantly, resulting in a reduced number of rachis branches and a smaller GNPP. In a *spa* and *lax1-1* (a weak allelic mutant of *lax1*) double mutant, the panicle becomes a linear structure lacking all rachis branches. In addition, *spa* or *lax1* has little effect on tillering, which is almost completely inhibited in the *spa lax1* double mutant [32]. Therefore, *SPA* positively regulates the GNPP by promoting PRB, SRB and lateral spikelet formation.

Short Panicle 1 (*SP1*) positively regulates panicle size, the length of the main rachis and rachis branches, the number of rachis branches and GNPP (Table 1). *SP1* encodes a putative transporter of the peptide transporter (PTR) family. SP1 contains a conserved PTR2 domain consisting of 12 transmembrane domains, and the SP1-GFP fusion protein was found to be localised in the plasma membrane. *SP1* is highly expressed in the phloem of the rachis branches of young panicles and may control panicle size by regulating rachis growth. SP1 has been suggested to be a nitrate transporter. However, transport of neither nitrate nor any other compound transported by PTR proteins has been detected, suggesting that SP1 requires other component(s) to function as a transporter, or that it transports unknown substrates [25]. Compared with the wild type, panicle elongation in *sp1* plants (loss of function of SP1) is defective, resulting in a shorter panicle. Moreover, the number of SRBs and the GNPP are significantly reduced as a result of the decreased number of PRBs [25]. Therefore, *SP1* positively regulates the GNPP by regulating the growth of the main rachis and rachis branches in rice.

DENSE AND ERECT PANICLE1 (*DEP1*) positively regulates the number of rachis branches and the GNPP (Table 1). *DEP1* is located at a major quantitative trait locus (QTL) that controls rice grain yield. *DEP1*, a key gene with multiple functions including the control of rice grain yield, was isolated from Shennong 265, a super rice variety in northeast China. The dominant allele at the *DEP1* locus is a gain-of-function mutation causing truncation of a phosphatidylethanolamine-binding protein-like domain protein [42]. This allele increases meristem activity, increasing the number of rachis branches (PRBs and SRBs) and the GNPP, consequently increasing grain yield by 15–20% [42]. This allele is common to many high-yield rice varieties planted in a large area of China, suggesting that *DEP1* has played an important role in increasing rice grain yield in China.

OsNAC2/OMTN2/Ostil1 positively regulates the panicle length, rachis branch number and GNPP (Table 1). *OsNAC2/OMTN2/Ostil1* encodes a NAC transcription factor. *OsNAC2* is located at the same locus as *OMTN2* and *Ostil1* [43–45]. The microRNA miR164b interacts with and lyses OsNAC2. Overexpressing OErN (mutation of OsNAC2), which is not lysed by miR164b, leads to increased panicle length, stem thickness and number of vascular bundles in stem and leaves, resulting in increased numbers of PRBs and SRBs and an increased GNPP and grain yield [46]. *IPA1* and *DEP1* are significantly upregulated in *OErN* plants, and regulation by *OsNAC2* of the number of rachis branches and the GNPP may be related to CK signalling. *OsNAC2-RNAi* plants, transgenic rice plants with low expression of *OsNAC2*, have a shorter panicle, thinner stem, decreased number of vascular bundles in the stem and leaf, decreased numbers of PRBs and SRBs, and a smaller GNPP and thus have a lower grain yield. In addition, overexpression of miR164b in the transgenic rice OE164 resulted in the fragmentation of OsNAC2. Compared with wild-type plants, OE164 had shorter panicles, lower GNPP and a lower grain yield [46].

Grain Number4-1 (*GN4-1*) positively regulates the number of rachis branches, and *GN4-1* has a marked effect on the GNPP (Table 1) [47]. Insertion of a near-isogenic line (NIL; *GN4-1*) from Wuyunjing 8 into Zhonghui 8006 increases the number of PRBs and SRBs, the GNPP and the grain yield. Compared with Zhonghui 8006, the expression levels of *OsCKX2* (cytokinin oxidase) and other cytokinin oxidase genes (*OsCKX1, OsCKX4, OsCKX7, OsCKX8, OsCKX9, OsCKX10* and *OsCKX11*) were decreased significantly in NIL (*GN4-1*), elevating the contents of CKs (zeatin, zeatin riboside) and IM activity and increasing the number of PRBs and SRBs and the GNPP in NIL (*GN4-1*) [4]. *GN4-1* from Wuyunjing 8 promotes CK accumulation in rice inflorescences and increases the GNPP by 17% [4]. Therefore, *GN4-1* regulates the GNPP by controlling the CK content in rice inflorescences.

Grain Number per Panicle1 (*GNP1*) positively regulates the number of SRBs and the GNPP (Table 1). A T-DNA insertion mutant (*gnp1-D*) with enhanced *GNP1* expression exhibits increased plant height, more rachis branches and a greater GNPP. Compared with the NIL-*GNP1*^{LT} (isogenic control) line, the total and solid grain numbers in the NIL-*GNP1*^{TQ} line increased by 56% and 28%, respectively. In the NIL-*GNP1*^{TQ} line, although there was no obvious increase in PRB number, the SRB number increased significantly, significantly increasing grain yield at multiple experimental sites in China [48]. *GNP1* may upregulate *KNOX* expression, inducing the expression of the CK biosynthesis gene *OsIPT*, and enhance IM activity by increasing the CK content and enhancing CK signalling. Moreover, *GNP1* reduces the contents of active GAs (GA_1 and GA_3) by promoting their inactivation via the upregulation of *GA2ox* expression, which may positively regulate SRB number and GNPP [48]. Therefore, *GNP1* may enhance IM activity by enhancing CK signalling and suppressing GA signalling, thus increasing the SRB number and GNPP [48].

Grain Number per Panicle Gene4 (*Gnp4*)/*LAX PANICLE2* (*LAX2*) encodes a nuclear protein that regulates the formation of the axillary meristem and positively regulates the number of SRBs and the GNPP (Table 1). *Gnp4* is located within a 10.7-kb region in the long arm of rice chromosome 4. There is no sequence difference between the mutant and the wild type in this region, but there are differences in cytosine methylation levels in the CpG island region of the candidate gene promoter [49]. LAX2 is a nuclear protein with a plant-specific conserved domain that interacts with LAX1 and regulates axillary meristem formation in rice [50]. *Gnp4* is located at the same locus as *LAX2* [49,50]. The SRB of the *gnp4* mutant does not produce spikelets, and the SRB number and GNPP of the *gnp4 lax1-1* double mutant are significantly reduced [49]. The spikelets in the panicles of the *lax2* mutant are sparse, and the SRB number and GNPP of the *lax2-1* mutant are significantly reduced, although there is no difference in PRB number between *lax2-1* and wild-type plants. The decreases in rachis branch number and the GNPP are greater in the *lax1 lax2* double mutant than in either single mutant [50]. Therefore, *Gnp4/LAX2* positively regulates the number of SRBs and the GNPP in rice.

PLANT ARCHITECTURE AND YIELD1 (*PAY1*) positively regulates the SRB number and the GNPP (Table 1). Zhao et al. constructed a mutant library by mutating YIL55 (infiltration system of wild-type rice) via EMS mutagenesis and investigated a mutant (*PAY1*) with an upright and compact architecture. YIL55 had a shorter plant height, more tillers, a larger tiller angle, a lower GNPP and lower grain yield [51]. By contrast, *PAY1* mutants had a taller plant height, fewer tillers, a smaller tiller angle, thicker culm, more SRBs, a larger GNPP and lower grain yield but exhibited no significant changes in PRB number. Auxin polar transport activity is weakened in *PAY1*, and so *PAY1* may regulate rice phenotype by influencing auxin polar transport and altering the distribution of endogenous indole-3-acetic acid (the most important auxin in higher plants) [51]. Compared with wild-type plants, *PAY1-OE* plants had a thicker culm, fewer tillers, taller plant height, no significant change in PRB number, more SRBs, a larger GNPP and a higher grain yield. By contrast, *PAY1-RNAi* plants had a thinner culm, more tillers, shorter plant height, fewer SRBs, a lower GNPP and a lower grain yield [51]. Therefore, *PAY1* positively regulates GNPP by increasing the number of SRBs in rice.

LAX1, IPA1/WFP/OsSPL14, MOC1, OSH1, SPA, SP1, DEP1, OsNAC2 and *GN4-1* positively regulate the number of rachis branches and the GNPP. Loss of function of these genes reduces the number of PRBs and SRBs and the grain yield, whereas their upregulation or gain of function increases PRB and SRB numbers and the grain yield. By contrast, although *GNP1, GNP4/LAX2* and *PAY1* positively regulate the number of SRBs and the GNPP, they do not regulate the number of PRBs. However, the mechanisms by which these genes regulate the number of rachis branches and the GNPP are unclear, and some have not been cloned yet. Therefore, the positive regulation by these genes of the number of rachis branches and the GNPP warrants further investigation. It may be possible to increase the number of rachis branches and the GNPP using GNPP-related genes, thus improving grain yield.

Table 1. GNPP-related genes and their regulation of GNPP. GNPP, grain number per panicle; Phase transition, from rachis branch meristem to spikelet meristem; PRB, primary rachis branch; SRB, secondary rachis branch; −, negative regulation; +, positive regulation.

Gene	RGAP Locus ID	Protein Products	PRB	SRB	Phase Transition	Spikelet Specialisation	Lateral Spikelet	Terminal Spikelet	GNPP	References
LAX1	LOC_Os01g61480	bHLH transcription factor	+	+	+		+	+	+	[26,32]
IPA1/WFP/OsSPL14	LOC_Os08g39890	SOUAMOSA PROMOTER BINDING PROTEIN-LIKE protein	+	+					+	[36,37]
MOC1	LOC_Os06g40780	GRAS-family nuclear protein	+	+					+	[24,38]
OSH1	LOC_Os03g51690	A protein product of 364 amino acids	+	+					+	[22,39–41]
SPA	Not reported	Small panicle	+	+			+		+	[32]
SP1	LOC_Os11g12740	A putative transporter of the peptide transporter (PTR) family	+	+					+	[25]
DEP1	LOC_Os09g26999	G protein gamma subunit	+	+					+	[42]
OsNAC2/OMTN2/Ostil1	LOC_Os04g38720	NAC transcription factor	+	+					+	[43–46]
GN4-1	Not reported	Grain number4-1	+	+					+	[4,47]
GNP1	LOC_Os03g63970	GA20-oxidase 1		+					+	[48]
Gnp4/LAX2	LOC_Os04g32510	A nuclear protein with a plant-specific conserved domain		+					+	[49,50]
PAY1	LOC_Os08g31470	Peptidase		+					+	[51]
LP/EP3	LOC_Os02g15950	F-box protein	−	−					−	[52–54]
DEP3/OspPLAIIIδ	LOC_Os06g46350	Patatin-related phospholipase A	−	−					−	[55,56]
PAP2/OsMADS34	LOC_Os03g54170	SEP-like MADS box transcription factor	−	−		+			−	[34,57–60]
An-1	LOC_Os04g28280	bHLH protein	−	−					−	[61]
PROG1	LOC_Os07g05900	A 161-amino-acid Cys_2-His_2 zinc-finger protein	−	−					−	[62–64]
DST	LOC_Os03g57240	Zinc-finger transcription factor	−	−					−	[8,65]
GN2	Not reported	Gain number2		−					−	[5]
APO1/SCM2	LOC_Os06g45460	F-box protein of 429 amino acids	+	+	−				+	[9,27,28]

Table 1. *Cont.*

Gene	RGAP Locus ID	Protein Products	PRB	SRB	Phase Transition	Spikelet Specialisation	Lateral Spikelet	Terminal Spikelet	GNPP	References
RCN1	LOC_Os11g05470	ATP-binding cassette transporter		+	–				+	[29]
RCN2	LOC_Os02g32950	ATP-binding cassette transporter		+	–				+	[29]
TAW1	LOC_Os10g33780	Nuclear protein	+	+	–	–			+	[30]
FZP/BFL1	LOC_Os07g47330	ERF transcription factor		–	+	+			+	[33,35,66]
FZP2	Not reported	Frizzy panicle 2		–	+	+			+	[31]
GN1a/OsCKX2	LOC_Os01g10110	Cytokinin oxidase/dehydrogenase	–	–					–	[17]
LOG	LOC_Os01g40630	Cytokinin riboside 50-monophosphate phosphoribohydro-lase	+	+					+	[18]
PYL1 *PYL4* *PYL6*	LOC_Os01g61210 LOC_Os03g18600 LOC_Os05g39580	ABA receptor protein	–	–					–	[11]
Ghd7	LOC_Os07g15770	CCT(CO, CO-LIKE and TIMING OF CAB1)	+	+					+	[67–69]
Ghd8/DTH8/OsHAP3H/LHD1	LOC_Os08g07740	HAP3 subunit of the HAP (heterotrimeric haem activator) complex	+	+					+	[70–73]
DTH7/Ghd7.1/OsPRR37	LOC_Os07g49460	A pseudo-response regulator protein	+	+					+	[74–77]
GAD1	LOC_Os08g37890	A cysteine-rich secretory peptide							+	[78]
NOG1	LOC_Os01g54860	Enoyl-CoA hydratase/isomerase							+	[79]

3.1.2. Negative Regulation of GNPP by Rachis Branch Development-Related Genes

LARGER PANICLE (*LP*)/*ERECT PANICLE3* (*EP3*) negatively regulates the number of rachis branches and the GNPP (Table 1). *LP* encodes an F-box protein rich in Kelch, and in situ hybridisation showed that *LP* is mainly expressed in the rachis primordium [52]. The F-box protein encoded by *EP3* may function as a subunit of E3 ubiquitin ligase in the recognition and degradation of specific substrates [53]. *EP3* and *LP* are located at the same locus [52–54]. LP interacts with SKP1-like protein to upregulate the expression of *OsCKX2* and decrease the CK level in rice inflorescence, leading to more PRBs and SRBs and a higher grain yield [52]. Compared to wild-type plants, the number of rachis branches, particularly of PRBs, increased significantly in an *lp* (loss of function of *LP*) mutant, increasing the GNPP and grain yield in *lp* mutant plants [52]. In addition, *lp* mutant plants are more lodging resistant compared to wild-type plants [52].

DENSE AND ERECT PANICLE3 (*DEP3*)/*OspPLAIIIδ* negatively regulates the number of rachis branches and the GNPP (Table 1). *DEP3* and *OspPLAIIIδ* are located at the same locus [55,56]. Compared with wild-type *DEP3* gene, the *dep3* mutant allele loses 408 bp at LOC_Os06g46350, including the back 47 bp of the region encoding the third exon and the front 361 bp of the untranslated region (UTR) of the 3′-end [55]. The panicles of wild-type plants typically begin to droop after the flowering stage, but the panicles of *dep3* mutants remain upright from the flowering stage to the fully mature stage. Moreover, there are differences in the number of vascular bundles and the vascular bundle size, and in other phenotypic characteristics (including panicle length, rachis branch length and culm thickness) between *dep3* and wild-type plants. The *dep3* mutant plants have more vascular bundles in the upmost internode, a smaller vascular bundle, shorter panicle, shorter rachis branch and thicker culm [55]. In addition, compared with wild-type plants, *dep3* mutant plants have more PRBs and SRBs, a larger GNPP and a higher grain yield [55]. These results show that *DEP3* negatively regulates the number of rachis branches and the GNPP, thus increasing the grain yield.

PANICLE PHYTOMER 2 (*PAP2*)/*OsMADS34* negatively regulates the number of rachis branches and the GNPP (Table 1). *PAP2* encodes an SEP-like MADS box transcription factor. PAP2 is expressed only in the inflorescence meristem, PRB meristem, lateral spikelet meristem, floret meristem, glumes and degenerated glumes during the panicle development stage [34]. The PRB number, SRB number and GNPP of a *pap2-1* (*PAP2* loss of function) mutant increased significantly compared with those of wild-type plants. The panicle size of *pap2-1* is slightly smaller than that of wild-type plants because panicle elongation in the mutant is inhibited, but plant height, tillering number and leaf number do not differ significantly between *pap2-1* and wild-type plants [34]. *OsMADS34*, also known as *PAP2*, is a specific SEP-like MADS box gene in gramineous plants that regulates panicle morphology by controlling the rachis number and GNPP [34,57,58]. *OsMADS34* is expressed in the root, stem, leaf, leaf sheath, panicle, glume and degenerated glume and strongly expressed in developing organs such as young panicles. OsMADS34 determines the grain size and rice grain yield. In *osmads34-t* mutants, the numbers of PRBs and SRBs are increased but the panicle length is reduced, and the GNPP is increased but the grain size and seed-setting rate are reduced [59]. Six MADS-box genes—*OsMADS50*, *OsMADS56*, *OsMADS22*, *OSMADS47*, *OsMADS55* and *OsMADS34*—have been detected in rice [60]. These MADS-box genes significantly increase the number of rachis branches, including PRBs and SRBs, by inhibiting the expression of *REDUCED CULM NUMBER 4*. In addition, knockout of *OsMADS50*, *OsMADS56*, *OsMADS22*, *OsMADS47* and *OsMADS55* in an *osmads34* mutant significantly increased the number of PRBs and SRBs and the GNPP [60]. These findings indicate that *PAP2*/*OsMADS34* negatively regulates the number of rachis branches and the GNPP in rice.

Awn-1 (*An-1*) encodes a bHLH protein that positively regulates awn length and negatively regulates rachis branch number and GNPP in rice (Table 1). During spikelet development, *An-1* is first expressed in two degenerate glumes and two empty glume primordia, next in lemma primordia and palea primordia, and finally in stamens and carpel primordia. *An-1* expression gradually increases at the tip of the lemma primordium and is significantly enhanced in the awn primordium from the sixth to the eighth stage, gradually decreasing thereafter [61]. The *An-1* allele was introduced into Guanglu-Ai-4, an *indica* rice without awns, yielding the near-isogenic line NIL-*An-1*. The awn length and grain length increased, but the number of rachis branches and the GNPP decreased significantly in NIL-*An-1* compared with wild-type plants. Upregulation of *An-1* expression in inflorescences decreases IM activity, thus reducing the number of rachis branches and the GNPP. *An-1-OX*, a transgenic plant with high expression of *An-1*, has fewer PRBs and SRBs and a lower GNPP. By contrast, the number of rachis branches and the GNPP are increased in *An-1-RNAi* plants, in which *An-1* expression is knocked down [61].

PROSTRATE GROWTH1 (*PROG1*) positively regulates prostate growth and negatively regulates the number of PRBs and SRBs and the GNPP in rice (Table 1). The semi-dominant gene *PROG1* is located between RM298 and RM481, the short-arm simple sequence repeat markers of chromosome 7 [62]. *PROG1* has been isolated and cloned by two research teams [62,63]. The *PROG1* cDNA is 833 bp long and encompasses a 486-bp open reading frame, a 147-bp 5′ UTR and a 200-bp 3′ UTR. *PROG1* encodes a 161-amino-acid Cys_2-His_2 zinc-finger protein, which is mainly expressed in the axillary meristem [62,64]. There is a base mutation in the coding region of the gene in cultivated rice that causes an amino acid substitution, which may be selected during artificial domestication [63]. During the evolution of rice, *PROG1* of wild rice evolved into *prog1* of cultivated rice, resulting in the loss of function of *PROG1*, which mediates not only the transition from prostrate growth to erect growth but also changes in panicle architecture—e.g., increasing the numbers of PRBs and SRBs, the GNPP and the grain yield of cultivated rice. One hundred eighty-two varieties of cultivated rice, including *indica* and *japonica* cultivars from 17 countries, carry identical mutations in the *prog1* coding region, suggesting that *prog1* has become fixed during artificial domestication in rice [62]. Therefore, *PROG1* negatively regulates the number of rachis branches and the GNPP in rice.

DROUGHT AND SALT TOLERANCE (*DST*) negatively regulates the CK content, reducing the number of rachis branches and the GNPP in rice (Table 1). DST is a zinc-finger transcription factor in rice that directly regulates the expression of the cytokinin oxidase-encoding gene *OsCKX2*, thus increasing the number of rachis branches, GNPP and grain yield [8,65]. DST is mainly expressed in the SAM, PRB, SRB and young spikelets of the developing panicle [8]. DST promotes the expression of *OsCKX2*, thus reducing the CK content, SAM activity, number of rachis branches and GNPP. DST is a transcription factor with a C_2H_2 zinc-finger domain, by which DST proteins bind to DST-binding sequence (DBS) elements [42]. DBS elements are present in the promoter region of *OsCKX2* and other *OsCKX* genes [8]. In the semi-dominant mutant *reg1*, a single base insertion in *DST* led to premature termination of protein translation, resulting in the loss of the transcriptional activation ability of DST, accompanied by decreased expression of *OsCKX2* and other *OsCKX* genes, an increased CK content in the IM, increased numbers of PRBs and SRBs, and an enhanced GNPP in rice [8].

Grain Number2 (*GN2*) negatively regulates the number of rachis branches and the GNPP in rice (Table 1). Chen et al. inserted *GN2*, a gene from the wild rice Yuanjiang (*O. rufipogon* Griff.), into *indica* Teqing and obtained an introgression line (YIL9). Compared with Teqing, the GNPP, panicle length, PRB length and SRB number, but not the PRB number, of YIL9 were reduced [5]. Compared with wild-type plants, *GN2-OE*, a transgenic plant overexpressing *GN2*, had a reduced GNPP, shorter panicles and fewer SRBs [5]. Therefore, *GN2* negatively regulates grain yield mainly by reducing the number of SRBs and the GNPP in rice.

The above findings indicate that *LP/EP3*, *DEP3*, *PAP2/OsMADS34*, *An-1*, *PROG1* and *DST* negatively regulate the number of PRBs and SRBs and the GNPP in rice. Mutants with loss of function of these genes have more PRBs and SRBs and a larger GNPP compared to wild-type plants. *GN2* does not regulate the number of PRBs, but it negatively regulates the number of SRBs and the GNPP. However, the molecular mechanisms by which these genes regulate the formation of rachis branches and spikelets warrant further study. Elucidation of the molecular regulatory mechanisms of these genes will enable the breeding of rice varieties with more rachis branches, a larger GNPP and higher grain yield using molecular markers and gene editing.

3.2. Phase Transition (Rachis Branch Meristem to Spikelet Meristem)-Related Genes and Their Regulation of GNPP

ABERRANT PANICLE ORGANIZATION 1 (*APO1*)/*SCM2* plays an important role during the phase transition from rachis branch meristem to spikelet meristem and positively regulates the number of rachis branches and the GNPP in rice (Table 1). *APO1*/*SCM2* contains two exons and encodes an F-box protein of 429 amino acids. *APO1* is mainly expressed in the SAM and lateral organ primordia. *APO1* regulates the timing of meristem transition, and loss of function of *APO1* resulted in premature spikelet formation and an extended period of lodicule and carpel formation in an *apo1* mutant [27]. In plants with *apo1-D*, a gain-of-function mutation of *APO1*, IM activity is prolonged, and conversion from the rachis branch meristem to the lateral spikelet meristem and from the rachis branch meristem to the terminal spikelet meristem is delayed, resulting in a larger IM, more PRBs and SRBs, and higher grain yield than in wild-type plants [9]. By contrast, in plants with *apo1*, a loss-of-function mutation of *APO1*, IM activity is shortened, resulting in a smaller IM, fewer PRBs and SRBs, and a decreased GNPP compared to wild-type plants [27]. In addition, the QTL containing *SCM2* controls culm thickness. Mapping cloning results showed that *SCM2* is equivalent to *APO1*, and that NILs carrying *SCM2* (NIL-*SCM2*) exhibited increased stem strength, more tillers and a larger GNPP, indicating that *SCM2* is pleiotropic. Although *SCM2* is a gain-of-function mutant of *APO1*, there are differences in panicle architecture, including the GNPP and spikelet shape, between *SCM2* and *APO1* plants [28]. Compared with wild-type plants, the number of rachis branches and the GNPP in NIL-*SCM2* plants are obviously increased. However, the GNPP of *APO1-OE* plants is not larger than that of wild-type plants, and the reason why overexpression of *APO1* does not increase the GNPP is unclear. Therefore, *APO1*/*SCM2* plays an important role in controlling the phase transition from rachis branch meristem to spikelet meristem as well as regulating the number of rachis branches and the GNPP in rice [28].

RCN1 and *RCN2* increase the number of rachis branches and the GNPP by delaying the phase transition from rachis branch meristem to spikelet meristem (Table 1). *RCN1* and *RCN2* are rice TERMINAL FLOWER 1 /CENTRORADIALIS-like homologs, which regulate rice plant architecture (tiller number, plant height, panicle architecture) mainly by regulating meristem phase transition [29]. In *35S::RCN1* and *35S::RCN2* transgenic rice plants, phase transition to the reproductive stage is delayed, and transgenic rice plants have more tillers and denser panicles. Observation of the panicle structure revealed that the transition from the rachis branch meristem to the spikelet meristem is delayed, leading to the generation of higher-order rachis branches. Although there is no significant difference in PRB number between transgenic rice (*35S::RCN1* and *35S::RCN2*) plants and wild-type plants, the number of higher-order rachis branches, including SRBs and tertiary rachis branches, increases significantly, increasing the GNPP in *35S::RCN1* and *35S::RCN2* plants [29]. These findings indicate that *RCN1* and *RCN2* increase the number of rachis branches and the GNPP by delaying the phase transition to the spikelet meristem.

TAWAWA1 (*TAW1*) positively regulates the number of rachis branches and the GNPP by regulating the phase transition of the meristem (Table 1). *TAW1* encodes a nuclear protein of unknown function and is highly expressed in the SAM, IM and rachis branch meristem. TAW1 regulates inflorescence development by extending IM activity and delaying the phase transition from rachis branch meristem to spikelet meristem [30]. In the dominant gain-of-function mutant *tawawa1-D*, IM activity is extended and spikelet specialisation is delayed, resulting in delayed IM abortion and prolonged rachis branch formation, thus increasing the number of rachis branches, including SRBs and terminal rachis branches, and the GNPP. By contrast, in *TAW1-RNAi* transgenic plants, the decreased TAW1 activity causes precocious IM abortion and spikelet formation, resulting in the generation of small inflorescences with reduced numbers of PRBs and SRBs and a decreased GNPP [30].

Therefore, *RCN1*, *RCN2* and *TAW1* positively regulate the number of rachis branches and the GNPP by regulating the phase transition from rachis branch meristem to spikelet meristem. Overexpression of these genes and their dominant gain-of-function mutation may delay the phase transition from rachis branch meristem to spikelet meristem and prolong the formation of higher-order rachis branches, including SRBs and terminal rachis branches, thus increasing the GNPP. In addition, compared with wild-type plants, the dominant gain-of-function mutant *apo1-D* and NIL-*SCM2* have more rachis branches and a larger GNPP. Therefore, rational use of these genes can enable the breeding of rice varieties with a large number of rachis branches, greater GNPP and high grain yield.

3.3. Spikelet-Specialisation-Related Genes and Their Regulation of GNPP

FRIZZY PANICLE (*FZP*)/*BFL1* positively regulates spikelet specialisation and GNPP (Table 1). *FZP* is a single-copy gene in rice, located in chromosome 7, that encodes an ERF transcription factor and is the rice ortholog of maize *BD1*. FZP is required to prevent the formation of an axillary meristem instead of a spikelet meristem and for the subsequent establishment of spikelet meristem identity [33]. In a *fzp* mutant, instead of proceeding to spikelet formation, axillary meristems are formed in the axils of rudimentary glumes and are either arrested or develop into higher-order rachis branches, such as SRBs and terminal branches. Therefore, although the *fzp* mutant has more SRBs and terminal rachis branches, it has fewer spikelets and a reduced GNPP. In addition, there is no significant difference in PRB number between *fzp-11* mutant and wild-type plants, but *fzp-11* mutants have more SRBs and a lower GNPP compared with wild-type plants [35]. Therefore, *FZP* may positively regulate the GNPP by promoting spikelet specialisation in rice. Additionally, *FZP* and *BFL1* are located at the same locus [66]. *BFL1* encodes a transcription factor with an EREBP/AP2 domain, and BFL1 is involved in mediating spikelet specialisation in rice. The *bfl1* mutant harbours a single *Ds* insertion in the upstream region of *BFL1*, and *Ds* insertion drastically reduces the *BFL1* transcript level in the *bfl1* mutant. Compared to the wild type, the *bfl1* mutant has a similar PRB number and more higher-order rachis branches but fewer spikelets and a reduced GNPP as a result of defective spikelet formation [66].

FRIZZY PANICLE 2 (*FZP2*) positively regulates spikelet specialisation and GNPP (Table 1). *FZP2* plays an important role in spikelet specialisation. In *fzp2*, a loss-of-function mutant, rachis branch meristem activity is prolonged, and the phase transition from rachis branch meristem to spikelet meristem is delayed, significantly increasing the number of rachis branches and inhibiting lateral and terminal spikelet formation, resulting in a decreased GNPP in rice [31].

The above results indicate that *FZP*/*BFL1* and *FZP2* regulate GNPP by regulating spikelet specialisation. Compared with wild-type plants, although the number of higher-order rachis branches (PRBs and terminal rachis branches) increases, the number of spikelets and GNPP decrease in *fzp* and *fzp2*. Therefore, *FZP*/*BFL1* and *FZP2* play a key role in balancing the numbers of rachis branches and spikelets, and the regulation of these genes' functions increases rice grain yield by balancing the number of rachis branches and spikelets.

4. Other GNPP-Related Genes and Their Regulation of GNPP

Grain Number1a (*GN1a*)/*Cytokinin Oxidase2* (*OsCKX2*) negatively regulate GNPP by reducing the CK content. The *Gn1a* locus is the main QTL affecting GNPP in rice. A QTL-*Gn1* was identified in the short arm of chromosome 1 using an *indica* × *japonica* cross (Habataki/Koshihikari). Further, 96 F2 plants produced by heterozygous NIL-*Gn1* (*Gn1*/*gn1*) plants were used to divide *Gn1* into two loci—*Gn1a* and *Gn1b*—with similar functions. *Gn1a* is located in the region (less than 2 cM) between R3192 and C12072S, and *Gn1b* is located upstream of *Gn1a* [17]. Additionally, using 13,000 F2 plants generated from hybrid plants of NIL-*Gn1a* (*Gn1a*/*gn1a*), *Gn1a* was confirmed to be located in the

6.3-kb region between 3A28 and 3A20, where there is only one open reading frame, namely *OsCKX2*. Complementary transformation experiments showed that *Gn1a* is *OsCKX2* [17]. Decreased expression or loss of function of *OsCKX2* increases the GNPP and grain yield of rice. *OsCKX2* is highly expressed in the leaf, stem, IM and spikelet, weakly expressed in the SAM and not expressed in the root or embryo. The SAM is responsible for the development of the aboveground organs, such as leaves, stems and flowers, after embryo transfer. CK plays a key role in maintaining SAM activity. Decreased *OsCKX2* expression results in the accumulation of CK in the IM, increasing the number of rachis branches and GNPP (Table 1), and ultimately increasing the grain yield [17].

LONELY GUY (*LOG*) positively regulates GNPP by promoting CK biosynthesis. *LOG* encodes a CK-activating enzyme that mediates the final step of CK biosynthesis. *LOG* is expressed at the apex of the SAM, indicating that CK activation occurs in a specific developmental region. CK promotes the development of the SAM. Loss of function of *LOG* results in premature termination of SAM activity, decreasing the number of rachis branches and GNPP. LOG may regulate SAM activity by controlling the concentration and spatial distribution of CK [18].

Pyrabactin Resistance-Like (*PYL*) positively regulates ABA signalling and negatively regulates GNPP in rice. PYLs are ABA receptors implicated in ABA signal transduction [11]. Mutations in ABA receptor genes can promote rice growth and increase grain yield [11]. Rice *PYL* genes are divided into two groups. Group I includes *PYL1–PYL6* and *PYL12*, and Group II includes *PYL7–PYL11* and *PYL13*. CRISPR/Cas9 has been used to edit *PYL* genes. Polygenic mutations in Group I can promote the growth of rice, but no such effect has been found for Group II. Compared with the wild type, a *pyL1/4/6* mutant exhibited enhanced growth and a higher grain yield in paddy fields with significantly increased panicle length, number of PRBs and SRBs, and GNPP [11]. Therefore, PYLs regulate rice panicle development, possibly in a manner dependent on the ABA signal transduction pathway.

Grain Number, Plant Height, and Heading Date7 (*Ghd7*) is a major QTL that simultaneously controls the GNPP, plant height and heading stage of rice [67]. *Ghd7* encodes a nuclear protein with a CTT (CO-like and TIMING OF CAB1) domain [67]. Proteins containing this structural domain, such as *Arabidopsis CONSTANS* (*CO*) and rice *Hd1*, are implicated in the regulation of flowering time [68,69]. In the photoperiod-regulated flowering pathway, Ghd7 is located upstream of Ehd1 and Hd3a and acts via the Ghd7-Ehd1-Hd3A pathway. Ghd7 does not affect the expression of *Hd1*, but it does affect the expression of *Ehd1* and *Hd3a*. Under long-day conditions, *Ghd7* expression is upregulated, *Hd3a* expression is inhibited, and the heading date is delayed, enabling rice to make full use of light and temperature, thus increasing panicle length, plant height, GNPP and grain yield. In temperate regions, the short growth period of rice weakens or eliminates the function of Ghd7 in rice varieties in these regions, reducing or avoiding the effect of delayed heading date on rice grain yield. Therefore, the main function of *Ghd7* is to prolong the differentiation period of rice panicles and increase panicle length, thus enhancing the production of PRBs and SRBs and the GNPP as well as increasing the grain yield [67].

Grain Number, Plant Height, and Heading Date8 (*Ghd8*)/*DTH8*/*OsHAP3H*/*LHD1* regulates GNPP by adjusting the photoperiodic pathway in rice. *Ghd8, DTH8, LHD1* and *OsHAP3H* are located at the same locus [70–73]. *Ghd8* encodes the HAP3 subunit of the HAP (heterotrimeric haem activator) complex [71]. *Ghd8* expression is maintained at a high level in meristems at all developmental stages under long-day conditions but is low under short-day conditions. Ghd8 has bidirectional regulatory effects on the Ehd1-Hd3a pathway. Under short-day conditions, Ghd8 upregulates *Ehd1, Hd3a* and *RFT1*, resulting in premature heading and flowering in rice. Under long-day conditions, Ghd8 inhibits the expression of these three genes and delays the heading and flowering dates of rice, thus increasing the number of rachis branches and GNPP. *Ghd8* may increase the number of tillers, PRBs and SRB as well as the GNPP of rice by upregulating the expression of *MOC1* [71].

DTH7/Ghd7.1/OsPRR37 is a pleiotropic gene that controls heading date, plant height and GNPP in rice (Table 1). *DTH7/Ghd7.1/OsPRR37* encodes a pseudo-response regulator protein, and its expression is regulated by photoperiod. *OsPRR37*, *Ghd7.1* and *DTH7* are located at the same locus [74–77]. Under long-day conditions, DTH7 acts downstream of photosensitive pigment B and inhibits the expression of the anthocyanin genes *Hd3a* and *Ehd1* in rice, thus delaying flowering [76] and inducing an increase in the number of rachis branches, GNPP and grain yield per plant. *Ghd7.1* encodes pseudo-response REGULATOR37 (OsPRR37), which has a CCT domain [77]. *OsPRR37* is strongly expressed in the leaf and panicle, especially in the meristems of young panicles. Similar to *Ghd7*, under long-day conditions, *Ghd7.1* does not affect the expression of *Hd1*, but it does affect the expression of *Ehd1* and *Hd3a* [77]. In *indica* rice Zhenshan 97B, an eight-base deletion in *OsPRR37* leads to premature heading, shorter plant height, fewer rachis branches and a smaller GNPP [77]. Therefore, *DTH7/Ghd7.1/OsPRR37* positively regulates GNPP via the photoperiodic pathway in rice.

GRAIN NUMBER, GRAIN LENGTH AND AWN DEVELOPMENT1 (*GAD1*) negatively regulates GNPP (Table 1). *GAD1*, located in the long arm of chromosome 8 of rice, encodes a cysteine-rich secretory peptide and has greater homology with the EPIDERMAL PATTERNING FACTOR-LIKE family of *Arabidopsis thaliana* [78]. The GAD1 protein has a signal peptide site at the N terminus, and the mature peptide has conserved cysteine residues at the C terminus. Common wild rice (W2014) has *GAD1*, but cultivated rice 93-11 harbours *gad1*. Wild rice typically has a lower GNPP, longer grains and long awns atop the grains. GAD1 may reduce the CK content by activating the expression of *DST* and *OsCKX2*, thus decreasing the GNPP of wild rice [78]. By contrast, a code-shifting mutation of *GAD1* in cultivated rice destroys the conserved cysteine structure, leading to loss of function of *GAD1*, thereby increasing the GNPP, shortening grain length and inhibiting awn development [78].

NUMBER OF GRAINS 1 (*NOG1*) positively regulates the GNPP in rice (Table 1). In 2017, Chinese scientists cloned the *NOG1* gene, which is involved in the regulation of GNPP in rice. The introduction of *NOG1* increased grain yield by 25.8% in the *NOG1*-deficient rice cultivar Zhonghua 17, and overexpression of *NOG1* further increased grain yield by 19.5% in the *NOG1*-containing variety Teqing. *NOG1*, which encodes an enoyl-CoA hydratase/isomerase, increases the grain yield of rice by enhancing GNPP without reducing the number of panicles per plant or grain weight. Furthermore, *NOG1* plays important roles in regulating jasmonic acid homeostasis and the β-oxidation of fatty acids, which may be associated with its regulation of GNPP [79]. The mechanism by which *NOG1* regulates GNPP warrants further investigation.

5. Conclusions

GNPP, a grain-yield component in rice, is a hot research topic for breeders and molecular biologists. Several rice GNPP-related genes have been cloned, and their regulation of GNPP has been investigated. *LAX1*, *IPA1/WFP/OsSPL14*, *MOC1*, *OSH1*, *SPA*, *SP1*, *DEP1*, *OsNAC2* and *GN4-1* positively regulate the number of rachis branches and GNPP, whereas *GNP1*, *GNP4/LAX2* and *PAY1* positively regulate the number of SRBs and GNPP but not the number of PRBs (Table 1). By contrast, *LP/EP3*, *DEP3*, *PAP2/OsMADS34*, *An-1*, *PROG1* and *DST* negatively regulate the number of PRBs and SRBs and the GNPP in rice. *GN2* does not regulate the number of PRBs, but it negatively regulates the number of SRBs and the GNPP. Moreover, *RCN1*, *RCN2*, *TAW1* and *APO1/SCM2* positively regulate the number of rachis branches and the GNPP by regulating the phase transition from rachis branch meristem to spikelet meristem. In addition, *FZP/BFL1* and *FZP2* regulate the GNPP by regulating spikelet specialisation via balancing the numbers of rachis branches and spikelets (Table 1). Further, other GNPP-related genes and their regulation of GNPP have been reported (Table 1). However, the mechanisms by which some GNPP-related genes regulate GNPP determination are unclear in rice, and some have not been cloned yet. Therefore, the molecular regulatory networks of GNPP determination in rice need to be investigated. These molecular regulatory networks can be uncovered by constructing mutants and using molecular, genetic, physiological and -omics techniques, as well as bioinformatics. Furthermore, molecular marker-assisted selection, transgene techniques, gene editing and genome selection enable directional modification of GNPP-related genes and the polymerisation of favourable GNPP-related genes, allowing the cultivation of rice varieties with higher GNPP and grain yields.

Author Contributions: Writing—original draft preparation, C.Y.; writing—review and editing, C.Y., Y.Z. and X.L.; supervision, Y.L.; funding acquisition, C.Y. All authors have read and agreed to the published version of the manuscript.

Funding: This research was funded by the Fundamental Research Funds for the Central Universities (No. 2562018PY076), the National Key Research and Development Program of China (No. 2016YFD0300102) and the Science and Technology Major Projects of Guangxi Province of China (GuiKeChuang 010106).

Conflicts of Interest: The authors declare that they have no conflict of interest. The funders had no role in the design of the study; in the collection, analyses, or interpretation of data; in the writing of the manuscript, or in the decision to publish the results.

Abbreviations

ABA	Abscisic acid
An-1	*Awn-1*
APO1	*ABERRANT PANICLE ORGANIZATION 1*
CK	Cytokinin
DEP1	*DENSE AND ERECT PANICLE1*
DEP3	*DENSE AND ERECT PANICLE3*
DST	*DROUGHT AND SALT TOLERANCE*
EP3	*ERECT PANICLE3*
FZP	*FRIZZY PANICLE*
FZP2	*FRIZZY PANICLE 2*
GA	Gibberellin
GAD1	*GRAIN NUMBER, GRAIN LENGTH AND AWN DEVELOPMENT1*
Ghd7	*Grain Number, Plant Height, and Heading Date7*
Ghd8	*Grain Number, Plant Height, and Heading Date8*
GN1a	*Grain Number1a*
GN2	*Grain Number2*
GN4-1	*Grain Number4-1*
GNP1	*Grain Number per Panicle1*
Gnp4	*Grain Number per Panicle Gene4*
GNPP	Grain number per panicle
HAP	Heterotrimeric haem activator
LOG	*LONELY GUY*
IM	Inflorescence meristem
IPA1	*Ideal Plant Architecture 1*
KNOX	*Knotted1-like Homeobox*
LAX	*LAX PANICLE*
LAX2	*LAX PANICLE2*
LP	*LARGER PANICLE*
MOC1	*MONOCULM1*
NOG1	*NUMBER OF GRAINS 1*
OsCKX2	*Cytokinin oxidase2*
OSH1	*Oryza sativa Homeobox1*
PAP2	*PANICLE PHYTOMER 2*
PAY1	*PLANT ARCHITECTURE AND YIELD1*
PROG1	*PROSTRATE GROWTH1*
PRR37	Pseudo-response REGULATOR37
PTR	Peptide transporter

PYL	*Pyrabactin Resistance-Like*
QTL	Quantitative trait locus
SAM	Shoot apical meristem
SRBs	Secondary rachis branches
SP1	*Short Panicle 1*
SPA	*Small Panicle*
SPL14	*SOUAMOSA PROMOTER BINDING PROTEIN-LIKE 14*
TAW1	*TAWAWA1*
UTR	Untranslated region
WFP	*WEALTHY FARMER'S PANICLE*

References

1. Demont, M.; Stein, A.J. Global value of GM rice: A review of expected agronomic and consumer benefits. *New Biotechnol.* **2013**, *30*, 426–436. [CrossRef] [PubMed]
2. Xing, Y.; Zhang, Q. Genetic and molecular bases of rice yield. *Annu. Rev. Plant Biol.* **2010**, *61*, 421–442. [CrossRef] [PubMed]
3. Lu, H.; Shi, Z. Molecular research progress of rice panicle development. *Plant Physiol. J.* **2013**, *49*, 111–121.
4. Zhou, Y.; Tao, Y.; Yuan, Y.; Zhang, Y.; Miao, J.; Zhang, R.; Yi, C.; Gong, Z.; Yang, Z.; Liang, G. Characterisation of a novel quantitative trait locus, *GN4-1*, for grain number and yield in rice (*Oryza sativa* L.). *Theor. Appl. Genet.* **2018**, *131*, 637–648. [CrossRef]
5. Chen, H.; Tang, Y.; Liu, J.; Tan, L.; Jiang, J.; Wang, M.; Zhu, Z.; Sun, X.; Sun, C. Emergence of a novel chimeric gene underlying grain number in rice. *Genetics* **2017**, *205*, 993–1002. [CrossRef] [PubMed]
6. Chen, L.; Bian, J.; Shi, S.; Yu, J.; Khanzada, H.; Wassan, G.M.; Zhu, C.; Luo, X.; Tong, S.; Yang, X.; et al. Genetic analysis for the grain number heterosis of a super-hybrid rice WFYT025 combination using RNA-Seq. *Rice* **2018**, *11*, 37. [CrossRef]
7. Zhu, K.; Tao, H.; Min, C.; Yang, Y. Research progress on rice panicle development. *Mol. Plant Breed.* **2015**, *13*, 2109–2117. [CrossRef]
8. Li, S.; Zhao, B.; Yuan, D.; Duan, M.; Qian, Q.; Tang, L.; Wang, B.; Liu, X.; Zhang, J.; Wang, J.; et al. Rice zinc finger protein DST enhances grain production through controlling *Gn1a/OsCKX2* expression. *Proc. Natl. Acad. Sci. USA* **2013**, *110*, 3167–3172. [CrossRef]
9. Ikeda-Kawakatsu, K.; Yasuno, N.; Oikawa, T.; Iida, S.; Nagato, Y.; Maekawa, M.; Kyozuka, J. Expression level of *ABERRANT PANICLE ORGANIZATION1* determines rice inflorescence form through control of cell proliferation in the meristem. *Plant Physiol.* **2009**, *150*, 736–747. [CrossRef]
10. Deveshwar, P.; Prusty, A.; Sharma, S.; Tyagi, A.K. Phytohormone-Mediated molecular mechanisms involving multiple genes and QTL govern grain number in rice. *Front. Genet.* **2020**, *11*, 586462. [CrossRef]
11. Miao, C.; Xiao, L.; Hua, K.; Zou, C.; Zhao, Y.; Bressan, R.A.; Zhu, J.-K. Mutations in a subfamily of abscisic acid receptor genes promote rice growth and productivity. *Proc. Natl. Acad. Sci. USA* **2018**, *115*, 6058–6063. [CrossRef] [PubMed]
12. Yoshida, A.; Ohmori, Y.; Kitano, H.; Taguchi-Shiobara, F.; Hirano, H.Y. Aberrant spikelet and panicle1, encoding a TOPLESS-related transcriptional co-repressor, is involved in the regulation of meristem fate in rice. *Plant J.* **2012**, *70*, 327–339. [CrossRef] [PubMed]
13. Li, Y.; Zhu, J.; Wu, L.; Shao, Y.; Wu, Y.; Mao, C. Functional divergence of *PIN1* paralogous genes in rice. *Plant Cell Physiol.* **2019**, *60*, 2720–2732. [CrossRef] [PubMed]
14. Malik, N.; Ranjan, R.; Parida, S.K.; Agarwal, P.; Tyagi, A.K. Mediator subunit OsMED14_1 plays an important role in rice development. *Plant J.* **2020**, *101*, 1411–1429. [CrossRef]
15. Gao, S.; Chu, C. Gibberellin metabolism and signaling: Targets for improving agronomic performance of crops. *Plant Cell Physiol.* **2020**, *61*, 1902–1911. [CrossRef]
16. Zhou, J.; Li, Z.; Xiao, G.; Zhai, M.; Pan, X.; Huang, R.; Zhang, H. CYP71D8L is a key regulator involved in growth and stress responses by mediating gibberellin homeostasis in rice. *Science* **2020**, *71*, 1160–1170. [CrossRef]
17. Ashikari, M.; Sakakibara, H.; Lin, S.; Yamamoto, T.; Takashi, T.; Nishimura, A.; Angeles, E.R.; Qian, Q.; Kitano, H.; Matsuoka, M. Cytokinin oxidase regulates rice grain production. *Science* **2005**, *309*, 741–745. [CrossRef]
18. Kurakawa, T.; Ueda, N.; Maekawa, M.; Kobayashi, K.; Kojima, M.; Nagato, Y.; Sakakibara, H.; Kyozuka, J. Direct control of shoot meristem activity by a cytokinin-activating enzyme. *Nature* **2007**, *445*, 652–655. [CrossRef]
19. Sun, L.; Zhang, Q.; Wu, J.; Zhang, L.; Jiao, X.; Zhang, S.; Zhang, Z.; Sun, D.; Lu, T.; Sun, Y. Two rice authentic histidine phosphotransfer proteins, OsAHP1 and OsAHP2, mediate cytokinin signaling and stress responses in rice. *Plant Physiol.* **2014**, *165*, 335–345. [CrossRef]
20. Hirose, N.; Makita, N.; Kojima, M.; Kamada-Nobusada, T.; Sakakibara, H. Overexpression of a type-A response regulator alters rice morphology and cytokinin metabolism. *Plant Cell Physiol.* **2007**, *48*, 523–539. [CrossRef]
21. Wuriyanghan, H.; Zhang, B.; Cao, W.-H.; Ma, B.; Lei, G.; Liu, Y.-F.; Wei, W.; Wu, H.-J.; Chen, L.-J.; Chen, H.W.; et al. The ethylene receptor ETR2 delays floral transition and affects starch accumulation in rice. *Plant Cell* **2009**, *21*, 1473–1494. [CrossRef] [PubMed]
22. Sentoku, N.; Sato, Y.; Kurata, N.; Ito, Y.; Kitano, H.; Matsuoka, M. Regional expression of the Rice *KN1*-Type homeobox gene family during embryo, shoot, and flower development. *Plant Cell* **1999**, *11*, 1651–1664. [CrossRef] [PubMed]

23. Sakamoto, T.; Sakakibara, H.; Kojima, M.; Yamamoto, Y.; Nagasaki, H.; Inukai, Y.; Sato, Y.; Matsuoka, M. Ectopic expression of KNOTTED1-like homeobox protein induces expression of cytokinin biosynthesis genes in rice. *Plant Physiol.* **2006**, *142*, 54–62. [CrossRef] [PubMed]
24. Li, X.; Qian, Q.; Fu, Z.; Wang, Y.; Xiong, G.; Zeng, D.; Wang, X.; Liu, X.; Teng, S.; Hiroshi, F.; et al. Control of tillering in rice. *Nature* **2003**, *422*, 618–621. [CrossRef] [PubMed]
25. Li, S.; Qian, Q.; Fu, Z.; Zeng, D.; Meng, X.; Kyozuka, J.; Maekawa, M.; Zhu, X.; Zhang, J.; Li, J.; et al. *Short panicle1* encodes a putative PTR family transporter and determines rice panicle size. *Plant J.* **2009**, *58*, 592–605. [CrossRef] [PubMed]
26. Oikawa, T.; Kyozuka, J. Two-step regulation of *LAX PANICLE1* protein accumulation in axillary meristem formation in rice. *Plant Cell* **2009**, *21*, 1095–1108. [CrossRef] [PubMed]
27. Ikeda, K.; Ito, M.; Nagasawa, N.; Kyozuka, J.; Nagato, Y. Rice *ABERRANT PANICLE ORGANIZATION 1*, encoding an F-box protein, regulates meristem fate. *Plant J.* **2007**, *51*, 1030–1040. [CrossRef] [PubMed]
28. Ookawa, T.; Hobo, T.; Yano, M.; Murata, K.; Ando, T.; Miura, H.; Asano, K.; Ochiai, Y.; Ikeda, M.; Nishitani, R.; et al. New approach for rice improvement using a pleiotropic QTL gene for lodging resistance and yield. *Nat. Commun.* **2010**, *1*, 132. [CrossRef] [PubMed]
29. Nakagawa, M.; Shimamoto, K.; Kyozuka, J. Overexpression of *RCN1* and *RCN2*, rice *TERMINAL FLOWER 1/CENTRORADIALIS* homologs, confers delay of phase transition and altered panicle morphology in rice. *Plant J.* **2002**, *29*, 743–750. [CrossRef] [PubMed]
30. Yoshida, A.; Sasao, M.; Yasuno, N.; Takagi, K.; Daimon, Y.; Chen, R.; Yamazaki, R.; Tokunaga, H.; Kitaguchi, Y.; Sato, Y.; et al. *TAWAWA1*, a regulator of rice inflorescence architecture, functions through the suppression of meristem phase transition. *Proc. Natl. Acad. Sci. USA* **2013**, *110*, 767–772. [CrossRef]
31. Komatsu, M.; Maekawa, M.; Shimamoto, K.; Kyozuka, J. The *LAX1* and *FRIZZY PANICLE 2* genes determine the inflorescence architecture of rice by controlling rachis-branch and spikelet development. *Dev. Biol.* **2001**, *231*, 364–373. [CrossRef] [PubMed]
32. Komatsu, K.; Maekawa, M.; Ujiie, S.; Satake, Y.; Furutani, I.; Okamoto, H.; Shimamoto, K.; Kyozuka, J. *LAX* and *SPA*: Major regulators of shoot branching in rice. *Proc. Natl. Acad. Sci. USA* **2003**, *100*, 11765–11770. [CrossRef]
33. Komatsu, M.; Chujo, A.; Nagato, Y.; Shimamoto, K.; Kyozuka, J. *FRIZZY PANICLE* is required to prevent the formation of axillary meristems and to establish floral meristem identity in rice spikelets. *Development* **2003**, *130*, 3841–3850. [CrossRef] [PubMed]
34. Kobayashi, K.; Maekawa, M.; Miyao, A.; Hirochika, H.; Kyozuka, J. *PANICLE PHYTOMER2* (*PAP2*), encoding a SEPALLATA subfamily MADS-box protein, positively controls spikelet meristem identity in rice. *Plant Cell Physiol.* **2010**, *51*, 47–57. [CrossRef] [PubMed]
35. Bai, X.; Huang, Y.; Mao, D.; Wen, M.; Zhang, L.; Xing, Y. Regulatory role of *FZP* in the determination of panicle branching and spikelet formation in rice. *Sci. Rep.* **2016**, *6*, 19022. [CrossRef] [PubMed]
36. Miura, K.; Ikeda, M.; Matsubara, A.; Song, X.-J.; Ito, M.; Asano, K.; Matsuoka, M.; Kitano, H.; Ashikari, M. *OsSPL14* promotes panicle branching and higher grain productivity in rice. *Nat. Genet.* **2010**, *42*, 545–549. [CrossRef] [PubMed]
37. Jiao, Y.; Wang, Y.; Xue, D.; Wang, J.; Yan, M.; Liu, G.; Dong, G.; Zeng, D.; Lu, Z.; Zhu, X.; et al. Regulation of *OsSPL14* by OsmiR156 defines ideal plant architecture in rice. *Nat. Genet.* **2010**, *42*, 541–544. [CrossRef]
38. Shao, G.; Lu, Z.; Xiong, J.; Wang, B.; Jing, Y.; Meng, X.; Liu, G.; Ma, H.; Liang, Y.; Chen, F.; et al. Tiller bud formation regulators MOC1 and MOC3 cooperatively promote tiller bud outgrowth by activating *FON1* expression in rice. *Mol. Plant* **2019**, *12*, 1090–1102. [CrossRef]
39. Postma-Haarsma, A.D.; Verwoert, I.I.G.S.; Stronk, O.P.; Koster, J.; Lamers, G.E.M.; Hoge, J.H.C.; Meijer, A.H. Characterization of the KNOX class homeobox genes *Oskn2* and *Oskn3* identified in a collection of cDNA libraries covering the early stages of rice embryogenesis. *Plant Mol. Biol.* **1999**, *39*, 257–271. [CrossRef]
40. Sato, Y.; Hong, S.K.; Tagiri, A.; Kitano, H.; Yamamoto, N.; Nagato, Y.; Matsuoka, M. A rice homeobox gene, *OSH1*, is expressed before organ differentiation in a specific region during early embryogenesis. *Proc. Natl. Acad. Sci. USA* **1996**, *93*, 8117–8122. [CrossRef]
41. Tsuda, K.; Ito, Y.; Sato, Y.; Kurata, N. Positive autoregulation of a *KNOX* gene is essential for shoot apical meristem maintenance in rice. *Plant Cell* **2011**, *23*, 4368–4381. [CrossRef] [PubMed]
42. Huang, X.; Qian, Q.; Liu, Z.; Sun, H.; He, S.; Luo, D.; Xia, G.; Chu, C.; Li, J.; Fu, X. Natural variation at the *DEP1* locus enhances grain yield in rice. *Nat. Genet.* **2009**, *41*, 494–497. [CrossRef] [PubMed]
43. Chen, X.; Lu, S.; Wang, Y.; Zhang, X.; Lv, B.; Luo, L.; Xi, D.; Shen, J.; Ma, H.; Ming, F. *OsNAC2* encoding a NAC transcription factor that affects plant height through mediating the gibberellic acid pathway in rice. *Plant J.* **2015**, *82*, 302–314. [CrossRef] [PubMed]
44. Fang, Y.; Xie, K.; Xiong, L. Conserved miR164-targeted NAC genes negatively regulate drought resistance in rice. *J. Exp. Bot.* **2014**, *65*, 2119–2135. [CrossRef] [PubMed]
45. Mao, C.; Ding, W.; Wu, Y.; Yu, J.; He, X.; Shou, H.; Wu, P. Overexpression of a NAC-domain protein promotes shoot branching in rice. *New Phytol.* **2007**, *176*, 288–298. [CrossRef] [PubMed]
46. Jiang, D.; Chen, W.; Dong, J.; Li, J.; Yang, F.; Wu, Z.; Zhou, H.; Wang, W.; Zhuang, C. Overexpression of miR164b-resistant OsNAC2 improves plant architecture and grain yield in rice. *J. Exp. Bot.* **2018**, *69*, 1533–1543. [CrossRef]
47. Deshmukh, R.; Singh, A.; Jain, N.; Anand, S.; Gacche, R.; Singh, A.; Gaikwad, K.; Sharma, T.; Mohapatra, T.; Singh, N. Identification of candidate genes for grain number in rice (*Oryza sativa* L.). *Funct. Integr. Genom.* **2010**, *10*, 339–347. [CrossRef]

48. Wu, Y.; Wang, Y.; Mi, X.-F.; Shan, J.-X.; Li, X.-M.; Xu, J.-L.; Lin, H.-X. The QTL *GNP1* encodes GA20ox1, which increases grain number and yield by increasing cytokinin activity in rice panicle meristems. *PLoS Genet.* **2016**, *12*, e1006386. [CrossRef]
49. Zhang, Z.; Li, J.; Yao, G.; Zhang, H.; Dou, H.; Shi, H.; Sun, X.; Li, Z. Fine mapping and cloning of the grain number per-panicle gene (*Gnp4*) on chromosome 4 in rice (*Oryza sativa* L.). *Agric. Sci. China* **2011**, *10*, 1825–1833. [CrossRef]
50. Tabuchi, H.; Zhang, Y.; Hattori, S.; Omae, M.; Shimizu-Sato, S.; Oikawa, T.; Qian, Q.; Nishimura, M.; Kitano, H.; Xie, H.; et al. *LAX PANICLE2* of rice encodes a novel nuclear protein and regulates the formation of axillary meristems. *Plant Cell* **2011**, *23*, 3276–3287. [CrossRef]
51. Zhao, L.; Tan, L.; Zhu, Z.; Xiao, L.; Xie, D.; Sun, C. *PAY1* improves plant architecture and enhances grain yield in rice. *Plant J.* **2015**, *83*, 528–536. [CrossRef] [PubMed]
52. Li, M.; Tang, D.; Wang, K.; Wu, X.; Lu, L.; Yu, H.; Gu, M.; Yan, C.; Cheng, Z. Mutations in the F-box gene *LARGER PANICLE* improve the panicle architecture and enhance the grain yield in rice. *Plant Biotechnol. J.* **2011**, *9*, 1002–1013. [CrossRef] [PubMed]
53. Yu, H.; Murchie, E.H.; González-Carranza, Z.H.; Pyke, K.A.; Roberts, J.A. Decreased photosynthesis in the *erect panicle 3* (*ep3*) mutant of rice is associated with reduced stomatal conductance and attenuated guard cell development. *J. Exp. Bot.* **2015**, *66*, 1543–1552. [CrossRef] [PubMed]
54. Piao, R.; Jiang, W.; Ham, T.-H.; Choi, M.-S.; Qiao, Y.; Chu, S.-H.; Park, J.-H.; Woo, M.-O.; Jin, Z.; An, G.; et al. Map-based cloning of the *ERECT PANICLE 3* gene in rice. *Theor. Appl. Genet.* **2009**, *119*, 1497–1506. [CrossRef]
55. Qiao, Y.; Piao, R.; Shi, J.; Lee, S.-I.; Jiang, W.; Kim, B.-K.; Lee, J.; Han, L.; Ma, W.; Koh, H.-J. Fine mapping and candidate gene analysis of dense and *erect panicle 3*, *DEP3*, which confers high grain yield in rice (*Oryza sativa* L.). *Theor. Appl. Genet.* **2011**, *122*, 1439–1449. [CrossRef]
56. Liu, G.; Zhang, K.; Ai, J.; Deng, X.; Hong, Y.; Wang, X. Patatin-related phospholipase A, pPLAIIIα, modulates the longitudinal growth of vegetative tissues and seeds in rice. *J. Exp. Bot.* **2015**, *66*, 6945–6955. [CrossRef]
57. Zahn, L.M.; Kong, H.; Leebens-Mack, J.H.; Kim, S.; Soltis, P.S.; Landherr, L.L.; Soltis, D.E.; de Pamphilis, C.W.; Ma, H. The evolution of the *SEPALLATA* subfamily of MADS-box genes: A preangiosperm origin with multiple duplications throughout angiosperm history. *Genetics* **2005**, *169*, 2209–2223. [CrossRef]
58. Gao, X.; Liang, W.; Yin, C.; Ji, S.; Wang, H.; Su, X.; Guo, C.; Kong, H.; Xue, H.; Zhang, D. The *SEPALLATA*-like gene *OsMADS34* is required for rice inflorescence and spikelet development. *Plant Physiol.* **2010**, *153*, 728–740. [CrossRef]
59. Zhang, Y.; Yu, H.; Liu, J.; Wang, W.; Sun, J.; Gao, Q.; Zhang, Y.; Ma, D.; Wang, J.; Xu, Z.; et al. Loss of function of *OsMADS34* leads to large sterile lemma and low grain yield in rice (*Oryza sativa* L.). *Mol. Breed.* **2016**, *36*, 147. [CrossRef]
60. Liu, C.; Teo, Z.W.N.; Bi, Y.; Song, S.; Xi, W.; Yang, X.; Yin, Z.; Yu, H. A conserved genetic pathway determines inflorescence architecture in *Arabidopsis* and rice. *Dev. Cell* **2013**, *24*, 612–622. [CrossRef]
61. Luo, J.; Liu, H.; Zhou, T.; Gu, B.; Huang, X.; Shangguan, Y.; Zhu, J.; Li, Y.; Zhao, Y.; Wang, Y.; et al. *An-1* encodes a basic helix-loop-helix protein that regulates awn development, grain size, and grain number in rice. *Plant Cell* **2013**, *25*, 3360–3376. [CrossRef]
62. Tan, L.; Li, X.; Liu, F.; Sun, X.; Li, C.; Zhu, Z.; Fu, Y.; Cai, H.; Wang, X.; Xie, D.; et al. Control of a key transition from prostrate to erect growth in rice domestication. *Nat. Genet.* **2008**, *40*, 1360–1364. [CrossRef]
63. Jin, J.; Huang, W.; Gao, J.-P.; Yang, J.; Shi, M.; Zhu, M.-Z.; Luo, D.; Lin, H.-X. Genetic control of rice plant architecture under domestication. *Nat. Genet.* **2008**, *40*, 1365–1369. [CrossRef]
64. Wang, Y.; Li, J. Rice, rising. *Nat. Genet.* **2008**, *40*, 1273–1275. [CrossRef]
65. Guo, T.; Lu, Z.-Q.; Shan, J.-X.; Ye, W.-W.; Dong, N.-Q.; Lin, H.-X. *ERECTA1* acts upstream of the OsMKKK10-OsMKK4-OsMPK6 cascade to control spikelet number by regulating cytokinin metabolism in rice. *Plant Cell* **2020**, *32*, 2763–2779. [CrossRef]
66. Zhu, Q.-H.; Hoque, M.S.; Dennis, E.S.; Upadhyaya, N.M. *Ds* tagging of *BRANCHED FLORETLESS 1* (*BFL1*) that mediates the transition from spikelet to floret meristem in rice (*Oryza sativa* L). *BMC Plant Biol.* **2003**, *3*, 6. [CrossRef] [PubMed]
67. Xue, W.; Xing, Y.; Weng, X.; Zhao, Y.; Tang, W.; Wang, L.; Zhou, H.; Yu, S.; Xu, C.; Li, X.; et al. Natural variation in *Ghd7* is an important regulator of heading date and yield potential in rice. *Nat. Genet.* **2008**, *40*, 761–767. [CrossRef] [PubMed]
68. Putterill, J.; Robson, F.; Lee, K.; Simon, R.; Coupland, G. The *CONSTANS* gene of *Arabidopsis* promotes flowering and encodes a protein showing similarities to zinc finger transcription factors. *Cell* **1995**, *80*, 847–857. [CrossRef]
69. Yano, M.; Katayose, Y.; Ashikari, M.; Yamanouchi, U.; Monna, L.; Fuse, T.; Baba, T.; Yamamoto, K.; Umehara, Y.; Nagamura, Y.; et al. *Hd1*, a major photoperiod sensitivity quantitative trait locus in rice, is closely related to the *Arabidopsis* flowering time gene *CONSTANS*. *Plant Cell* **2000**, *12*, 2473–2484. [CrossRef]
70. Wei, X.; Xu, J.; Guo, H.; Jiang, L.; Chen, S.; Yu, C.; Zhou, Z.; Hu, P.; Zhai, H.; Wan, J. *DTH8* suppresses flowering in rice, influencing plant height and yield potential simultaneously. *Plant Physiol.* **2010**, *153*, 1747–1758. [CrossRef]
71. Yan, W.-H.; Wang, P.; Chen, H.-X.; Zhou, H.-J.; Li, Q.-P.; Wang, C.-R.; Ding, Z.-H.; Zhang, Y.-S.; Yu, S.-B.; Xing, Y.-Z.; et al. A major QTL, *Ghd8*, plays pleiotropic roles in regulating grain productivity, plant height, and heading date in rice. *Mol. Plant* **2011**, *4*, 319–330. [CrossRef] [PubMed]
72. Dai, X.; Ding, Y.; Tan, L.; Fu, Y.; Liu, F.; Zhu, Z.; Sun, X.; Sun, X.; Gu, P.; Cai, H.; et al. *LHD1*, an allele of *DTH8/Ghd8*, controls late heading date in common wild rice (*Oryza rufipogon*). *J. Integr. Plant Biol.* **2012**, *54*, 790–799. [CrossRef] [PubMed]
73. Thirumurugan, T.; Ito, Y.; Kubo, T.; Serizawa, A.; Kurata, N. Identification, characterization and interaction of *HAP* family genes in rice. *Mol. Genet. Genom.* **2008**, *279*, 279–289. [CrossRef]

74. Murakami, M.; Matsushika, A.; Ashikari, M.; Yamashino, T.; Mizuno, T. Circadian-associated rice pseudo response regulators (*OsPRRs*): Insight into the control of flowering time. *Biosci. Biotech. Biochem.* **2005**, *69*, 410–414. [CrossRef] [PubMed]
75. Koo, B.-H.; Yoo, S.-C.; Park, J.-W.; Kwon, C.-T.; Lee, B.-D.; An, G.; Zhang, Z.; Li, J.; Li, Z.; Paek, N.-C. Natural variation in *OsPRR37* regulates heading date and contributes to rice cultivation at a wide range of latitudes. *Mol. Plant* **2013**, *6*, 1877–1888. [CrossRef]
76. Gao, H.; Jin, M.; Zheng, X.-M.; Chen, J.; Yuan, D.; Xin, Y.; Wang, M.; Huang, D.; Zhang, Z.; Zhou, K.; et al. *Days to heading 7*, a major quantitative locus determining photoperiod sensitivity and regional adaptation in rice. *Proc. Natl. Acad. Sci. USA* **2014**, *111*, 16337–16342. [CrossRef]
77. Yan, W.; Liu, H.; Zhou, X.; Li, Q.; Zhang, J.; Lu, L.; Liu, T.; Liu, H.; Zhang, C.; Zhang, Z.; et al. Natural variation in *Ghd7.1* plays an important role in grain yield and adaptation in rice. *Cell Res.* **2013**, *23*, 969–971. [CrossRef]
78. Jin, J.; Hua, L.; Zhu, Z.; Tan, L.; Zhao, X.; Zhang, W.; Liu, F.; Fu, Y.; Cai, H.; Sun, X.; et al. *GAD1* encodes a secreted peptide that regulates grain number, grain length, and awn development in rice domestication. *Plant Cell* **2016**, *28*, 2453–2463. [CrossRef]
79. Huo, X.; Wu, S.; Zhu, Z.; Liu, F.; Fu, Y.; Cai, H.; Sun, X.; Gu, P.; Xie, D.; Tan, L.; et al. *NOG1* increases grain production in rice. *Nat. Commun.* **2017**, *8*, 1497. [CrossRef]

International Journal of
Molecular Sciences

MDPI

Article

OsBRKq1, Related Grain Size Mapping, and Identification of Grain Shape Based on QTL Mapping in Rice

Jae-Ryoung Park [1,2,†], Dany Resolus [3,†] and Kyung-Min Kim [1,2,*]

1. Division of Plant Biosciences, School of Applied Biosciences, College of Agriculture and Life Science, Kyungpook National University, Daegu 41566, Korea; icd92@naver.com
2. Coastal Agriculture Research Institute, Kyungpook National University, Daegu 41566, Korea
3. Department of Crop Production, Faculté d'Agronomie et de Médecine Vétérinaire (FAMV), Université d'Etat D'Haiti (UEH), National Road number 1, Damien, Port-au-Prince 1441, Haiti; resolusdany@yahoo.fr

* Correspondence: kkm@knu.ac.kr; Tel.: +82-53-950-5711

† These authors contributed equally to this manuscript.

Abstract: The world population is growing rapidly, and food shortage remains a critical issue. Quantitative trait locus (QTL) mapping is a statistical analytical method that uses both phenotypic and genotypic data. The purpose of QTL mapping is to determine the exact gene location for various complex traits. Increasing grain weight is a way to increase yield in rice. Genes related to grain size were mapped using the Samgang/Nagdong double haploid (SNDH) populations. Grain sizes were diversely distributed in SNDH 113 populations, and *OsBRKq1* was detected on chromosome 1 in an analysis of QTL mapping that used 1000 grain weight, grain length, and grain width. *OsBRKq1* exhibited high sequence similarity with the brassinosteroid leucine-rich repeat-receptor kinases of *Arabidopsis thaliana* and *Zea mays*. It was also predicted to have a similar function because of its high homology. *OsBRKq1* interacts with various grain-size control genes. Among the SNDH populations, the analysis of the relative expression level during the panicle formation stage of *OsBRKq1* in panicles of SNDH117, which has the largest grain size, and SNDH6, which has the smallest grain size, the relative expression level was significantly increased in SNDH117 panicles. SNDH populations have been advancing generations for 10 years; various genetic traits have been fixed and are currently being used as bridging parents. Therefore, the stable expression level of *OsBRKq1* was confirmed via QTL mapping. In the future, *OsBRKq1* can be effectively used to increase the yield of rice and solve food problems by increasing the size of seeds.

Keywords: rice; QTL; food shortage; yield; grain size; *OsBRKq1*

Citation: Park, J.-R.; Resolus, D.; Kim, K.-M. *OsBRKq1*, Related Grain Size Mapping, and Identification of Grain Shape Based on QTL Mapping in Rice. *Int. J. Mol. Sci.* **2021**, *22*, 2289. https://doi.org/10.3390/ijms22052289

Academic Editor: Kiyosumi Hori

Received: 29 January 2021
Accepted: 23 February 2021
Published: 25 February 2021

1. Introduction

Rice (*Oryza sativa* L.) is an important part of human food consumption worldwide [1,2]. In fact, rice is an essential crop that is consumed by more than half of the world's population [3–5]. In Asia alone, more than two billion people obtain around 70% of their energy intake from rice and its derivatives [6]. The total area used to plant rice worldwide is estimated at 154 million hectares per year, and its production is around 600 million tons [7]. This production level is sufficient to feed the current population; however, over an extended period, this equilibrium may be reversed because of demographic explosion. The current rice production should be increased by at least 40% by 2030 to meet the growing demand of the population [8]. To achieve this, early-maturing and high-yield rice varieties that can adapt to agro-ecological conditions need to be developed. Therefore, it is necessary to combine traditional breeding techniques with modern biotechnology tools to meet the projected production demand.

This study is relevant because it advocates the use of quantitative trait loci (QTLs) related to rice yield and grain size, to develop new varieties of rice. The QTLs include genes that contribute to the formation and composition of quantitative traits [9,10]. QTL analysis is a statistical method that considers two types of information: phenotypic data (particularly trait measurements) and genotypic data (usually molecular markers) [11]. The purpose of QTL analysis is to explain the genetic basis of the variation in complex traits [12,13]. QTL analysis allows researchers in fields like agriculture to link complex phenotypes to specific regions of chromosomes [14]. Yield is a complex agronomic trait that is regulated by several genes. In the case of rice, it is characterized by three main components: the filled grain number per panicle, the number of panicles per plant, and the 1000 grain weight. In recent decades, significant progress has been made in the cloning of key QTLs that can control grain yield and its components. Complex agronomic traits from segregated population studies involving recombinant lines, F_2 and progeny, crossover populations, and doubled haploids (DHs) have been reported since the 1980s [15]. A total of 2060 QTLs related to rice yield and its components had been identified up to March 2014 (http://www.gramene.org (accessed on 12 October 2020)). Although this progress has facilitated a better understanding of the regulatory mechanisms of rice production, the findings of QTL mapping and gene cloning did not significantly improve rice yield under field conditions, and the application of QTLs to the practice of selection was minimal [16]. This could be attributed to the dependence of quantitative traits on agro-ecological conditions. However, there is evidence that the 1000 grain weight is an agronomic trait that plays a predominant role in the adaptation of rice to agro-ecological conditions (i.e., sea-sons and cultivated areas), to achieve an ideal yield. Regulating the 1000 grain weight is one of the major objectives of rice breeding programs. The association between yield and its components and the grain characteristics should be investigated.

The food shortage problem is a serious debate around the world. Therefore, in this study, seed characteristics like grain length, grain width, and 1000 grain weight were investigated for QTL mapping related to grain size. Using these, QTL mapping related to grain size was performed. More specifically, it aimed to identify QTLs that could affect these traits and construct a framework linkage map to detect candidate genes associated with the variation in these agronomic traits. In additionm the QTL regions and open reading frames (ORFs) detected in this study can be effectively used to breed crops with large grain sizes.

2. Results

2.1. Phenotype Evaluation

The association of between grain length, grain width, and 1000 grain weight was evaluated by QTL mapping. These data were obtained through 2 years of research (Figure 1 and Supplementary Table S1). In 2018, the grain length of the DH population (7.56 ± 0.34 mm) was closer to that of Samgang (7.60 ± 0.15 mm) than to that of Nagdong (7.36 ± 0.11 mm). Similarly, in 2019, the grain length of the DH population (7.53 ± 0.37 mm) was closer to that of Samgang (7.53 ± 0.12 mm) than to that of Nagdong (7.39 ± 0.13 mm). However, in 2018, the grain width of the DH population (2.77 ± 0.07 mm) was greater than that of the parents Nagdong (2.73 ± 0.02 mm) and Samgang (2.76 ± 0.03 mm). Similarly, in 2019, the grain width of the DH population (2.73 ± 0.05 mm) was greater than that of the parents Nagdong (2.75 ± 0.04 mm) and Samgang (2.71 ± 0.05 mm). In 2018, the 1000 grain weight of the DH population was moderate, with an average value of 22.89 g and a standard deviation of 3.10 g. The 1000 grain weight of Nagdong was 25.71 g with a standard deviation of 0.69 g, whereas that of Samgang was 21.57 g with a standard deviation of 0.38 g. In 2019, the 1000 grain weight of the DH population was moderate, with an average value of 23.58 g and a standard deviation of 2.15 g. The 1000 grain weight of Nagdong was 26.11 g with a standard deviation of 0.72 g, whereas that of Samgang was 22.01 g with a standard deviation of 0.27 g. Supplementary Table S2 presents the correlation between grain length, grain width, and 1000 grain weight. The three seed characteristics showed high correlations (Figure 2 and Supplementary Table S2). The frequency distributions of heading date, grain length, grain width, and 1000 grain weight in all Samgang/Nagdong double haploid (SNDH) populations showed continuous changes close to a normal distribution; therefore, they represented quantitative traits regulated by one or more genes (Figure 3). Because all traits were normally distributed, QTL mapping can be performed using these data.

Figure 1. Mature paddy grains and brown rice gains of the SNDH 6 and SNDH 117 rice lines. Grain size varied in the SNDH (Samgang/Nagdong double haploid) line. The SNDH6 line had the smallest grain size, whereas SNDH117 had the largest grain size. When comparing the grain size before and after the chaff removal, the grain size of SNDH117 was the largest and the grain size of SNDH6 was the smallest. (**a**) Rough rice without removing the chaff. (**b**) Brown rice with chaff removed.

2.2. QTL (Quantitative Trait Locus) Mapping for Grain Characteristics

QTL (quantitative trait locus) mapping using the two-year phenotypic findings and analysis is presented in Supplementary Table S3 and Figure 4. A total of 850 SSR markers were used to construct the SNDH genetic map. Of these, 222 SSR markers with polymorphism in Samgang and Nagdong were finally selected and used to construct the SNDH genetic map. The total length of the genetic map constructed with 222 SSR markers was 2082.4 cM, and the interval between markers is 9.4 cM. These SSR markers are evenly distributed across 12 chromosomes. QTL mapping was performed using the phenotypic data collected in 2018 and 2019 to determine the QTL related to the grain length. Grain length, grain width, and 1000 grain weight were termed qGl, qGw, and qTgw, respectively. Two of these data were subjected to genetic mapping and QTL analysis for 2 years, and a total of 10 QTLs were obtained. In 2018, two QTLs (qGl1 and qGl1-1) were detected on chromosome 1, whereas in 2019, another two QTLs (qGl1-2 and qGl8) were identified on chromosomes 1 and 8, respectively. The qGl1 in 2018 was located at the RM575–RM1287 marker on chromosome 1, and yielded an LOD score of 2.8. The qGl1-1 (2018) was located at the s1021–s1024 marker on chromosome 1, and yielded an LOD score of 3.9. The qGl1-2 in 2019 was located at the s1024–s1026 marker on chromosome 1 and yielded an LOD score of 2.9. Lastly, the qGl8 (2019) was located at the s8017–s8018 marker on chromosome

8, and yielded an LOD score of 2.7. In addition, the QTL related to grain width (qGw), which can affect seed size, was found at a position similar to that detected for the grain length QTL. The qGw1 from 2018 was located at the s1026–s1030 marker on chromosome 1 and yielded an LOD score of 3.8. The qGw1-1 from 2019 was located at the s1021–s1024 marker on chromosome 1 and yielded a LOD score of 2.5. Meanwhile, the qGw1-2 (2019) was located at the s1028–s1030 marker on chromosome 1 and yielded an LOD score of 2.7. However, qGw1 (2018) originated from the allele of Samgang, and qGw 1-2 (2019) originated from the allele of Nagdong. Finally, the QTL region related to 1000 grain weight (qTgw) was searched. qTgw1 (2018) was located between s1024 and s1026 on chromosome 1 and yielded an LOD score of 2.4. The qTgw1-1 (2019) was located between s2030 and RM450 on chromosome 1 and yielded an LOD score of 3.3. The qGw2 (2019) was located at the s2027-RM450 marker on chromosome 2 and yielded an LOD score of 2.3. However, the qTgw in 2018 was derived from the allele of Nagdong, and the qTgw in 2019 was derived from the allele of Samgang (Supplementary Table S3). Over 2 years, one grain length, one grain width, and two 1000 grain weight QTLs were detected at approximately the same region.

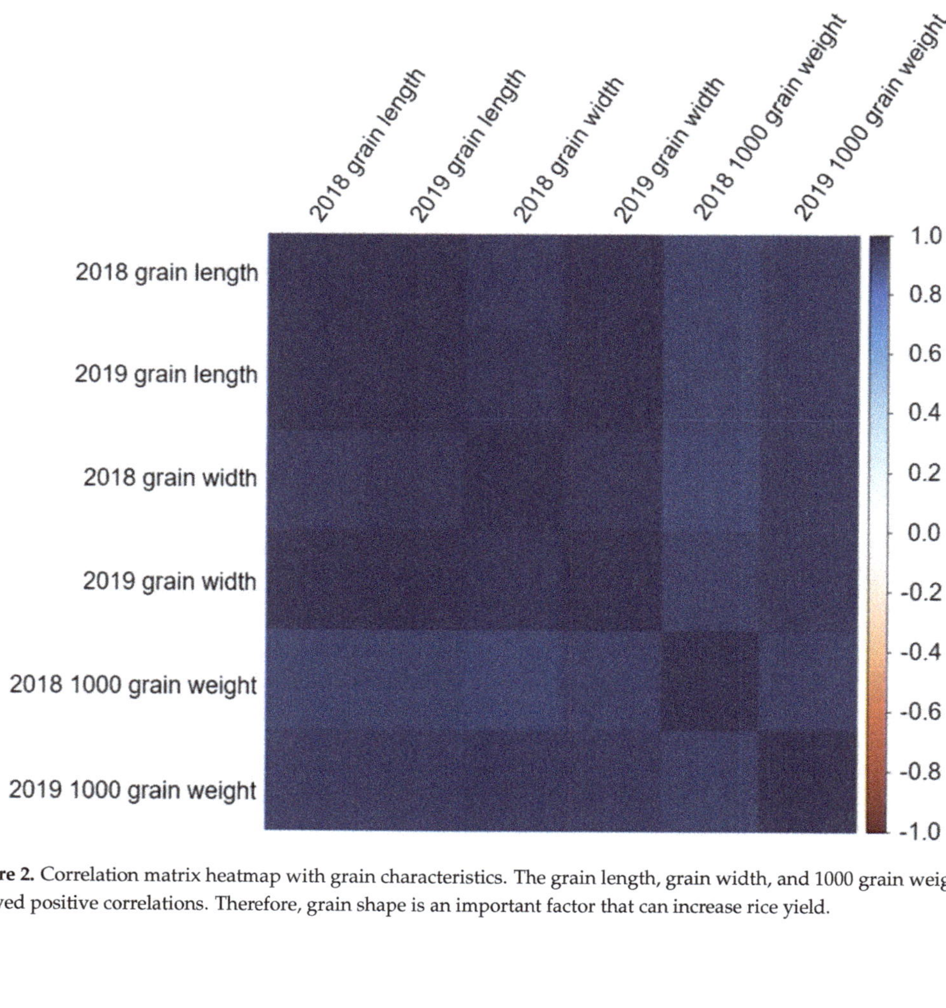

Figure 2. Correlation matrix heatmap with grain characteristics. The grain length, grain width, and 1000 grain weight all showed positive correlations. Therefore, grain shape is an important factor that can increase rice yield.

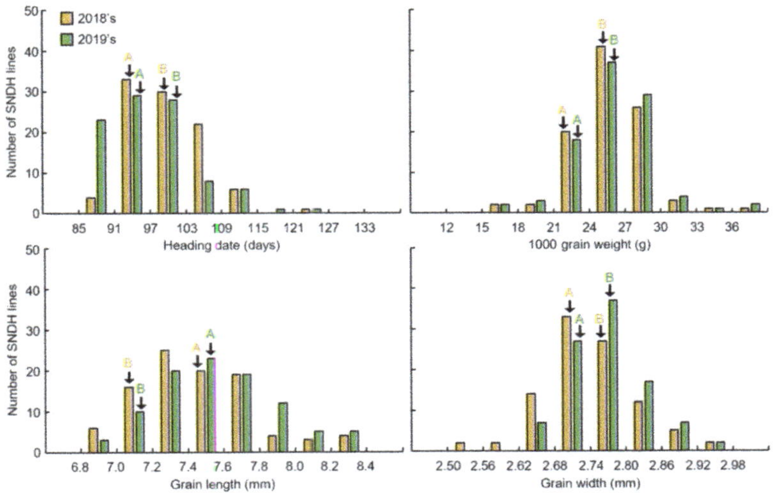

Figure 3. Frequency distribution for heading date, 1000 grain weight, grain length, and grain width in SNDH lines. Because all traits showed a normal distribution, the investigated traits can be considered quantitative traits. (**A**) Samgang; (**B**) Nagdong.

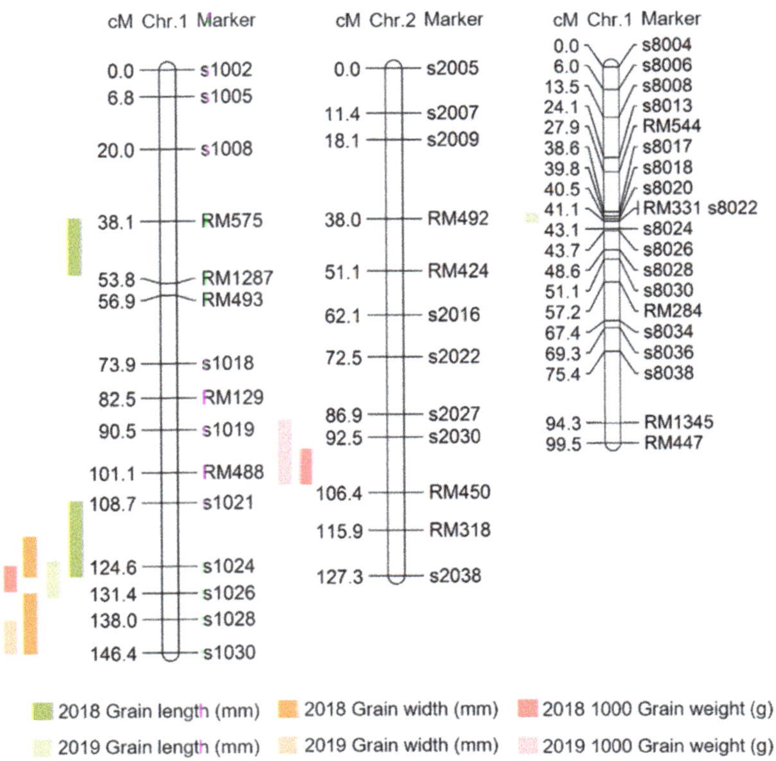

Figure 4. Chromosomal location of quantitative trait loci (QTLs) associated with grain size in the SNDH line. The QTLs mapped to chromosome 1, chromosome 2, and chromosome 8. Moreover, QTLs for all traits related to grain size were commonly located at s1024–s1026 on chromosome 1.

2.3. Searching for Candidate Genes Related to Grain Characteristic

The two-year, grain size-related chromosome 1 QTL mapping result was intensively searched for QTLs related to grain length, grain width, and 1000 grain weight. Among these regions, s1024–s1028 on chromosome 1 and s2030–RM450 on chromosome 2 were commonly detected regions related to grain size for the 2 years. Therefore, we searched for candidate genes related to grain shape, centering on s1024–s1028 on chromosome 1 and s2030–RM450 on chromosome 2. An analysis of simple sequence repeat (SSR) markers from NCBI in the s1024–s1028 region on chromosome 1 and the s2030–RM450 region on chromosome 2 revealed 18 ORFs (Figure 5). Among them, six ORFs (*LOC_Os01g52050*, *LOC_Os01g48444*, *LOC_Os01g53880*, *LOC_Os01g61690*, *LOC_Os02g42310*, and *LOC_Os02g46260*) were responsible for the functions related to grain shape (Supplementary Table S4). These ORFs corresponded to the brassinosteroid receptor kinase, auxin-responsive protein, and serine carboxypeptidase family protein.

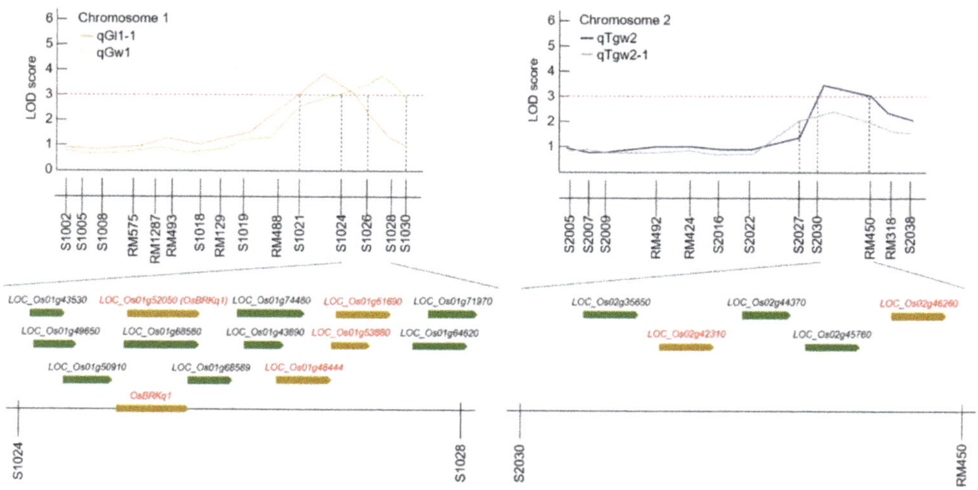

Figure 5. Quantitative trait locus (QTL) analysis and physical mapping related to grain-size candidate genes. After 2 years of QTL mapping, s1024–s1028 on chromosome 1 and s2030–RM450 on chromosome 2 were mapped consecutively. There were various candidate genes in this region, of which *OsBRKq1* on chromosome 1 was screened as a grain-size-associated gene.

2.4. Phylogenetic Tree and Homology Sequence Analyses

Among the candidate genes identified in the NCBI analysis, the sequence of *LOC_Os01g52050 (OsBRKq1)* was similar to that of the brassinosteroid LRR receptor kinase (BRL). Therefore, gene homology was compared and analyzed between *OsBRKq1* and brassinosteroid receptor kinases in *Arabidopsis thaliana*, *Oryza sativa*, and *Zea mays*. BLAST analysis showed that *OsBRKq1* had 68% sequence similarity with brassinosteroid insensitive 1-associated receptor kinase 1 of *Arabidopsis thaliana*. Phylogenetic tree analysis revealed that *OsBRKq1* had a very similar BRL sequence in *Zea mays*, in addition to BRL1 in *Arabidopsis thaliana*, and belonged to the same group (Figure 6a). Moreover, the results of the domain search for protein sequences showed very high similarity to *Arabidopsis thaliana*, *Oryza sativa*, and *Zea mays* (Figure 6c). Also, we used the OsBRKq1 domain to identify functional partners to predict the protein interactions of OsBRKq1. *OsBRKq1* interacted with 10 proteins (OS03T0132800-01, OS10T0571300-01, BSK3, BSL1, DWARF4, SERK1, BKI1, OSJ_11117, HRD3, and OS9) (Figure 6b).

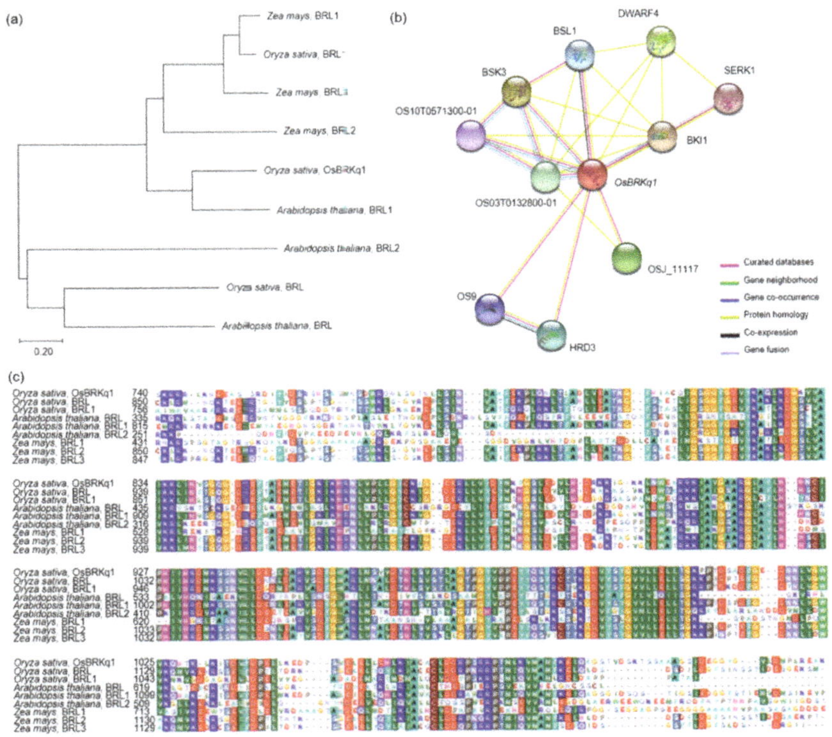

Figure 6. Sequence analysis of *OsBRKq1*. (**a**) Analysis of the relationships of the *OsBRKq1* gene and analysis of a homologous gene, using a phylogenetic tree that was constructed using 1000 bootstrap replicates. (**b**) Various proteins interacted with *OsBRKq1*. Brassinosteroids (BRs) regulate the growth and development all of plants. BRs regulate various pathways, such as the development of grain size, ethylene biosynthesis, and proton secretion into the cell wall. All of these proteins are related to the BR biosynthesis pathway. (**c**) Conserved domain of the protein sequences of homologous genes of *OsBRKq1*; very high similarity was observed in *Oryza sativa*, *Arabidopsis thaliana*, and *Zea mays*.

2.5. Analysis of the Relative Expression Levels of Candidate Genes Related to Grain Size

Grain length, grain width, and 1000 grain weight related QTL mapping was performed and detected six candidate genes for grain size. The relative expression level of these candidate genes was checked in the SNDH6, which had the smallest grain size, and the SNDH117, which had the largest grain size. The relative expression levels of candidate genes related to grain size at the panicle formation stage were analyzed. *OsBRKq1* exhibited a significant difference in relative expression level in SNDH117 at the 1% level compared with SNDH6 on the 20th, 15th, and 10th days before heading. On the fifth day before heading, the relative expression level of *OsBRKq1* was high in SNDH117, with a significant difference at the 5% level. However, after heading, there was no significant difference in the expression of this gene between SNDH6 and SNDH117. *LOC_Os01g48444* exhibited a significant difference at the 1% level on the 20th and 5th days before heading, and its relative expression level was high in SNDH117. For *LOC_Os01g53880*, the relative expression level was high in SNDH117, with a significant difference at the 1% level on the 20th and 5th days before heading; in addition, the relative expression level detected in SNDH117 was high, with a significant difference at the 5% level on the tenth day before heading. *LOC_Os01g61690* had a high relative expression level in SNDH117 at the 5% level on the 15th day before heading. *LOC_Os02g42310* exhibited a significant difference at the 1% level on the 15th day before heading, and its relative expression level in SNDH117 was high. *LOC_Os02g46260* showed a high relative expression level in SNDH117, with a significant difference at the level of 1% at 25, 15, 10, and 5 days before heading; immediately after heading; and 5 days after heading (Figure 7). In addition, the comparison of the relative expression levels of leaf candidate genes revealed no significant differences among all candidate genes in SNDH6 and SNDH117 (Figure 8). However, the analysis of the relative expression levels of the candidate genes in panicles identified a significant difference between SNDH6 and SNDH117 (Figure 8). In particular, *OsBRKq1* showed a significant difference at the 1% level and a higher level of expression in SNDH117, which had a larger grain size. *LOC_Os01g48444*, *LOC_Os01g61690*, and *LOC_Os02g46260* showed significant differences at the 5% level between SNDH6 and SNDH117. In contrast, *LOC_Os01g53880* and *LOC_Os02g42310* showed no significant differences between SNDH6 and SNDH117.

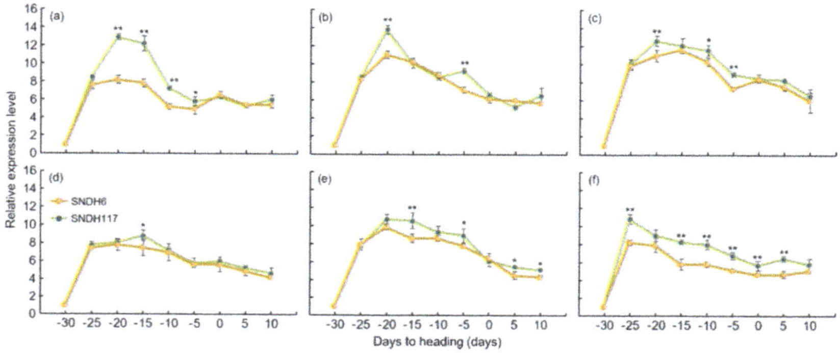

Figure 7. Analysis of the relative expression level of candidate genes related to grain size at the spikelet differentiation stage. Relative expression levels were analyzed at intervals of 5 days from 30 days before heading to 10 days after heading. The panicle formation stage starts from 30 days before heading, and grain size is determined at this time. The relative expression levels of candidate genes were analyzed in SNDH6 and SNDH117. * significant at the 0.05 level; ** significant at the 0.01 level. (**a**) *OsBRKq1*, (**b**) *LOC_Os01g48444*, (**c**) *LOC_Os01g53880*, (**d**) *LOC_Os01g61690*, (**e**) *LOC_Os02g42310*, and (**f**) *LOC_Os02g46260*.

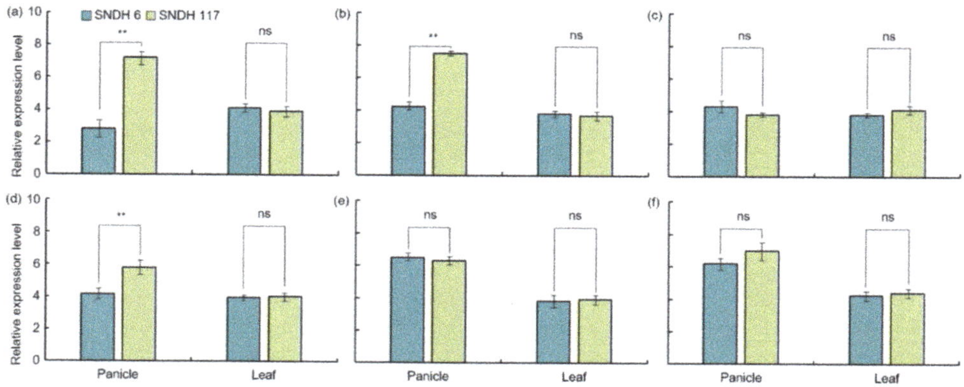

Figure 8. Comparison of relative gene expression levels in the leaves and panicles of the SNDH 6 and SNDH 117 lines, which had the smallest and largest grains, respectively, among the SNDH lines. There were no significant differences in any of the candidate genes in leaves. However, in the panicles, *LOC_Os01g48444* and *LOC_Os01g61690* were significantly different ($p < 0.05$) between the SNDH 6 and SNDH 117 lines. Moreover, *LOC_Os01g52050* was significantly different ($p < 0.01$) between the SNDH 6 and SNDH 117 lines in the panicles. ^{ns} no significant; ** significant at the 0.01 level. (**a**) *OsBRKq1*, (**b**) *LOC_Os01g48444*, (**c**) *LOC_Os01g53880*, (**d**) *LOC_Os01g61690*, (**e**) *LOC_Os02g42310*, and (**f**) *LOC_Os02g46260*.

3. Discussion

Grain length and grain width are related to 1000 grain weight. Here, the correlations between grain length, grain width, and 1000 grain weight were strongly positive (Supplementary Table S2). Moreover, as all values were close to 1, the grain length and grain width were closely related to the 1000 grain weight. Thus, studying the length and width of the grains can increase the 1000 grain weight, which leads to yield increase. Therefore, the traits investigated in this study were all related to 1000 grain weight and are important factors for increasing rice yield.

Here, QTLs for grain length, grain width, and 1000 grain weight were investigated using DH lines from a cross between Samgang and Nagdong, and were shown to be located on chromosome 1. To improve the accuracy of QTL mapping for grain length, grain width, and 1000 grain weight, 2 years of phenotypic data were examined, and the chromosomal location of QTLs was determined. A total of 10 QTLs were detected from the data collected in 2018 and 2019, and these QTLs were uniformly distributed on chromosomes 1, 2, and 8. The QTLs (qGl, qGw) related to the grain length and width, which determine the shape of the seeds, were all located at the s1028–s1030 marker on chromosome 1. The LOD scores for qGl (2018) and qGl (2019) were 3.9 and 2.9, respectively, while the LOD scores for qGw (2018) and qGw (2019) were 3.8 and 2.3, respectively. Wang et al. [17] conducted a QTL mapping of genes related to the grain size and shape of rice. They identified QTLs related to the grain shape at different locations on chromosome 1. This may be attributed to the different numbers of groups and markers used in that experiment. In this research, two-year data were used, and a high confidence level was observed because of the similar location and high LOD scores. The locations at which the LOD score exceeded 3.0 in at least one of the 2 years were mapped. All two-year data were mapped to the same marker position. Therefore, the gene that determines the shape of rice seeds was located at the s1028-s1030 marker on chromosome 1 of rice. qTgw1 (2018) and qTgw1 (2019) were located at the s1024–S1026 marker on chromosome 1, based on the QTL mapping of the 1000 grain weight. In addition, as the QTL position for the qTgw was close to the QTL position for the grain length and width, the weight of the seed may be affected by the shape of the seed.

The candidate genes identified in this study are all genes associated with increased seed size. A total of 18 candidate genes were detected in the QTL mapping, six of which were related to grain size. *OsBRKq1* exhibited sequence similarity with the brassinosteroid LRR receptor kinase. Brassinosteroids transcriptionally modulate seeds' developmental pathways by binding to multiple seed development promoters, and affect seed size and shape through integuments, endosperm, and embryo development [18]. *LOC_Os01g48444* and *LOC_Os01g53830* exhibit sequence similarity with auxin-responsive protein IAA. Sun et al. [19] reported that the auxin response factor regulated a pathway upstream of auxin signaling, and was related to plant organ size and seed size. In addition, grain size increased when the auxin response factor was overexpressed [20]. *LOC_Os01g61690*, *LOC_Os02g42310*, and *LOC_Os02g46260* have serine carboxypeptidase-like sequences. Serine carboxypeptidase increases grain size by increasing cell proliferation and expansion in spikelet hulls [21]. In addition, serine carboxypeptidase affects the signaling of brassinosteroids and influences grain-size regulation [22]. The analysis of the relative expression levels of all of these candidate genes showed an absence of significant differences in the leaves. However, the analysis of the relative expression level of grain-size-related candidate genes in the panicles revealed a significant difference in *OsBRKq1* expression at the 1% level between the lines with the largest and smallest grain size. *LOC_Os01g48444*, *LOC_Os01g61690*, and *LOC_Os02g46260* also showed a significant difference in grain size at the 5% level between these two lines. Moreover, the size of the panicle was completely determined by the panicle formation stage. Therefore, in this study, the panicle formation stage was subdivided and sampled every 5 days, and the relative expression levels of all candidate genes were analyzed. *OsBRKq1* expression exhibited a significant difference in the initial stage of the spikelet differentiation stage, and its relative expression level was high in SNDH117. In SNDH6 and SNDH117, sequencing and alignment analyses were performed to determine whether the relative expression level of *OsBRKq1* was due to sequence variation; however, there was no difference in their sequences (data not shown). Grain size is considered the most important among the panicle formation stage [23]. Therefore, *OsBRKq1* detected in this study can be effectively used to control grain size. In particular, *OsBRKq1* is similar to systemin receptor SR160 precursor, and functions as a brassinosteroid LRR receptor kinase. Jiang et al. [24] explains that brassinosteroid genes are very important for seed control in *Arabidopsis*. In addition, not only brassinosteroids but also embryonic development plays an important role in seed size determination [25,26]. Dong et al. [27] also reported that embryonic development is important for seed size determination, and in that study, genes related to embryonic development were mapped onto chromosome 1, as in the present study.

The analysis of the interaction of *OsBRKq1* with intracellular proteins revealed that it interacts with a total of 10 proteins (OS03T0132800-01, OS10T0571300-01, BSK3, BSL1, DWARF4, SERK1, BKI1, OSJ_11117, HRD3, and OS9). First, *OsBRKq1* responds to brassinosteroids by binding to the brassinosteroid LRR receptor kinase, and regulates signals involved in plant development, cell kidney, and flowering promotion. *OsBRKq1* regulates grain size by interacting with BRI1, which is a negative regulator of brassinosteroid signal transduction. BSK3 is a serine/threonine protein kinase that acts as a positive regulator of the downstream brassinosteroid signal transduction of BRI1, and SERK1 acts as a positive regulator of somatic embryogenesis and the defense reaction of the LRR receptor kinase. DWARF4 catalyzes the hydroxylation step in brassinosteroid biosynthesis, while BSL1, OS10T0571300-01, and OS03T0132800-01 catalyze the reaction via protein phosphorylation. *OsBRKq1* interacts with OS9, HRD3, and OSJ_11117, which are proteins associated with quality control of the endoplasmic reticulum and endoplasmic-reticulum-associated degradation, in addition to the brassinosteroid biosynthetic reaction.

In this study, grain size-related QTL mapping was performed, which led to the search for various candidate genes at s1024–s1028 on chromosome 1 and s2030–RM450 on chromosome 2. Moreover, proteins with domains similar to those of these candidate genes have been reported previously as being associated with grain size. Ying et al. [28] mapped a major QTL for grain length and width on chromosome 3. Furthermore, Yosida et al. [29] mapped a rice grain-size-related QTL on chromosome 11. Thus, some studies have speculated that the differences in QTL mapping for the same trait may be due to differences in the materials used in the experiment, as well as genetic and environmental factors [30]. The present research identified the *OsBRKq1* gene, which was expressed in the spikelet differentiation stage and affected grain size. *OsBRKq1* is a putative causal gene, and can be effectively used to increase rice yield and investigate unknown grain size regulatory pathways. The results of this study may provide important information for increasing seed size and improving rice yield.

4. Materials and Methods

4.1. Rice Materials and Field Experiment

The rice Samgang/Nagdong double haploid (SNDH) line used here was the result of a cross between Samgang and Nagdong. This is a DH population composed of 113 lines. Samgang is an indica elite cultivar characterized by a heavy panicle and a high yield. In contrast, Nagdong is a japonica cultivar with good quality and many panicles per plant [31]. This study was conducted at the experimental field of Kyungpook National University in Gunwi. The planting density was 30 × 15 cm per plant. One row was transplanted for each line, and there were 23 plants in each row. Therefore, 23 plants were transplanted into one line. The N-P_2O_5-K_2O level was 9.0/4.5/5.7 kg/10a; phosphoric acid and obscured fertilizer were used as the full base, and nitrogenous fertilizers were used with a base of 70% and a powder ratio of 30%. Herbicide and insecticide spraying was performed according to the standard rice cultivation method of the Rural Development Administration.

4.2. Phenotypic Evaluation

The 1000 grain weight associated with rice yield was investigated. In addition, the 1000 grain weight of the seeds was related to the shape of the seed; thus, the grain length and grain width of the seeds were measured together. Grains of the SNDH population 45 days after heading were harvested. After harvesting the SNDH line to investigate grain-size-related traits, ripening rice was carefully selected, and the grain length and width were measured using a caliper (caliper CD-15CP, Mitutoyo Corp., Kawasaki-shi, Kanagawa, Japan). To determine 1000 grain weight, 1000 grains of brown rice were repeatedly measured (five times), and the average value was used. The 1000 grain weight, seed length, and seed width were used for both the 2018 and 2019 examinations, for data reliability. The characteristics of grain size investigated in this research were repeatedly investigated (10 plants each) and statistically analyzed using the SPSS program (IBM SPSS Statistics, version 22, Armonk, NY, USA).

4.3. DNA Extraction

Leaf samples of the Samgang, Nagdong, and DH populations were collected and DNA was extracted. Healthy leaves (100 mg) of each line were placed in a 2 mL e-tube (MCT-200-C, AXYGEN, AZ, USA) with 3 mm beads, rapidly cooled with liquid nitrogen, and diluted with TissueLyser (TissueLyser II, QIAGEN, Hilden, Germany) for grinding. After adding 700 µL of DNA extraction buffer (2% CTAB (Cetyltrimethylammonium bromide), 0.1 M Tris pH 8.0, 1.4 M NaCl, and 1% PVP (Polyvinylpyrrolidone)) to the ground leaf sample, it was incubated for 20 min in a constant temperature bath at 65 °C. After the reaction, 700 µL of PCI (phenol/chloroform/isoamylalcohol = 25:24:1) was added. The sample was kept inverted at room temperature for 20 min, then centrifuged at 14,000 rpm for 10 min. After centrifugation, the supernatant was transferred to a 1.5 mL e-tube (microcentrifuge tubes, Sorenson, Murray, NY, USA). Subsequently, 350 µL of isopropanol was added to the separated supernatant, and the tube was inverted for 5 min and incubated at −72 °C for 2 h. After dissolving at room temperature, the sample was centrifuged at 13,000 rpm for 10 min, the supernatant was discarded, and the pellet was washed with 70% ethanol and dried. After drying, the DNA was dissolved in 20 µL of ddH_2O, and the DNA concentration of each sample was adjusted to 20 ng/µL using a NanoDrop 2000 Spectrophotometer (ND-2000; Nanodrop, Waltham, MA, USA) for QTL analysis.

4.4. SNDH Gene Mapping

A total of 222 SSR markers were obtained from the Rural Development Administration and used to map the QTLs related to the heading date, seed length and width, and 1000 grain weight. PCR was performed in a total volume of 12 µL, which included 20 ng/µL genomic DNA, 200 µM dNTP (deoxynucleoside triphosphate)s, 0.1 U of Taq polymerase (RR001A, TaKaRa, Seoul, Korea), and 10 pmol primer. PCR (C1000, BioRad, Hercules, CA, USA) amplification was performed using pre-denaturation at 95 °C for 5 min, followed by 35 cycles of denaturation at 95 °C for 30 s, annealing at 55 °C for 30 s, and extension at 72 °C for 1 min; and a final extension at 72 °C for 7 min. The PCR products were separated using a vertical electrophoresis apparatus (NA-1114; NIHON EIDO Co. Ltd, Tokyo, Japan) at 350 V for 1 h and 30 min after loading in an 8% natural acrylamide gel. The gel was then stained with EtBr (Ethidium bromide), to confirm polymorphism.

4.5. QTL Analysis

Win QTL Cart 2.5 version was used for QTL analysis in this study. The effective use of this program requires multiple factors, such as the genome distance between markers, the label of the markers, the number of chromosomes, the genotypic data, and the number of target traits. After inserting all the data required by Win QTL Cart version 2.5, composite interval mapping at an LOD threshold of 2.5 was performed considering the entire genome [32,33].

4.6. Candidate Gene Information Analysis

The QTL analysis approximated the locations of the genes involved in 1000 grain weight, grain length, and grain width. Once the gaps between the markers were known, numerous genes in these regions were identified on the Rapdb (https://rapdb.dna.affrc.go.jp/ (accessed on 12 September 2020)) and Rice X pro (http://ricexpro.dna.affrc.go.jp/ (accessed on 15 September 2020)) websites. Candidate genes were selected from a number of genes obtained through these sites. Moreover, all sequences located between the two flanking markers at both ends of the QTL interval were downloaded, which corresponded to the QTL range. Each open reading frame (ORF) was sorted and analyzed according to function, and the shape associated with the target was filtered out. The Simple Modular Architecture Research Tool (http://smart.embl-heidelberg.de/ (accessed on 15 September 2020)) was used to predict the protein interactions of the selected candidate genes. In addition, NCBI (http://www.ncbi.nlm.nih.gov (accessed on 18 September 2020)) and BioEdit 7.0 (https://bioedit.software.informer.com/7.0/ (accessed on 18 September 2020)) were used for homology sequence analysis and gene sequence analysis.

4.7. Analysis of Relative Expression Levels Related to Candidate Genes for Grain Size

To check the relative expression levels of candidate genes, sampling was performed at the spikelet differentiation stage. The spikelet differentiation stage started at 30 days before heading, and sampling was performed every five days up to 10 days after heading. RNA was extracted from a sampled spikelet. Total RNA was extracted using the RNeasy plant mini kit (QIAGEN, Hilden, Germany) from the leaves and panicles of the SNDH6 (with the smallest grain size) and SNDH117 (with the largest grain size) lines, and 1 μg of RNA was used as a template for cDNA synthesis using transcriptase (Invitrogen, Carlsbad, CA, USA) and an oligodt primer. A qRCRBIO cDNA Synthesis kit (cat no. PB30.11-10, PCRBIOSYSTEM, Wayne, PA, USA) was used for cDNA synthesis. Quantitative real-time PCR was performed using an Eco Real-Time PCR system (Illmina, San Diego, CA, USA). *OsActin* was used as a control, each reaction was run in triplicate, and the mean and standard deviation were calculated.

4.8. Statistical Analysis

All heading date-, 1000 grain weight-, grain width-, and grain length-determining experiments in the SNDH lines were replicated at least five times each year, and all data were analyzed using the SPSS program (IBM SPSS Statistics, version22, Armonk, NY, USA).

5. Conclusions

Grain size is a very important factor for increasing rice yield. In this research, QTL mapping for grain length, grain width, and 1000 grain weight (which is related to grain size) was performed over a period of 2 years. The frequency distribution table for each trait revealed that all traits showed a normal distribution; moreover, it was found that they were quantitative traits, which implies continuous variation. Also, these three characteristics—grain length, grain width, and 1000 grain weight—are highly positively correlated. As a result of QTL mapping for 2 years, s1024–s1028 on chromosome 1 and s2030–RM450 on chromosome 2 were mapped at the same region. Six grain-size-related candidate genes (one brassinosteroid LRR receptor kinase, two auxin-responsive proteins, and three serine carboxypeptidases) were detected in the s1024–s1028 and s2030–RM450 regions. Among the SNDH lines, the relative expression levels of candidate genes in SNDH117, which has the largest grain size, and SNDH 6, which has the smallest grain size, were confirmed in leaves and panicles. In the leaves, none of the candidate genes showed significant differences. However, in the panicles, *OsBRKq1*, *LOC_Os01g48444*, and *LOC_Os01g61690* exhibited significant differences between SNDH 6 and SNDH 117. In particular, *OsBRKq1* showed a significant difference at the level of 1% between SNDH 6 and SNDH 117. This was the largest difference in relative expression level among candidate genes related to grain size; therefore, we examined *OsBRKq1* and found that it interacts with 10 proteins

(OS03T0132800-01, OS10T0571300-01, BSK3, BSL1, DWARF4, SERK1, BKI1, OSJ_11117, HRD3, and OS9). All of these proteins are related to cell lodge and somatic embryogenesis. In addition, the homology analysis of *OsBRKq1* revealed that it had a very similar domain to the brassinosteroid LRR receptor kinase (BRL) of *Arabidopsis thaliana*, *Oryza sativa*, and *Zea mays*. Therefore, it is predicted that *OsBRKq1* will exhibit a similar function to that of the BRL proteins. *OsBRKq1*, which was newly discovered through this research, can be effectively used for the breeding of rice varieties with improved yield via an increase in grain size.

Supplementary Materials: The following are available online at https://www.mdpi.com/1422-0067/22/5/2289/s1, Table S1: The grain length, grain width and 1000-grain weight from the 113 SNDH lines, Table S2: Analysis of correlation between grain length, grain width, and 1000-grain weight, Table S3: QTL related to the grain length, grain width, and 1000-grain weight in the Samgang/Nagdong DH population, Table S4: Candidate genes related to grain length, grain width, and 1000 grain weight.

Author Contributions: Conceptualization, J.-R.P. and D.R.; methodology, J.-R.P.; formal analysis, J.-R.P. and D.R.; investigation, J.-R.P. and D.R.; writing—original draft preparation, J.-R.P. and D.R.; writing—review and editing, J.-R.P.; supervision, K.-M.K. All authors have read and agreed to the published version of the manuscript.

Funding: None.

Data Availability Statement: The data presented in this study are available on request from the corresponding author.

Acknowledgments: This work was supported by a grant from the New Breeding Technologies development Program (project no. PJ014793012021), Rural Development Administration, Republic of Korea.

Conflicts of Interest: The authors declare no conflict of interest.

References

1. Bruno, E.; Choi, Y.S.; Chung, I.K.; Kim, K.M. QTLs and analysis of the candidate gene for amylose, protein, and moisture con-tent in rice (*Oryza sativa* L.). *3 Biotech.* **2017**, *7*, 40. [CrossRef] [PubMed]
2. Maraseni, T.N.; Deo, R.C.; Qu, J.; Gentle, P.; Neupane, P.R. An international comparison of rice consumption behaviours and greenhouse gas emissions from rice production. *J. Clean. Prod.* **2018**, *172*, 2288–2300. [CrossRef]
3. Bodie, A.R.; Micciche, A.C.; Atungulu, G.G.; Rothrock, M.J., Jr.; Ricke, S.C. Current trends of rice milling byproducts for agricultural applications and alternative food production systems. *Front. Sustain. Food Syst.* **2019**, *3*, 47. [CrossRef]
4. Charoenthaikij, P.; Chaovanalikit, A.; Uan-On, T.; Waimaleongora-ek, P. Quality of different rice cultivars and factors influencing consumer willingness-to-purchase rice. *Int. J. Food Sci. Technol.* **2020**. [CrossRef]
5. Hafez, E.H.; Abou El Hassan, W.H.; Gaafar, I.A.; Seleiman, M.F. Effect of gypsum application and irrigation intervals on clay saline-sodic soil characterization, rice water use efficiency, growth, and yield. *J. Agric. Sci.* **2015**, *7*, 208–219. [CrossRef]
6. Wang, X.; Pang, Y.; Zhang, J.; Tao, Y.; Feng, B.; Zheng, T. Genetic background effects on QTL and QTL × environment interaction for yield and its component traits as revealed by reciprocal introgression lines in rice. *Crop J.* **2014**, *2*, 345–357. [CrossRef]
7. Prince, S.J.; Beena, R.; Gomez, S.M.; Senthivel, S.; Babu, R.C. Mapping consistent rice (*Oryza sativa* L.) yield QTLs under drought stress in target rainfed environments. *Rice* **2015**, *8*, 53. [CrossRef]
8. Rath, S.; Mishra, R.K.; Mishra, A.K.; Sarangi, K.K. Entropy and its relation with human population and rice production: A review. *Entropy* **2019**, *SP5*, 226–229.
9. Dixit, S.; Singh, A.; Sandhu, N.; Bhandari, A.; Vikram, P.; Kumar, A. Combining drought and submergence tolerance in rice: Marker-assisted breeding and QTL combination effects. *Mol. Breed.* **2017**, *37*, 1–12. [CrossRef]
10. Kearsey, M.J. The principles of QTL analysis (a minimal mathematics approach). *J. Exp. Bot.* **1998**, *49*, 1619–1623. [CrossRef]
11. Al-Ashkar, I.; Alderfasi, A.; Ben Romdhane, W.; Seleiman, M.F.; El-Said, R.A.; Al-Doss, A. Morphological and genetic diversity within salt tolerance detection in eighteen wheat genotypes. *Plants* **2020**, *9*, 287. [CrossRef]
12. Maurer, M.J.; Sutardja, L.; Pinel, D.; Bauer, S.; Muehlbauer, A.L.; Ames, T.D.; Arkin, A.P. Quantitative trait loci (QTL)-guided metabolic engineering of a complex trait. *ACS Synth. Biol.* **2017**, *6*, 566–581. [CrossRef] [PubMed]
13. Shabir, G.; Aslam, K.; Khan, A.R.; Shahid, M.; Manzoor, H.; Noreen, S.; Arif, M. Rice molecular markers and genetic mapping: Current status and prospects. *J. Integr. Agric.* **2017**, *16*, 1879–1891. [CrossRef]
14. Miles, C.; Wayne, M. Quantitative trait locus (QTL) analysis. *Nat. Educ.* **2008**, *1*, 208.

15. Liang, G.U.O.; Zhang, Z.H.; Zhuang, J.Y. Quantitative trait loci for heading date and their relationship with genetic control of yield traits in rice (*Oryza sativa*). *Rice Sci.* **2013**, *20*, 1–12.
16. Li, G.X.; Chen, A.H.; Liu, X.; Wang, W.Y.; Ding, H.F.; Li, J.; Liu, W.; Li, S.S.; Yao, F.Y. QTL detection and epistasis analysis for heading date using single segment substitution lines in rice (*Oryza sativa* L.). *J. Integr. Agric.* **2014**, *13*, 2311–2321. [CrossRef]
17. Wang, S.; Wu, K.; Yuan, Q.; Liu, X.; Liu, Z.; Lin, X.; Fu, X. Control of grain size, shape and quality by OsSPL16 in rice. *Nat. Genet.* **2012**, *44*, 950–954. [CrossRef]
18. Jiang, W.B.; Lin, W.H. Brassinosteroid functions in Arabidopsis seed development. *Plant Signal. Behav.* **2013**, *8*, e25928. [CrossRef] [PubMed]
19. Sun, Y.; Wang, C.; Wang, N.; Jiang, X.; Mao, H.; Zhu, C.; Xu, Z. Manipulation of Auxin Response Factor 19 affects seed size in the woody perennial Jatropha curcas. *Sci. Rep.* **2017**, *7*, 1–12. [CrossRef] [PubMed]
20. Schruff, M.C.; Spielman, M.; Tiwari, S.; Adams, S.; Fenby, N.; Scott, R.J. The AUXIN RESPONSE FACTOR 2 gene of Arabidopsis links auxin signalling, cell division, and the size of seeds and other organs. *Development* **2006**, *133*, 251–261. [CrossRef]
21. Li, N.; Li, Y. Signaling pathways of seed size control in plants. *Curr. Opin. Plant Biol.* **2016**, *33*, 23–32. [CrossRef] [PubMed]
22. Xu, C.; Liu, Y.; Li, Y.; Xu, X.; Xu, C.; Li, X.; Zhang, Q. Differential expression of GS5 regulates grain size in rice. *J. Exp. Bot.* **2015**, *66*, 2611–2623. [CrossRef] [PubMed]
23. Guan, Y.; Li, G.; Chu, Z.; Ru, Z.; Jiang, X.; Wen, Z.; Zhang, G.; Wang, Y.; Zhang, Y.; Wei, W. Transcriptome analysis reveals important candidate genes involved in grain-size formation at the stage of grain enlargement in common wheat cultivar "Bainong 4199". *PLoS ONE* **2019**, *14*, e0214149. [CrossRef] [PubMed]
24. Jiang, W.B.; Huang, H.Y.; Hu, Y.W.; Zhu, S.W.; Wang, Z.Y.; Lin, W.H. Brassinosteroid regulates seed size and shape in Ara-bidopsis. *J. Plant Physiol.* **2013**, *162*, 1965–1977. [CrossRef]
25. Berger, F.; Grini, P.E.; Schnittger, A. Endosperm: An integrator of seed growth and development. *Curr. Opin. Plant Biol.* **2006**, *9*, 664–670. [CrossRef]
26. Zhou, Y.; Zhang, X.; Kang, X.; Zhao, X.; Zhang, X.; Ni, M. Short hypocotyl under blue1 associates with MINISEED3 and HAIKU2 promot-ers in vivo to regulate Arabidopsis seed development. *Plant Cell* **2009**, *21*, 106–117. [CrossRef] [PubMed]
27. Dong, Y.; Tsuzuki, E.; Kamiunten, H.; Terao, H.; Lin, D. Mapping of QTL for embryo size in rice. *Crop Sci.* **2003**, *43*, 1068–1071. [CrossRef]
28. Ying, J.Z.; Ma, M.; Bai, C.; Huang, X.H.; Liu, J.L.; Fan, Y.Y.; Song, X.J. TGW3, a major QTL that negatively modulates grain length and weight in rice. *Mol. Plant* **2018**, *11*, 750–753. [CrossRef] [PubMed]
29. Yoshida, S.; Ikegami, M.; Kuze, J.; Sawada, K.; Hashimoto, Z.; Ishii, T.; Kamijima, O. QTL analysis for plant and grain characters of sake-brewing rice using a doubled haploid population. *Breed. Sci.* **2002**, *52*, 309–317. [CrossRef]
30. Ayalew, H.; Liu, H.; Liu, C.; Yan, G. Identification of early vigor QTLs and QTL by environment interactions in wheat (*Triticum eastivum* L.). *Plant Mol. Biol. Rep.* **2018**, *36*, 399–405. [CrossRef]
31. Qin, Y.; Kim, S.M.; Sohn, J.K. Detection of main-effect QTLs, epistatic QTLs and QE interactions for grain appearance of brown rice (*Oryza sativa* L.). *J. Crop Sci. Biotechnol.* **2008**, *11*, 151–156.
32. Boopathi, N.M. QTL Analysis. In *Genetic Mapping and Marker Assisted Selection*; Springer: Singapore, 2020; pp. 253–326.
33. Wang, S.; Basten, C.J.; Zeng, Z.B. *Windows QTL Cartographer 2.5*; Department of Statistics: Raleigh, NC, USA, 2007.

International Journal of
Molecular Sciences

MDPI

Article

CRISPR/Cas9 Guided Mutagenesis of *Grain Size 3* Confers Increased Rice (*Oryza sativa* L.) Grain Length by Regulating Cysteine Proteinase Inhibitor and Ubiquitin-Related Proteins

Babar Usman [1,†], Neng Zhao [1,†], Gul Nawaz [1], Baoxiang Qin [1], Fang Liu [1], Yaoguang Liu [2,*] and Rongbai Li [1,*]

1 State Key Laboratory for Conservation and Utilization of Subtropical Agro-Bioresources, College of Agriculture, Guangxi University, Nanning 530004, China; babarusman119@gmail.com (B.U.); nengzhao@st.gxu.edu.cn (N.Z.); gulnawazmalik@yahoo.com (G.N.); bxqin@gxu.edu.cn (B.Q.); fangliu@gxu.edu.cn (F.L.)

2 State Key Laboratory for Conservation and Utilization of Subtropical Agricultural Bioresources, South China Agricultural University, Guangzhou 510642, China

* Correspondence: ygliu@scau.edu.cn (Y.L.); lirongbai@126.com (R.L.); Tel.: +86-20-8528-1908 (Y.L.); +86-136-0009-4135 (R.L.)

† These authors contributed equally to this work.

Abstract: Clustered Regularly Interspaced Short Palindromic Repeats (CRISPR)/CRISPR-associated protein (Cas9)-mediated genome editing has become an important way for molecular breeding in crop plants. To promote rice breeding, we edited the *Grain Size 3* (*GS3*) gene for obtaining valuable and stable long-grain rice mutants. Furthermore, isobaric tags for the relative and absolute quantitation (iTRAQ)-based proteomic method were applied to determine the proteome-wide changes in the *GS3* mutants compared with wild type (WT). Two target sites were designed to construct the vector, and the Agrobacterium-mediated method was used for rice transformation. Specific mutations were successfully introduced, and the grain length (GL) and 1000-grain weight (GWT) of the mutants were increased by 31.39% and 27.15%, respectively, compared with WT. The iTRAQ-based proteomic analysis revealed that a total of 31 proteins were differentially expressed in the *GS3* mutants, including 20 up-regulated and 11 down-regulated proteins. Results showed that differentially expressed proteins (DEPs) were mainly related to cysteine synthase, cysteine proteinase inhibitor, vacuolar protein sorting-associated, ubiquitin, and DNA ligase. Furthermore, functional analysis revealed that DEPs were mostly enriched in cellular process, metabolic process, binding, transmembrane, structural, and catalytic activities. Pathway enrichment analysis revealed that DEPs were mainly involved in lipid metabolism and oxylipin biosynthesis. The protein-to-protein interaction (PPI) network found that proteins related to DNA damage-binding, ubiquitin-40S ribosomal, and cysteine proteinase inhibitor showed a higher degree of interaction. The homozygous mutant lines featured by stable inheritance and long-grain phenotype were obtained using the CRISPR/Cas9 system. This study provides a convenient and effective way of improving grain yield, which could significantly accelerate the breeding process of long-grain japonica parents and promote the development of high-yielding rice.

Keywords: rice; genome editing; homozygous; yield; proteomics

Citation: Usman, B.; Zhao, N.; Nawaz, G.; Qin, B.; Liu, F.; Liu, Y.; Li, R. CRISPR/Cas9 Guided Mutagenesis of *Grain Size 3* Confers Increased Rice (*Oryza sativa* L.) Grain Length by Regulating Cysteine Proteinase Inhibitor and Ubiquitin-Related Proteins. *Int. J. Mol. Sci.* **2021**, *22*, 3225. https://doi.org/10.3390/ijms22063225

Academic Editors: Kiyosumi Hori and Matthew Shenton

Received: 2 March 2021
Accepted: 20 March 2021
Published: 22 March 2021

1. Introduction

Rice is one of the most important crops globally and a staple food for more than half of the world's population. Increasing rice production plays an extremely important role in the stability of the world economy and the development of human society. Since the 1960s, China's rice yield has been stagnant for a long time after two major leaps in dwarf and hybrid rice breeding. At present, with population increase, arable land reduction, environmental pollution, and frequent extreme weather disasters, rice production is facing severe challenges [1]. It is urgent to apply new technologies to break the bottleneck of rice production. Rice yield is a complex agronomic trait mainly determined by the effective panicle number (PN), grain number per panicle (GNPP), and 1000-grain weight (GWT), all of which are typical quantitative traits. In recent years, the demand for high-yield and quality rice has increased, especially for long-grain rice. In the process of conventional rice variety improvement, the aggregation of excellent genes is mainly achieved through hybridization and backcrossing. However, conventional breeding has the disadvantages of a long breeding cycle and low efficiency. Through the development of specific gene chips and functional molecular markers, the function of yield-related genes has been extensively explored, and rice breeding is planned to cultivate new rice varieties with high and stable yields [2]. The completion of rice genome sequencing, development of functional genomics, proteomics, bioinformatics, and the widespread use of next-generation molecular marker technology has laid an important foundation for rice yield improvement [3]. Recently developed gene-editing technologies can bring new improvements to the rapid development of rice varieties with improved grain yield.

Mutants are important materials for studying gene function as well as for breeding programs. Traditional artificial mutations are usually produced by ethyl methanesulfonate (EMS) mutagenesis, radiation mutagenesis, and (transgene-DNA) T-DNA insertion. However, these methods cause random mutations genome wide, which are difficult to detect and sometimes deleterious. Through transgenic technologies, genetically modified (GM) crops can be given a variety of beneficial traits, such as insect resistance, herbicide tolerance, stress resistance, improved nutritional value, and grain yield. There is still a big gap between GM rice and large-scale commercialization. The technical reason is that it is difficult to completely remove the T-DNA fragments. At present, the Clustered Regularly Interspaced Short Palindromic Repeats/CRISPR-associated protein 9 (CRISPR/Cas9) emerged as a new generation of gene-editing technology, and it is widely used in crop breeding and functional genomics research [4–9]. Recent studies have proved that CRISPR/Cas9 can generate T-DNA-free mutants with inheritable mutations [10–12].

Researchers generally believe that grain length (GL), grain width (GWD), and aspect ratio are generally controlled by multiple genes. Studies have also shown that the genes that control various traits of grain type also have complementary and additive effects [13]. Many quantitative trait loci (QTLs) related to grain size have been identified in rice, such as the negative regulatory factors *GW2*, *GS9*, *qW5/GW5*, and *TGW6* [14–18], and positive regulators including *GL2*, *GL3.1*, *GS5*, *GL7*, *GLW7*, and *GW8* [19–22]. *Grain Size 3* (*GS3*) is the first gene to be cloned to regulate grain yield, and its loss-of-function mutations result in enhanced grain yield [15,23]. The negative regulator genes related to grain yield are suitable for carrying out knockout offspring to obtain slender kernels. It is difficult to increase yield through traditional breeding methods, so the employment of modern technologies is necessary to achieve the required grain production in a timely manner. Currently, rice is one of the most successful crops for CRISPR/Cas9 applications. Studies have shown that through CRISPR/Cas9-mediated genome editing, the probability of obtaining homozygous or biallelic genes is as high as 90% in T_0 generation [24]. Previous studies suggest that mutations in the *GS3* gene results in increased GL and GWT, and the overall yield improvement [25]. Zhao et al. (2018) [16] used CRISPR/Cas9 technology to generate mutations in the first exon of *GS9*, which resulted in a frameshift mutation, disrupted *GS9* normal expression, and increased GL. Lowder et al. (2015) [26] designed sgRNAs mediated by U6 and U3 promoters respectively according to the sequences of rice target genes *OsYSA* and *OsROC5* and obtained albino seedling and curled leaf phenotype. CRISPR/Cas9-based mutations in three homoeologs of cytochrome *P450* genes and *OsBADH2* resulted in increased grain yield and aroma [10]. After *OsPYL9* was mutated by CRISPR/Cas9, the drought tolerance and grain yield of rice were improved significantly [11]. A recent study has shown that CRISPR/Cas9 can successfully introduce homozygous mutations in *GW8*, resulting in increased grain yield [12]. These studies suggest that CRISPR/Cas9 technology has set a precedent in rice gene editing, and at the same time, it also provides a new technical approach and method for yield improvement.

With the rapid development of proteomics, quantitative study of protein changes has become one of the important contents of proteomics research. Relative and absolute quantitative isotope labeling (iTRAQ) technology combined with tandem mass spectrometry and multidimensional liquid chromatography is the latest technology. The tools for qualitative and quantitative protein research with better effects have been widely used in the proteomics research of rice [10–12]. Recently, iTRAQ technology has been used to compare the proteomic changes of CRISPR/Cas9 rice mutants, and differentially expressed proteins (DEPs) have been successfully screened [27–31]. These studies suggest that iTRAQ technology is helpful to study the changes in plant protein expression because of mutations. Therefore, with the continuous improvement of genome and plant protein databases, there will be more room for the application of iTRAQ technology in CRISPR/Cas9 generated mutant plants.

In this study, using CRISPR/Cas9 technology, *GS3* mutants were successfully generated, and comparative proteomic analysis was performed to reveal the changes proteome-wide. Mutants exhibited increased grain size without any change in other agronomic traits. Proteomics screening found that multiple identified proteins were differentially regulated, and mutant plants showed enhanced grain yield. In short, this work suggests that *GS3* mutants hold great potential in rice breeding to improve grain yield.

2. Results

2.1. Construction of CRISPR/Cas9 Expression Vector

According to the "Golden Gate" cloning method, the guided RNA (gRNA) expression cassette with two targets was connected to the pYLCRISPR/Cas9Pubi-H vector backbone (Figure 1A). The amplification of the sgRNA expression cassette for the first and second targets (T1 and T2) was verified by the overlapping polymerase chain reaction (PCR). The CRISPR/Cas9 binary vector was effectively built, and both sgRNA sequences were confirmed in the vector (Figure 1B) by using the SP-L1 and SP-R (Table S1) primers.

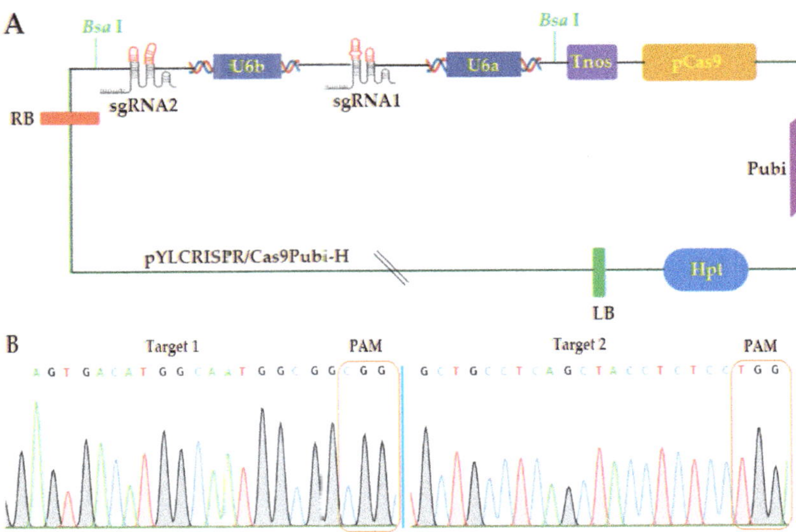

Figure 1. (**A**) Schematic representation of pYLCRISPR/Cas9 vector construction and (**B**) sequencing peak map of both target sites assembled in vector. sgRNA, single guided RNA; LB, left border; mpCas9, Cas9 protein; U6, rice U6 promoter; HPT, hygromycin phosphotransferase gene; Bsa, cutting sites; Pubi, maize ubiquitin promoter; Tnos: gene terminator, RB, right border; PAM, protospacer adjacent motif.

2.2. Obtaining Mutant Plants and Genotyping

In total, we treated 75 calli with transformed *A. tumefaciens* and attained 20 plantlets. We extracted the corresponding genomic DNA from each mutant plant, and the specific primers *HPTF*/R (Table S1) were used to identify whether the T-DNA regions were successfully integrated. The results showed that 15 tissue cultured plantlets were transgenic positive.

Sequencing results revealed that among the 15 plantlets, 12 plantlets were edited, representing an editing efficiency of 80%. According to the decoding of sequencing results, four types of plantlets were obtained with no editing, homozygous editing, monoallelic heterozygous editing, and biallelic heterozygous editing. By counting the different types of edits, we found that there were four WT, four homozygous, three biallelic heterozygous, and four monoallelic heterozygous plantlets for the first target in T_0 generation. The editing results of the second target revealed that there were three WT, five homozygous, two biallelic heterozygous, and five mono-allelic heterozygous plantlets in the T_0 generation.

Two mutant lines (GXU27-4 and GXU27-9) showed homozygous mutations for both target sites. GXU27-4 exhibited 25 bp and 6 bp deletions at the first and second target positions, respectively. GXU27-9 presented 9 bp and 19 bp deletions at the first and second target locations, respectively (Figure 2). Deletion and insertions with at least one nucleotide were achieved successfully.

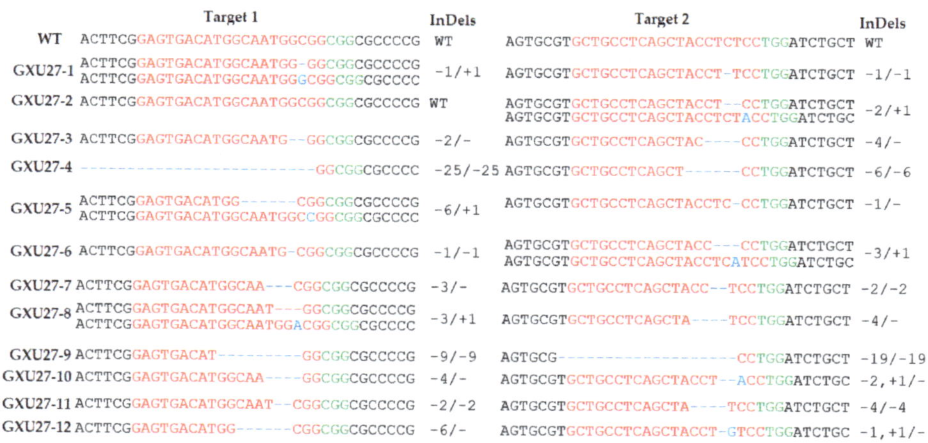

Figure 2. Sequence alignment of both target sites and information about deletions and insertions in all T_0 mutant lines. The targeted sequence is highlighted in red, and the protospacer adjacent motif (PAM) sequence in green. Deletions and insertions are represented by blue hyphens and uppercase letters, respectively.

Using the Cas9F/R specific primers (Table S2), we amplified the DNA of 20 T_1 mutant plants for the five most likely positions with the higher off-target ranking. The sequencing results revealed that there were no off-target effects found against sgRNA1 and sgRNA2 in selected putative loci (Table S3).

2.3. Screening of T-DNA-Free Mutant Plants in the T_1 Generation and Segregation Analysis

To obtain mutant plants without T-DNA components, the progeny of the T_0 generation was evaluated. A total of 27 mutants of the T_1 generation were screened for T-DNA fragments using Cas9F/R primers. Those mutants that were amplified to the 600 bp fragment length were considered T-DNA positive, whereas mutants not amplified were considered T-DNA-free. Results showed that 15 plants were not amplified to Cas9-specific primers (Figure S1). The T-DNA-free mutants appeared with a frequency of 60%.

Transmissions of the targeted mutations induced by CRISPR/Cas9 were investigated by the self-fertilization of T_0 mutants and subjected to segregation analyses. The T_1 progeny of homozygous plants (GXU27-4) exhibited homozygosity for both target sites with the same mutations. These results indicate that homozygous mutations were stably transmitted from T_0 to T_1 generation for all target sites. We observed the segregation pattern of a monoallelic heterozygous mutation in GXU27-3 for the first target site and GXU27-8 for the second target site. The T_1 progeny (GXU27-3 and GXU27-8) of T_0 monoallelic heterozygous lines was segregated according to Mendelian inheritance and resulted in homozygous and heterozygous mutations and WT plants. The inheritance pattern of biallelic mutations was studied using GXU27-1 for the first target and GXU27-2 for the second target. The progeny of the biallelic mutants also followed the classic Mendelian inheritance (1:2:1) (Table S4). Therefore, gene-edited plants that do not contain T-DNA components can be obtained in the T_1 generation, and the mutation characteristics of these plants remain consistent with their T_0 generation.

2.4. Investigation of Agronomic Traits

Two homozygous (GXU27-4 and GXU27-9) and one monoallelic heterozygous line (GXU27-3) from T_0, T_1, and T_2 generations were tested for agronomic traits evaluation. The results showed that the grain size of all mutants was significantly increased than that of the WT, whereas there was not any change in other agronomic traits was observed (Table 1; Figure 3). The GL of mutants was increased from 8.6 to 11.3 mm compared with WT plants. The results showed that the GWT of the mutants was also increased significantly compared with the corresponding WT plants. At the same time, the plant height (PH), PN, flag leaf length (FLL), flag leaf width (FLW), GNPP, and GWD of all *GS3* mutants showed a non-significant difference compared to WT (Table 1). The T_1 and T_2 generation showed consistent results with the T_0 generation, which clearly showed that mutations were passed to the next generation successfully.

Figure 3. (**A**) Plant type and (**B**) grain phenotype of wild type (WT) and mutants (GXU27-3, GXU27-4, and GXU27-9) in the T_1 generation.

Table 1. Increased yield of mutant lines in T_0, T_1, and T_2 generations.

Generation	Genotypes	PH	PN	FLL	FLW	GNPP	GL	GWD	GWT
T_0	WT	145.3 ± 3.6	9.5 ± 1.6	60.2 ± 2.3	2.6 ± 0.3	148.6 ± 0.9	8.7 ± 0.2	2.9 ± 0.1	31.3 ± 1.3
	GXU27-3	144.6 ± 4.2 ns	9.7 ± 2.2 ns	59.9 ± 3.1 ns	2.5 ± 0.5 ns	147.4 ± 0.8 ns	11.1 ± 0.3 *	3.0 ± 0.3 ns	39.8 ± 1.2 *
	GXU27-4	146.4 ± 5.6 ns	9.5 ± 1.4 ns	60.3 ± 2.2 ns	2.6 ± 0.4 ns	146.5 ± 0.7 ns	10.8 ± 0.5 *	2.9 ± 0.2 ns	39.1 ± 1.6 *
	GXU27-9	145.9 ± 4.8 ns	9.7 ± 1.8 ns	59.8 ± 2.5 ns	2.7 ± 0.3 ns	149.2 ± 0.8 ns	10.9 ± 0.2 *	2.9 ± 0.1 ns	39.0 ± 1.4 *
T_1	WT	146.5 ± 2.8	9.6 ± 1.5	61.3 ± 2.6	2.5 ± 0.2	149.4 ± 0.6	8.8 ± 0.3	3.0 ± 0.1	32.1 ± 1.1
	GXU27-3	145.4 ± 3.4 ns	9.3 ± 1.9 ns	60.2 ± 2.4 ns	2.7 ± 0.2 ns	148.5 ± 0.9 ns	11.3 ± 0.2 *	3.1 ± 0.1 ns	39.7 ± 1.4 *
	GXU27-4	147.3 ± 4.2 ns	9.3 ± 1.2 ns	59.9 ± 2.5 ns	2.4 ± 0.5 ns	148.2 ± 0.5 ns	10.9 ± 0.3 *	3.0 ± 0.2 ns	39.3 ± 1.0 *
	GXU27-9	145.8 ± 3.5 ns	9.5 ± 1.4 ns	60.1 ± 2.2 ns	2.6 ± 0.4 ns	149.6 ± 0.4 ns	10.6 ± 0.4 *	2.9 ± 0.2 ns	39.2 ± 1.2 *

Table 1. *Cont.*

Generation	Genotypes	PH	PN	FLL	FLW	GNPP	GL	GWD	GWT
T_2	WT	144.6 ± 2.8	9.8 ± 1.3	59.9 ± 2.4	2.7 ± 0.2	148.2 ± 0.6	8.6 ± 0.3	2.9 ± 0.2	31.6 ± 1.2
	GXU27-3	144.8 ± 3.7 ^{ns}	9.6 ± 2.0 ^{ns}	60.1 ± 3.2 ^{ns}	2.6 ± 0.3 ^{ns}	145.4 ± 0.6 ^{ns}	11.2 ± 0.4 *	3.0 ± 0.1 ^{ns}	39.6 ± 1.5 *
	GXU27-4	145.9 ± 3.5 ^{ns}	9.8 ± 1.6 ^{ns}	60.0 ± 2.3 ^{ns}	2.6 ± 0.4 ^{ns}	147.5 ± 0.5 ^{ns}	10.9 ± 0.3 *	3.0 ± 0.1 ^{ns}	39.3 ± 1.2 *
	GXU27-9	145.3 ± 4.2 ^{ns}	9.9 ± 1.7 ^{ns}	59.9 ± 2.4 ^{ns}	2.5 ± 0.2 ^{ns}	149.3 ± 0.7 ^{ns}	10.7 ± 0.2 *	2.9 ± 0.3 ^{ns}	39.1 ± 1.3 *

WT (wild type); PH (plant height) cm; PN (panicle numbers); FLL (flag leaf length) cm; FLW (flag leaf width) cm; GNPP (grain number per panicle); GL (grain length) mm; GWD (grain width) mm; GWT (1000-grain weight) g. Five independent plants were used to collect data from three replicates (n = 5). For grain phenotyping, five grains from each plant were selected randomly. * and ^{ns} denote the significant and non-significant differences (Student's *t*-test, $p < 0.01$), respectively.

2.5. Proteomic Data Outcome

The box and whisker plot showed a clear difference between WT and mutant plants' proteomic data (Figure 4A). The count data distribution plot also showed a clear difference among WT and mutant plants' expression data, whereas there was no difference between the replicates (Figure 4B). The significant difference between WT and GXU27-1 was found in two-dimensional t-Distributed Stochastic Neighbor Embedding (t-SNE) graph. The points representing the replicates from both samples were near to each other while there was a significant difference among the points sample point (Figure 4C). Using iTRAQ labeled proteomics, 574,173 total spectra, 66,313 matched spectra, and a total of 26,986 peptides were detected in the six samples tested. Peptides were searched through the UniProt database, and finally, 4743 proteins were identified and quantified (Figure 4D; Supplementary File S1).

The results of differential analysis of protein expression levels showed that there were 20 proteins up-regulated and 11 proteins down-regulated, and 4712 proteins were not differentially expressed (Figure 4E). The proteins related to cysteine synthase (A2ZMY2, Q5JNB0, and B8AJV7) cysteine proteinase inhibitor (Q0JNR2, A0A0A7EQF3, and P20907), ubiquitin (A2XEA1), Vacuolar protein sorting-associated protein (Q10NQ3, A2 × 377, A0A0E0GXY5, and Q8H8K1), DNA damage-binding protein 1 (Q6L4S0), DNA ligase (Q7XD67), and some other were differentially regulated (Table 2; Supplementary File S1).

Table 2. List of some important differentially expressed proteins (DEPs).

Protein ID	Protein Names	log2 FC	Regulation
A2ZMY2	Cysteine synthase	1.66	Up
Q5JNB0	Cysteine synthase	1.55	Up
B8AJV7	Cysteine synthase	1.46	Up
Q0JNR2	Cysteine proteinase inhibitor 12	−2.38	Down
A0A0A7EQF3	Cysteine proteinase inhibitor	−2.66	Down
P20907	Cysteine proteinase inhibitor 2	−2.30	Down
A2XEA1	Ubiquitin	−1.50	Down
Q10NQ3	Vacuolar protein sorting-associated protein 9A	2.87	Up
A2X377	Vacuolar protein sorting-associated protein 29	3.12	Up
A0A0E0GXY5	Vacuolar protein sorting-associated protein 41 homolog	3.09	Up
Q8H8K1	Putative vacuolar sorting receptor protein	2.13	Up
Q6L4S0	DNA damage-binding protein 1	1.86	Up
Q7XD67	DNA ligase	3.83	Up

2.6. Functional Assignment and Pathway Analysis

The results of the significant enrichment of gene ontology (GO) function and Kyoto Encyclopedia of Genes and Genomes (KEGG) pathway analysis are shown in Figure 5. GO analysis is mainly divided into three parts: molecular function (MF), biological process (BP), and cellular component (CC). It is often used to provide functional classification labels of DEPs and background knowledge of gene function research.

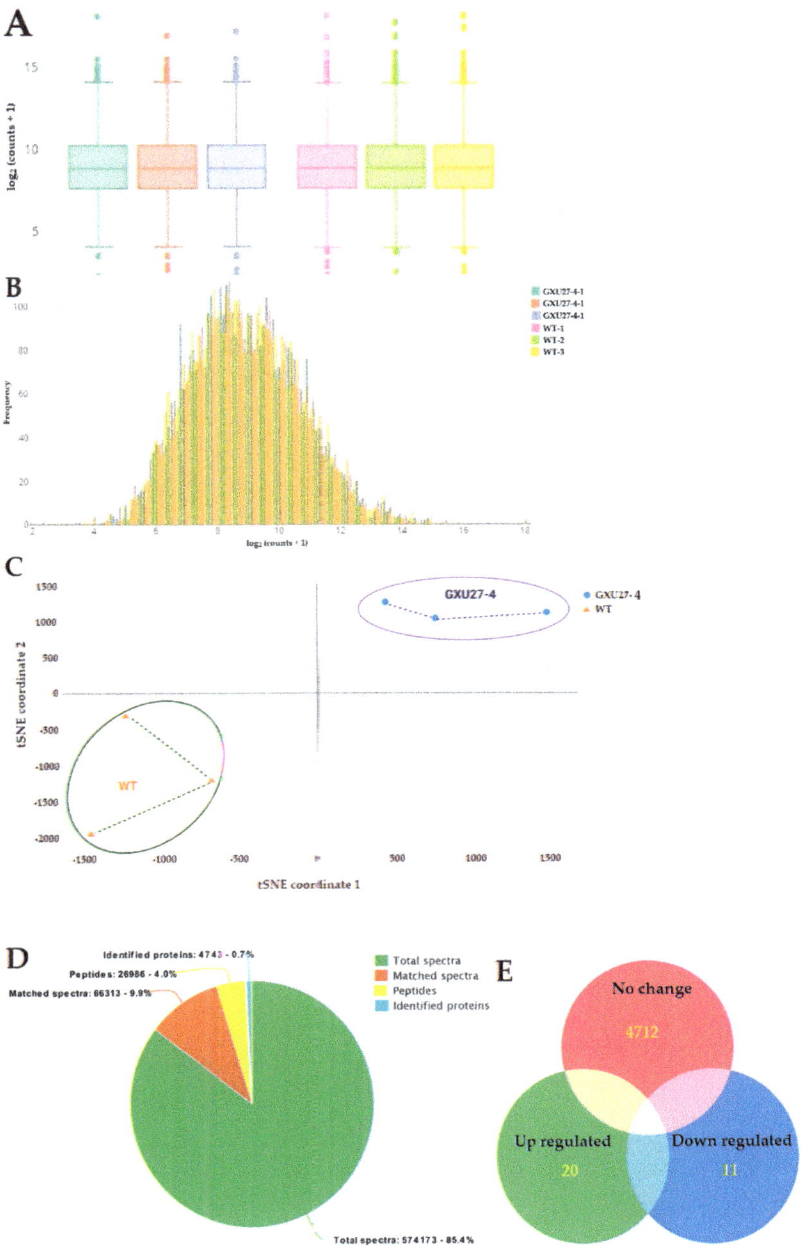

Figure 4. Basic information about the outcome from the proteomic analysis. (**A**) Representation of count data distribution with log2 values in a Box and Whisker plot. (**B**) Histogram showing the frequency of count data distribution of transformed data (**C**) Two-dimensional t-Distributed Stochastic Neighbor Embedding (t-SNE) graph representing the statistical difference between wild type (WT) and mutant line (GXU27-4) (**D**) Information about total spectra, identified matched spectra, peptide, and identified proteins and (**E**) Venn diagram showing the distribution of differentially expressed proteins (DEPs).

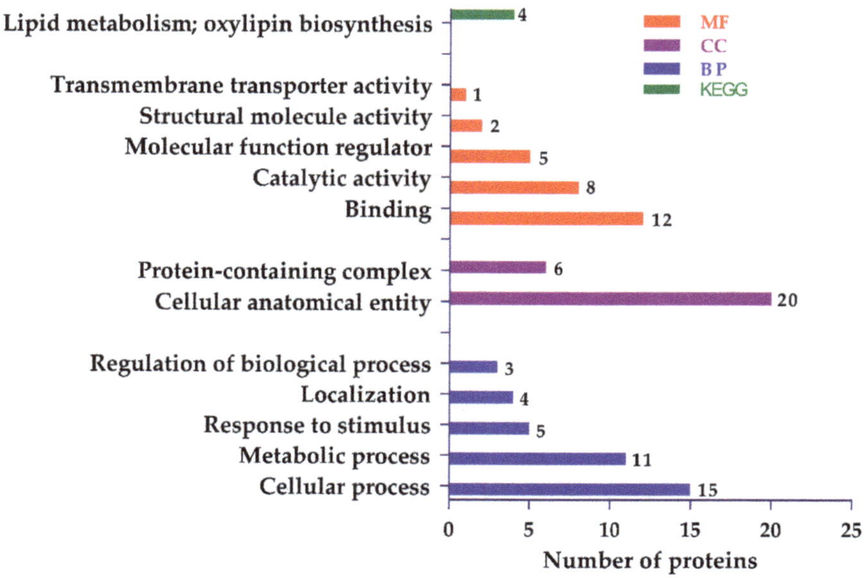

Figure 5. Gene ontology (GO) classification and pathway assignment of differentially expressed proteins (DEPs).

The results of significant GO enrichment showed that from the perspective of BP, the DEPs were significantly enriched (p-value ≤ 0.05) in the cellular process, metabolic process, response to stimulus, localization, and regulation of the BP. Regarding CC, the GO functional attributes were significantly enriched in cellular anatomical entity and the protein-containing complex. Finally, regarding the MF perspective, the DEPs were significantly enriched in binding, catalytic activity, molecular function regulator, structural molecule activity, and transmembrane transporter activity. The result of significant enrichment of the KEGG pathway showed that DEPs were only enriched in lipid metabolism and oxylipin biosynthesis.

2.7. Functional Interaction Networks of the Differentially Expressed Proteins (DEPs)

The Search Tool for the Retrieval of Interacting Genes/Proteins (STRING) database was used to find the protein interactions. After retrieving proteins with the highest connectivity from the projected network, higher interaction was found between Q6L4S0 (DNA damage-binding protein 1), Q8H936 (JUN-activation-domain-binding protein 1), Q7XN7 (ubiquitin-40S ribosomal protein S27a-2), Q9ARZ9 (ubiquitin-40S ribosomal protein S27a-1), Q7XD47 (putative ubiquitin/ribosomal protein S27a fusion protein), A0A0P0X6U8 (putative ubiquitin/ribosomal protein CEP52), A0A0P0X005 (ubiquitin family protein), A0A0P0X0E0 (pentameric polyubiquitin-like), Q7XN78 (polyubiquitin 3 Ubiquitin-related Ubiquitin, and A0A0P0VF30 (polyubiquitin) (Figure 6) with a degree higher than 12 (Supplementary File S1). Some proteins including P20907 (cysteine proteinase inhibitor 2), Q0JNR2 (cysteine proteinase inhibitor 12), Q10Q47 (putative cysteine proteinase inhibitor 7), Q10Q46 (cysteine proteinase inhibitor 6), Q0DS16 (PP1/PP2A phosphatases pleiotropic regulator PRL1), and Q5Z4U6 (putative C2H2 zinc-finger protein) showed a degree value of less than 3 and no or poor interaction with other proteins. The above results revealed that the DNA-damage binding proteins and ubiquitin ribosomal proteins were found to be highly interactive, whereas cysteine proteinase inhibitor proteins showed poor or no interaction.

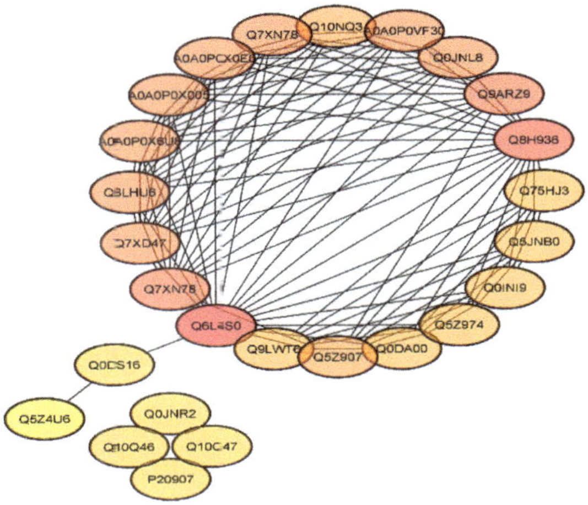

Figure 6. Protein–protein interaction (PPI) network of differentially expressed proteins (DEPs). The nodes with redder color showed a higher degree of interaction, whereas the nodes out of the circle showed less or no interaction.

2.8. *Verification of mRNA Expression Patterns of GS3 and Differentially Expressed Proteins (DEPs)*

The mRNA expression patterns of *GES3* and five representative DEPs coding genes are shown in Figure 7. RT-qPCR was performed to assess the expression level of target genes expression level in mutant plants and WT. The rice *actin* gene was used as internal control. The RT-qPCR results exhibited that the expression of *GS3* was significantly suppressed in mutant plants compared to WT (Figure 7A). To validate the proteomic data, we selected the five genes associated with DEPs. In total, two key genes encoding down-regulated proteins including *Os01g0270100* and *GLUP6*, and three genes encoding up-regulated proteins including *Os01g0978100*, *Os06g0698859*, and *Os06g0698859* were chosen. The protein expression pattern obtained by iTRAQ differential labeling was consistent with the trend of mRNA expression pattern in RT-qPCR (Figure 7B), indicating that the result of identifying and quantifying the proteome by iTRAQ differential labeling combined with high-performance liquid chromatography is reliable. The primers used for RT-qPCR are enlisted in Table S5.

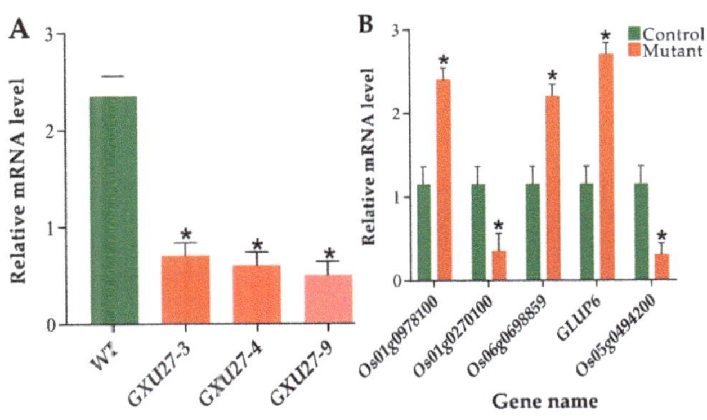

Figure 7. RT-qPCR-based assessment of *GS3* expression and validation of proteomic experiment. (**A**) The expression level of *GS3* in wild-type (WT) and mutant plants. (**B**) Expression analysis of five selected genes associated with differentially expressed proteins (DEPs). * Denotes a significant difference, Student's *t*-test, $p \leq 0.01$, $n = 3$.

3. Discussion

Gene-editing technology has emerged in recent years, which uses sequence-specific nucleases (SSNs) to introduce or delete bases at specific sites in DNA to generate high-frequency-induced mutations in target genes. It has attracted the attention of most researchers because of its broad application prospects in the fields of medicine, agriculture, and industry. Compared with conventional breeding, CRISPR/Cas9 technology is widely used in rice germplasm improvement due to its simple operation, high editing efficiency, and low cost [11,12,28,31], which greatly shortens the breeding cycle. CRISPR/Cas9 has been widely used for crop genetic improvement, and various vectors have been developed which are used to edit the genome of monocots and dicots [24]. The mining and utilization of key functional genes is an important way to improve crop yield and quality, but obtaining mutants with loss of function is a prerequisite for the identification of genetic functions of genes. Traditional genetic research strategies, such as EMS mutagenesis, T-DNA insertion mutation, transposon insertion mutation, and RNA interference greatly limit the functional identification of target genes due to the random nature of mutation sites. The CRISPR/Cas gene-editing technology has the advantages of high efficiency, and it can knock out almost any functional gene. Therefore, it has shown great application prospects in the field of research on the utilization of key functional genes for crop improvement. Currently, rice is one of the most successful crops for CRISPR/Cas9 applications. Studies have shown that in CRISPR/Cas9-mediated T_0 mutants, the probability of obtaining homozygous or biallelic mutations is highest [10–12].

The editing results of the *GS3* gene showed that the *Cas9* system with a U6 promoter had high efficiency. The editing efficiency of target one was 73%, and the editing efficiency of the second target was 80%. Currently, CRISPR/Cas9 is widely used in rice, and an editing efficiency of more than 80% has been achieved [30]. Higher editing efficiency makes it easy to get a variety of different types of mutations. In this study, the deletions were occurred frequently, whereas the rate of base insertion was relatively low. The GL and GWT were increased by 31.39% and 27.15%, respectively. The present results show that the use of CRISPR/Cas9 technology can breed long-grain and high-yielding varieties and accelerate the creation of long-grain germplasm resources.

To identify the proteome-wide changes between mutants and WT, a comparative iTRAQ-proteomics analysis was performed, and DEPs were screened successfully. We found that proteins related to cysteine synthase, cysteine proteinase inhibitor, ubiquitin, vacuolar protein sorting-associated, and DNA ligase were differentially regulated in mutant plants. The GO analysis also showed that DEPs were enriched in cellular process, metabolic process, binding, transmembrane, structural and catalytic activities. The KEGG analysis found that the DEPs were only enriched in lipid metabolism and oxylipin biosynthesis. In the PPI network, the proteins related to DNA damage-binding, JUN-activation-domain-binding, vacuolar protein sorting-associated, ubiquitin-40S ribosomal, and cysteine proteinase inhibitor showed the highest interaction.

The plant ubiquitin/proteasome system is the main way of protein degradation in cells and plays an important role in the process of plant growth and development, morphogenesis, and disease resistance [32]. Recent studies have shown that some pathogenic bacteria can simulate the host plant ubiquitin/protease system components [33]. The ubiquitin/proteasome pathway is mainly composed of ubiquitin-activating enzyme E1, ubiquitin-conjugating enzyme E2, ubiquitin-protein ligase E3, proteasome, and deubiquitinating enzymes (DUBs). Studies have shown that ubiquitination plays an important role in plant growth and development and the response of plants to biotic and abiotic stress [33–35]. Previous studies revealed that ubiquitin ligase is a functional partner of the Ribosome Quality Control Complex (RQC), which forms a stable complex with 60S ribosomal subunits containing stalled polypeptides and triggers their degradation. It also causes dissociation of the ribosome into the 40S and 60S subunits and translation-stress signaling pathway from the ribosome dependent on the RQC members [36]. Ubiquitin has been characterized as a mark for the degradation of proteins and regulates endogenous proteins lacking any folding defect [37–39]. Ubiquitin can control various cellular processes by altering protein localization, inducing structural changes, and regulating protein interactions [40–43]. These different functions have been observed in all signaling pathways, such as DNA repair, endocytosis, kinase regulation, transcriptional and translational control [43–46]. The Arabidopsis E3 ubiquitin ligase positively regulates the cytosol's protein levels, probably before importing pre-proteins into chloroplast during the formation of functional chloroplasts [47]. Misfolded proteins exported to the cytosol are subsequently ubiquitinated by E1 ubiquitin-activating enzymes, E2 ubiquitin-conjugating enzymes, and E3 ubiquitin ligases for their eventual degradation through the 26S proteasome [48,49]. The cytosolic protein recognizes specific sequence motifs and promotes pre-protein degradation by the 26S proteasome through interaction with the C-terminus of the E3 ubiquitin ligase Hsc70-interacting protein in Arabidopsis [50].

The results indicate that *GS3* mutations may function in the ubiquitination pathway. Interestingly, among the four grain-weight related genes identified previously through map-based cloning of QTLs, three of them encoding proteins that are possibly associated with the ubiquitination types of machinery. *GW2* has been shown to encode a cytosolic RING-type protein with E3 ubiquitin ligase activity [14], and *GW5/qSW5* encodes a nuclear protein that physically interacts with a polyubiquitin [51,52], whereas the protein encoded by *GS3* contains a putative cysteine-rich domain of the tumor necrosis factor receptor (TNFR), which likely colocalizes with ubiquitin in human cells [53]. These data together raise the possibility that *GS3*, *GW2*, and *GW5/qSW5* act through the same ubiquitination pathways to determine grain size and grain weight in rice.

Cysteine is a superfamily of secreted proteins that are widely found in plants and animals. Its primary function is to interact with proteases and control various physiological traits of organisms. These proteins are involved in preventing the functional protein from being degraded by exogenous proteases. The cysteine protease inhibitor superfamily can be divided into three families (I, II, and III) according to the amino acid sequence [54–56]. Cysteine proteins are recognized as the major enzymes for the catabolism of the majority of reserve proteins in seeds. Studies have shown that plant cysteine proteinase inhibitor proteins play an important role in terms of insect feeding inhibitors, resulting in abnormal development or death of insects [57–61]. In addition, cysteine protease inhibitors are also involved in plant growth and development, such as seed germination and maturation, seedling growth, fruit ripening, and programmed cell death [62–65] and improving plant abiotic stress tolerance [66–68]. At present, cysteine protease inhibitors have been cloned from a variety of plants, such as *Arabidopsis thaliana* [57], rice [69], cotton [70], and tobacco [71]. Vacuolar protein sorting-associated (VPS) proteins are the main component of endosomal sorting and transport complex II and play an important role in the ubiquitin-mediated degradation of membrane proteins in the multivesicular pathway. In rice, *OsVPS22* is very important for seedling viability and grain filling [72]. Lipid metabolism plays a vital role in plant reproductive development and is also involved in biosynthesis and transport [73,74]. Lipid metabolism is also crucial for callus formation, female gametophyte development, spikelet development, and flowering [75–79]. The defects in lipid metabolism-related genes are lethal to embryo or seed development in various plants [80–82]. Lipid-derived jasmonic acids (JA) genes have been demonstrated to play critical roles in anther dehiscence and pollen maturation in Arabidopsis and rice. Loss functions of genes, such as *dad1*, *fad3/fad7/fad8*, *opr3*, and *aos* in Arabidopsis, and *osjar1* in rice, generally lead to defects in anther dehiscence, filament elongation, and pollen maturation [83–87]. Several JA biosynthetic genes, *Defective in Anther Dehiscence1/Extra Glume 1 (OsDAD1/EG1)*, *Allene Oxide Cyclase gene (OsAOC)*, *Jasmonic Acid Carboxyl Methyltransferase gene (OsJMT)*, and *Open Glume 1 (OsOG1)/OsOPR7* in rice, are required for rice spikelet development and flowering [77,88–90]. Therefore, it might increase grain yield per plant by proper manipulation of these genes in rice. Notably, the mechanism of lipid metabolism underlying *GS3* mutations and reproductive development processes needs to be further elucidated.

We can attribute that functional crosstalk maybe exists between the differentially regulated proteins. Thus, different functional changes generated by the mutations analyzed here could directly and/or indirectly contribute to the mutant's phenotype. Alternatively, the relatively increased level of expression of these proteins may be advantageous in grain development. The differential expression of these DEPs may be triggered by *GS3* mutations. It is speculated that *GS3* mutations may regulate the expression of vacuolar protein sorting-associated, putative ubiquitin, and cysteine proteinase inhibitor proteins, and then play a positive regulatory role in the process of grain development. Therefore, the molecular mechanism of the differential response of these proteins in *GS3* mutants deserves further attention. The author believes that using different bioinformatic approaches and gene functional studies, it is possible to further understand the molecular regulation network mechanism of rice grain development to select parents with improved yield.

4. Materials and Methods

4.1. Test Materials

This study used the japonica rice variety TP309 as the test material. WT and mutant plants were grown in the experimental area of Guangxi University (45° N latitude) and Hainan (19° N latitude) and planted with a row spacing of 25 cm × 25 cm under natural conditions. The promoters (U6a and U6b) and pYLCRISPR/Cas9Pubi-H binary vector used in the experiment were kindly provided by Liu Yaoguang, South China Agricultural University. In this study, the CRISPR/Cas9 construct was carefully designed with higher specificity and low off-target score, and the *GS3* gene with the expectation to produce a null mutation was edited. Moreover, the iTRAQ-based proteomic analysis was also performed to assess the effect of mutations on the whole proteome. The schematic representation of the entire workflow of generation and analysis of targeted mutated plants is described in Figure 8A.

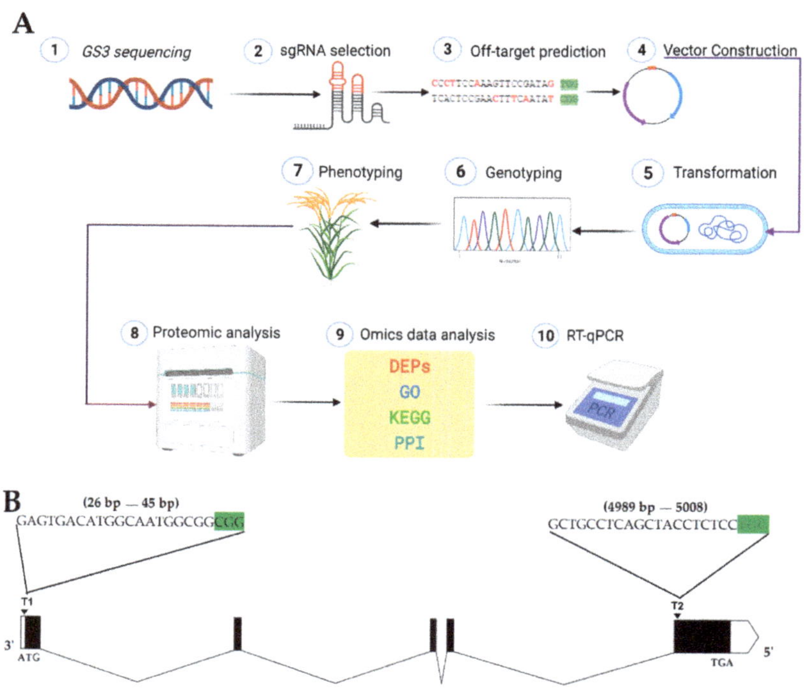

Figure 8. (**A**) Schematic diagram of the procedure for CRISPR/Cas9-based generation of mutant plants and analysis of mutations. Two sgRNAs were selected using the CRISPR-GE online web-based tool, and vector was constructed.

Agrobacterium-mediated transformation was performed, and T_0 plants were regenerated. Later generations were produced by self-pollination, and genotyping was performed using target-specific primers. The phenotypic data of mutant and wild-type (WT) plants were recorded and further analyzed. The proteomic analysis was also performed, and RT-qPCR was performed to assess the *GS3* expression level and validate the proteomic data. (**B**) Diagram of *GS3* gene and positions of both target sites. ATG is the start codon; TGA is the stop codon; green highlighted CGG and TGG are the PAM sequences; the white boxes at extreme left and right represent the untranslated (UTR) regions, the black boxes represents the exons; black lines in between the exon regions represent the intron regions, T1 and T2 represent target 1 and target 2, respectively.

4.2. gRNA Design, Vector Construction, and Transformation

The gene sequence of *GS3* (*Os03g0407400*) was retrieved through NCBI (https://www.ncbi.nlm.nih.gov/) [91] and using the CRISPR-GE online website (http://skl.scau.edu.cn/) [92], a pair of targets and their linker primers were designed in the first and last exon (Figure 8B; Table S6). The structure of both sgRNAs was developed using CRISPR-P ver 2.0 (http://crispr.hzau.edu.cn/CRISPR2/) [93] (Figure S2). Referring to Ma et al. (2015) [94], the vector was contrasted. The connected vector was transferred into DH5α by the heat shock method. The positive colonies were selected by colony PCR using specific primers (SPL1/SP-R), and the product was directly sequenced. The plasmid that was detected correctly was transferred into EHA105 Agrobacterium competent cells by heat shock. EHA105 by electroporation and rice transformation was performed according to Hiei et al. (1994) [95].

4.3. Genotyping, Phenotyping, and Screening of T-DNA-free Plants

At the seedling stage of the T_0 plants, the CTAB method was used to extract the genomic DNA from the leaves, and the target-specific primers (GS3T1F/R and GS3T2F/R) were used for genotyping. The sequencing results were detected using the degenerate sequence decoding (DSDecode) method [96] to analyze the mutation types and frequency. The leaf DNA of the transgenic T_0 generation plants was extracted by the CTAB method and stored in a refrigerator at −20 °C. The designed hygromycin primer HPTF/R and carrier primer Cas9F/R were used to screen T-DNA-free plants. The PCR system was as follows: 2× *Taq* Master Mix 7.5 µL, forward and reverse primers (10 µ mol/L) 0.5µL each, 1 µL DNA, 5.5 µL ddH_2O. The PCR program was as follows: 1 min and 30 s at 94 °C; 30 s at 94 °C; 30 s at 57 °C; 30 s at 72 °C; 5 min at 72 °C; and 5 min at 10 °C. The PCR products were run on gel electrophoresis using 2% agarose gel and stained with ethidium bromide solution. The plants not showing bands were considered as T-DNA-free plants. WT, T_0, T_1, and T_2 generation plants were planted in field conditions, and the agronomic data were recorded. The PH, FLL, and GLW were measured at the maturity stage. The seeds were dried, and data for GL, GWD, and GWT were measured from randomly selected grains.

4.4. Protein Extraction and Quality Inspection

The three g of rice leaf samples from WT and mutant plant of GXU27-4 were taken and ground into a powder with liquid nitrogen and transferred to a 15 mL centrifuge tube. Protein extraction buffer (8 mol/L urea, 0.1% SDS, 1 mmol/L PMSF, 1 mmol/L DTT) was added to the sample and shaken for 3 h at room temperature on an automatic vortex mixer (Thermo Fisher Scientific, Shanghai, China). After vortexing, the sample was centrifuged for 15 min at 4 °C, 14,000 r/min, and the supernatant was transferred to a new 15 mL centrifuge tube. Pre-cooled acetone was added to the 6 times volume of a sample and precipitated overnight at −20 °C. The precipitated overnight sample was centrifuged at 4 °C and 12,000× *g* for 15 min, and the supernatant was removed. The precipitate was vacuum dried at 4 °C to obtain the total protein dry powder, which was stored at −80 °C for later use.

A total of 50 mg of dry protein powder was taken and lysis solution (containing 8 mol/L urea, 2 mol/L thiourea, mass fraction 4% CHAPS, pH 3–10 volume fraction 0.5% ampholyte, 50 mmol/L dithiothreitol (DTT), and 1.0 mmol/L after phenylmethanesulfonyl fluoride (PMSF)) was added, dissolved, and centrifuged at 4 °C, 12,000× *g* for 15 min.

The supernatant (total protein solution) was transferred to a 1.5 mL centrifuge tube and SDS-PAGE electrophoresis, and Coomassie brilliant blue staining were used to detect the integrity of the total protein. The concentration of protein samples was determined by the Bradford protein assay.

4.5. Protein Digestion and iTRAQ Labeling

The protein sample was first reduced and alkylated and then digested with trypsin enzyme. One hundred μg of protein was taken from each sample, and after incubating with 10 mmol/L dithiothreitol (DTT) at 37 °C for 1 h, and 55 mmol/L iodoacetamide at room temperature for 1 h, 3.3 μg of pancreatic protein enzyme digestion was carried out at 37 °C for 12 h. At the end of digestion, 100 μL of formic acid with a volume fraction of 1% was added to terminate the enzymatic hydrolysis reaction and then vacuum-dried to obtain a dry powder of protein peptides. The dry powder of protein peptides was dissolved in 8 mol/L urea (containing 0.1% SDS) and 500 mmol/L triethylammonium bicarbonate in water and the 8-plex iTRAQ kit of AB Sciex was used for labeling. The 6 tubes of labeling reagents (110, 111, 112, 113, 114, and 115) in the kit were diluted with 50 μL of isopropanol and mixed with the corresponding protein-peptide dry powder reconstitution samples and placed at room temperature. The three WT and GXU27-4 samples were labeled with iTRAQ reagent.

4.6. Strong Cation Exchange Column Classification and Mass Spectrometry Detection of iTRAQ Labeled Samples

Six iTRAQ-labeled samples were mixed in equal amounts and then graded using a C18 strong cation column (SEC). The peptide mixture was loaded onto SEC in equilibration buffer A (containing pH 2.55, 5 mmol/L KH_2PO_4, 20% acetonitrile, and H_3PO_4 by volume) for 25 min. Then, 300 μL of the labeled sample mixture was taken and used SEC equilibration buffer B (containing pH 2.75, 5 mmol/L $KH\ _2PO_4$, 20% acetonitrile by volume, 600 mmol/L KCl, $H\ _3PO_4$) to dilute 7 times, and orthophosphoric acid was added to adjust the pH value to 2.5. The sample was centrifuged to take the supernatant for gradient elution with the eluent flow rate of 0.2 mL/min. In the full scan, the top 20 precursor ions of the ion intensity were selected after fragmentation in the higher-energy collisional dissociation (HCD) mode with a standard collision energy of 30 eV, and then, the secondary mass spectrometry sequence was determined to report the ion. The rest of the procedure was followed according to Wang et al. (2014) [97].

4.7. Proteomic Data Analysis and Functional Annotation

The mass spectrometry results were imported into the Proteome Discoverer software to search the Uniprot database (http://www.uniprot.org/) for the functional annotation of the protein. According to the protein expression counted by the software Proteome Discoverer, the DEPs analysis was performed. The DEPs were screened between the WT and mutant lines with a fold change (FC) of 1.2 and $p < 0.05$. The online GO database (http://geneontology.org/) and the KEGG database (http://www.genome.jp/kegg/pathway) were searched using the KOBAS software according to the protein-coding gene ID. STRING database (version 10.0) (https://string-db.org/) was searched for PPI network for all up and down-regulated proteins. Then, the network was visualized by Cytoscape version 3.8.0.

4.8. Verification of mRNA Expression of GS3 and Protein-Coding Genes

The expression pattern of *GS3* was analyzed in WT and mutants' plants. To verify the reliability of the iTRAQ data, 10 representative DEPs were selected from different functional classifications. According to their coding genes, the mRNA sequences were retrieved from China Rice Data Center (http://www.ricedata.cn/), and RTqPCR primers were designed using qPrimerDB (https://biodb.swu.edu.cn/qprimerdb/). The RT-qPCR analyses were conducted, and the gene expression was calculated by using the $2^{-\Delta\Delta Ct}$ (cycle threshold) process as described previously [98].

4.9. Statistical Analyses

Statistical analysis ($p < 0.05$) was completed using SPSS 16.0 Statistical Software Program and GraphPad Prism (version 7.0, GraphPad Software Inc. San Diego, CA, USA) was used to display the graphs.

5. Conclusions

In this study, *GS3* mutants were generated using the CRISPR/Cas9 system and iTRAQ-based quantitative proteomic analysis was performed. The proteome-wide characterization of mutants provides new understandings, and the targeted genome editing also facilitated the identification DEPs and pathways that may be involved in rice grain development. The differential response of proteins related to cysteine synthase, cysteine proteinase inhibitor, ubiquitin, and DNA ligase in *GS3* mutants will provide a reference and sharp focus in rice molecular breeding. Therefore, the molecular mechanism of the up-regulation of these proteins' expression in *GS3* mutants deserves further attention. Genome editing facilitated pathway-level study, and DEPs found in this study might be directly or indirectly involved in rice grain development and can be analyzed further to reveal their functional role. The mutants with different mutations represent population diversity, which is the main driving force of breeding work. This study provides a strategy for the rapid introduction of genetic diversity in the process of crop breeding and practical significance.

Supplementary Materials: The following are available online at https://www.mdpi.com/1422-0067/22/6/3225/s1.

Author Contributions: Conceptualization, B.U. and N.Z.; Data curation, G.N.; Formal analysis, B.U., G.N., B.Q. and F.L.; Funding acquisition, R.L.; Investigation, B.U., G.N. and B.Q.; Methodology, B.U., N.Z., G.N. and Y.L.; Project administration, R.L.; Resources, R.L. and Y.L.; Software, B.U., N.Z., G.N., B.Q. and F.L.; Supervision, R.L.; Validation, F.L. and R.L.; Visualization, R.L. and Y.L.; Writing—original draft, B.U. and N.Z.; Writing—review and editing, B.U. and R.L. All authors have read and agreed to the published version of the manuscript.

Funding: This study was supported by the State Key Laboratory for Conservation and Utilization of Subtropical Agro-bioresources (SKLWSA-a201914).

Institutional Review Board Statement: Not applicable.

Informed Consent Statement: Not applicable.

Data Availability Statement: Raw data can be provided to researchers interested on request to the corresponding author.

Acknowledgments: We are grateful to Mohsin Niaz, Mudassar Abbas, and Umair Khalid for their support to make this study meaningful. The authors sincerely acknowledge the services of Jiwei Li (Shanghai Life Genes Biotechnology Co. Pvt. Ltd.; Website: www.lifegenes.com.cn) and assistance to perform comparative proteomic data analysis.

Conflicts of Interest: The authors declare that they have no competing professional or personal interests that might have influenced the presentation of the work described in this manuscript.

Abbreviations

CRISPR	Clustered regularly interspaced short palindromic repeats
Cas9	CRISPR-associated protein 9
GWT	1000-grain weight
GL	Grain length
iTRAQ	Isobaric tags for relative and absolute quantitation
DEPs	Differentially expressed proteins
KEGG	Kyoto Encyclopedia of Genes and Genomes
GO	Gene Ontology

References

1. Godfray, H.C.J.; Beddington, J.R.; Crute, I.R.; Haddad, L.; Lawrence, D.; Muir, J.F.; Pretty, J.; Robinson, S.; Thomas, S.M.; Toulmin, C. Food security: The challenge of feeding 9 billion people. *Science* **2010**, *327*, 812–818. [CrossRef]
2. Zhou, X.; Bai, X.; Xing, Y. A rice genetic improvement boom by next generation sequencing. *Curr. Issues Mol. Biol.* **2018**, *27*, 109–126. [CrossRef] [PubMed]
3. Rao, Y.; Li, Y.; Qian, Q. Recent progress on molecular breeding of rice in China. *Plant Cell Rep.* **2014**, *33*, 551–564. [CrossRef] [PubMed]
4. Wang, M.; Mao, Y.; Lu, Y.; Wang, Z.; Tao, X.; Zhu, J.K. Multiplex gene editing in rice with simplified CRISPR-Cpf1 and CRISPR-Cas9 systems. *J. Integr. Plant Biol.* **2018**, *60*, 626–631. [CrossRef]
5. Cai, Y.; Chen, L.; Liu, X.; Guo, C.; Sun, S.; Wu, C.; Jiang, B.; Han, T.; Hou, W. CRISPR/Cas9-mediated targeted mutagenesis of *GmFT2a* delays flowering time in soya bean. *Plant Biotechnol. J.* **2018**, *16*, 176–185. [CrossRef]
6. Qi, W.; Zhu, T.; Tian, Z.; Li, C.; Zhang, W.; Song, R. High-efficiency CRISPR/Cas9 multiplex gene editing using the glycine tRNA-processing system-based strategy in maize. *BMC Biotechnol.* **2016**, *16*, 1–8. [CrossRef]
7. Zhang, S.; Zhang, R.; Song, G.; Gao, J.; Li, W.; Han, X.; Chen, M.; Li, Y.; Li, G. Targeted mutagenesis using the Agrobacterium tumefaciens-mediated CRISPR-Cas9 system in common wheat. *BMC Plant Biol.* **2018**, *18*, 1–12. [CrossRef]
8. Wang, P.; Zhang, J.; Sun, L.; Ma, Y.; Xu, J.; Liang, S.; Deng, J.; Tan, J.; Zhang, Q.; Tu, L. High efficient multisites genome editing in allotetraploid cotton (*Gossypium hirsutum*) using CRISPR/Cas9 system. *Plant Biotechnol. J.* **2018**, *16*, 137–150. [CrossRef]
9. Liu, G.; Li, J.; Godwin, I.D. Genome editing by CRISPR/Cas9 in sorghum through biolistic bombardment. In *Sorghum*; Springer: Berlin/Heidelberg, Germany, 2019; pp. 169–183. [CrossRef]
10. Usman, B.; Nawaz, G.; Zhao, N.; Liu, Y.; Li, R. Generation of high yielding and fragrant rice (*Oryza sativa* L.) Lines by CRISPR/Cas9 targeted mutagenesis of three homoeologs of cytochrome P450 gene family and *OsBADH2* and transcriptome and proteome profiling of revealed changes triggered by mutations. *Plants* **2020**, *9*, 788. [CrossRef]
11. Usman, B.; Nawaz, G.; Zhao, N.; Liao, S.; Liu, Y.; Li, R. Precise Editing of the *OsPYL9* Gene by RNA-Guided Cas9 Nuclease Confers Enhanced Drought Tolerance and Grain Yield in Rice (*Oryza sativa* L.) by Regulating Circadian Rhythm and Abiotic Stress Responsive Proteins. *Int. J. Mol. Sci.* **2020**, *21*, 7854. [CrossRef]
12. Usman, B.; Nawaz, G.; Zhao, N.; Liao, S.; Qin, B.; Liu, F.; Liu, Y.; Li, R. Programmed Editing of Rice (*Oryza sativa* L.) *OsSPL16* Gene Using CRISPR/Cas9 Improves Grain Yield by Modulating the Expression of Pyruvate Enzymes and Cell Cycle Proteins. *Int. J. Mol. Sci.* **2021**, *22*, 249. [CrossRef] [PubMed]
13. Zhou, Q.Y.; An, H.; Zhang, Y.; Shen, F.C. Study on heredity of morphological characters of rice grain. *J. Southwest Agric. Univ.* **2000**, *22*, 102–104.
14. Song, X.J.; Huang, W.; Shi, M.; Zhu, M.Z.; Lin, H.X. A QTL for rice grain width and weight encodes a previously unknown RING-type E3 ubiquitin ligase. *Nat. Genet.* **2007**, *39*, 623–630. [CrossRef] [PubMed]
15. Fan, C.; Xing, Y.; Mao, H.; Lu, T.; Han, B.; Xu, C.; Li, X.; Zhang, Q. GS3, a major QTL for grain length and weight and minor QTL for grain width and thickness in rice, encodes a putative transmembrane protein. *Theor. Appl. Genet.* **2006**, *112*, 1164–1171. [CrossRef]
16. Zhao, D.S.; Li, Q.F.; Zhang, C.Q.; Zhang, C.; Yang, Q.Q.; Pan, L.X.; Ren, X.Y.; Lu, J.; Gu, M.H.; Liu, Q.Q. *GS9* acts as a transcriptional activator to regulate rice grain shape and appearance quality. *Nat. Commun.* **2018**, *9*, 1–14. [CrossRef] [PubMed]
17. Liu, J.; Chen, J.; Zheng, X.; Wu, F.; Lin, Q.; Heng, Y.; Tian, P.; Cheng, Z.; Yu, X.; Zhou, K. *GW5* acts in the brassinosteroid signalling pathway to regulate grain width and weight in rice. *Nat. Plants* **2017**, *3*, 1–7. [CrossRef] [PubMed]
18. Ishimaru, K.; Hirotsu, N.; Madoka, Y.; Murakami, N.; Hara, N.; Onodera, H.; Kashiwagi, T.; Ujiie, K.; Shimizu, B.I.; Onishi, A. Loss of function of the IAA-glucose hydrolase gene *TGW6* enhances rice grain weight and increases yield. *Nat. Genet.* **2013**, *45*, 707–711. [CrossRef] [PubMed]
19. Qi, P.; Lin, Y.S.; Song, X.J.; Shen, J.B.; Huang, W.; Shan, J.X.; Zhu, M.Z.; Jiang, L.; Gao, J.P.; Lin, H.X. The novel quantitative trait locus *GL3.1* controls rice grain size and yield by regulating Cyclin-T1; 3. *Cell Res.* **2012**, *22*, 1666–1680. [CrossRef] [PubMed]
20. Li, Y.; Fan, C.; Xing, Y.; Jiang, Y.; Luo, L.; Sun, L.; Shao, D.; Xu, C.; Li, X.; Xiao, J. Natural variation in *GS5* plays an important role in regulating grain size and yield in rice. *Nat. Genet.* **2011**, *43*, 1266–1269. [CrossRef]
21. Wang, Y.; Xiong, G.; Hu, J.; Jiang, L.; Yu, H.; Xu, J.; Fang, Y.; Zeng, L.; Xu, E.; Xu, J. Copy number variation at the *GL7* locus contributes to grain size diversity in rice. *Nat. Genet.* **2015**, *47*, 944–948. [CrossRef]
22. Si, L.; Chen, J.; Huang, X.; Gong, H.; Luo, J.; Hou, Q.; Zhou, T.; Lu, T.; Zhu, J.; Shangguan, Y. *OsSPL13* controls grain size in cultivated rice. *Nat. Genet.* **2016**, *48*, 447–456. [CrossRef]
23. Mao, H.; Sun, S.; Yao, J.; Wang, C.; Yu, S.; Xu, C.; Li, X.; Zhang, Q. Linking differential domain functions of the *GS3* protein to natural variation of grain size in rice. *Proc. Natl. Acad. Sci. USA* **2010**, *107*, 19579–19584. [CrossRef] [PubMed]
24. Ma, X.; Zhu, Q.; Chen, Y.; Liu, Y.G. CRISPR/Cas9 platforms for genome editing in plants: Developments and applications. *Mol. Plant* **2016**, *9*, 961–974. [CrossRef]
25. Peng, B.; Li, J.; Kong, D.Y.; He, L.L.; Li, M.G.; Nassirou, T.Y.; Song, X.H.; Peng, J.; Jiang, Y.; Sun, Y.F. Genetic Improvement of Grain Quality Promoted by High and New Technology in Rice. *J. Agric. Sci.* **2019**, *11*. [CrossRef]
26. Lowder, L.G.; Zhang, D.; Baltes, N.J.; Paul, J.W.; Tang, X.; Zheng, X.; Voytas, D.F.; Hsieh, T.F.; Zhang, Y.; Qi, Y. A CRISPR/Cas9 toolbox for multiplexed plant genome editing and transcriptional regulation. *Plant Physiol.* **2015**, *169*, 971–985. [CrossRef] [PubMed]

27. Liao, S.; Qin, X.; Luo, L.; Han, Y.; Wang, X.; Usman, B.; Nawaz, G.; Zhao, N.; Liu, Y.; Li, R. CRISPR/Cas9-Induced Mutagenesis of *Semi-Rolled Leaf1, 2* Confers Curled Leaf Phenotype and Drought Tolerance by Influencing Protein Expression Patterns and ROS Scavenging in Rice (*Oryza sativa* L.). *Agronomy* **2019**, *9*, 728. [CrossRef]
28. Han, Y.; Teng, K.; Nawaz, G.; Feng, X.; Usman, B.; Wang, X.; Luo, L.; Zhao, N.; Liu, Y.; Li, R. Generation of semi-dwarf rice (*Oryza sativa* L.) lines by CRISPR/Cas9-directed mutagenesis of *OsGA20ox2* and proteomic analysis of unveiled changes caused by mutations. *3 Biotech* **2019**, *9*, 1–17. [CrossRef] [PubMed]
29. Nawaz, G.; Usman, B.; Peng, H.; Zhao, N.; Yuan, R.; Liu, Y.; Li, R. Knockout of *Pi21* by CRISPR/Cas9 and iTRAQ-Based Proteomic Analysis of Mutants Revealed New Insights into *M. oryzae* Resistance in Elite Rice Line. *Genes* **2020**, *11*, 735. [CrossRef]
30. Nawaz, G.; Han, Y.; Usman, B.; Liu, F.; Qin, B.; Li, R. Knockout of *OsPRP1*, a gene encoding proline-rich protein, confers enhanced cold sensitivity in rice (*Oryza sativa* L.) at the seedling stage. *3 Biotech* **2019**, *9*, 1–18. [CrossRef]
31. Han, Y.; Luo, D.; Usman, B.; Nawaz, G.; Zhao, N.; Liu, F.; Li, R. Development of high yielding glutinous cytoplasmic male sterile rice (*Oryza sativa* L.) lines through CRISPR/Cas9 based mutagenesis of *Wx* and *TGW6* and proteomic analysis of anther. *Agronomy* **2018**, *8*, 290. [CrossRef]
32. Hershko, A. The ubiquitin system for protein degradation and some of its roles in the control of the cell division cycle. *Cell Death Diff.* **2005**, *12*, 1191–1197. [CrossRef]
33. Dielen, A.S.; Badaoui, S.; Candresse, T.; German-Retana, S. The ubiquitin/26S proteasome system in plant–pathogen interactions: A never-ending hide-and-seek game. *Mol. Plant Pathol.* **2010**, *11*, 293–308. [CrossRef] [PubMed]
34. Stone, S.L. The role of ubiquitin and the 26S proteasome in plant abiotic stress signaling. *Front. Plant Sci.* **2014**, *5*, 135. [CrossRef]
35. Marino, D.; Peeters, N.; Rivas, S. Ubiquitination during plant immune signaling. *Plant Physiol.* **2012**, *160*, 15–27. [CrossRef] [PubMed]
36. Brandman, O.; Stewart-Ornstein, J.; Wong, D.; Larson, A.; Williams, C.C.; Li, G.W.; Zhou, S.; King, D.; Shen, P.S.; Weibezahn, J. A ribosome-bound quality control complex triggers degradation of nascent peptides and signals translation stress. *Cell* **2012**, *151*, 1042–1054. [CrossRef] [PubMed]
37. Hershko, A.; Ciechanover, A.; Heller, H.; Haas, A.L.; Rose, I.A. Proposed role of ATP in protein breakdown: Conjugation of protein with multiple chains of the polypeptide of ATP-dependent proteolysis. *Proc. Natl. Acad. Sci. USA* **1980**, *77*, 1783–1786. [CrossRef]
38. Hershko, A.; Heller, H.; Elias, S.; Ciechanover, A. Components of ubiquitin-protein ligase system. Resolution, affinity purification, and role in protein breakdown. *J. Biol. Chem.* **1983**, *258*, 8206–8214. [CrossRef]
39. Johnson, E.S.; Bartel, B.; Seufert, W.; Varshavsky, A. Ubiquitin as a degradation signal. *EMBO J.* **1992**, *11*, 497–505. [CrossRef]
40. Hicke, L.; Riezman, H. Ubiquitination of a yeast plasma membrane receptor signals its ligand-stimulated endocytosis. *Cell* **1996**, *84*, 277–287. [CrossRef]
41. Hurst, J.H.; Dohlman, H.G. Dynamic ubiquitination of the mitogen-activated protein kinase kinase (MAPKK) Ste7 determines mitogen-activated protein kinase (MAPK) specificity. *J. Biol. Chem.* **2013**, *288*, 18660–18671. [CrossRef]
42. Jentsch, S.; McGrath, J.P.; Varshavsky, A. The yeast DNA repair gene RAD6 encodes a ubiquitin-conjugating enzyme. *Nature* **1987**, *329*, 131–134. [CrossRef] [PubMed]
43. Mayer, J.; Layfield, R.; Dhananjayan, S.C.; Ismail, A.; Nawaz, Z. Ubiquitin and control of transcription. *Essays Biochem.* **2005**, *41*, 69–80. [CrossRef]
44. An, H.; Harper, J.W. Ribosome abundance control via the ubiquitin–proteasome system and autophagy. *J. Mol. Biol.* **2020**, *432*, 170–184. [CrossRef] [PubMed]
45. Back, S.; Gorman, A.W.; Vogel, C.; Silva, G.M. Site-specific K63 ubiquitinomics provides insights into translation regulation under stress. *J. Prot. Res.* **2018**, *18*, 309–318. [CrossRef] [PubMed]
46. Sung, M.K.; Porras-Yakushi, T.R.; Reitsma, J.M.; Huber, F.M.; Sweredoski, M.J.; Hoelz, A.; Hess, S.; Deshaies, R.J. A conserved quality-control pathway that mediates degradation of unassembled ribosomal proteins. *eLife* **2016**, *5*, e19105. [CrossRef]
47. Wei, J.; Qiu, X.; Chen, L.; Hu, W.; Hu, R.; Chen, J.; Sun, L.; Li, L.; Zhang, H.; Lv, Z. The E3 ligase AtCHIP positively regulates Clp proteolytic subunit homeostasis. *J. Exp. Bot.* **2015**, *66*, 5809–5820. [CrossRef]
48. Tsai, B.; Ye, Y.; Rapoport, T.A. Retro-translocation of proteins from the endoplasmic reticulum into the cytosol. *Nat. Rev. Mol. Cell Biol.* **2002**, *3*, 246–255. [CrossRef]
49. Ye, Y.; Meyer, H.H.; Rapoport, T.A. The AAA ATPase Cdc48/p97 and its partners transport proteins from the ER into the cytosol. *Nature* **2001**, *414*, 652–656. [CrossRef]
50. Lee, S.; Lee, D.W.; Lee, Y.; Mayer, U.; Stierhof, Y.D.; Lee, S.; Jürgens, G.; Hwang, I. Heat shock protein cognate 70-4 and an E3 ubiquitin ligase, CHIP, mediate plastid-destined precursor degradation through the ubiquitin-26S proteasome system in Arabidopsis. *Plant Cell* **2009**, *21*, 3984–4001. [CrossRef] [PubMed]
51. Shomura, A.; Izawa, T.; Ebana, K.; Ebitani, T.; Kanegae, H.; Konishi, S.; Yano, M. Deletion in a gene associated with grain size increased yields during rice domestication. *Nat. Genet.* **2008**, *40*, 1023–1028. [CrossRef] [PubMed]
52. Weng, J.; Gu, S.; Wan, X.; Gao, H.; Guo, T.; Su, N.; Lei, C.; Zhang, X.; Cheng, Z.; Guo, X. Isolation and initial characterization of *GW5*, a major QTL associated with rice grain width and weight. *Cell Res.* **2008**, *18*, 1199–1209. [CrossRef]
53. Todd, I.; Radford, P.M.; Draper-Morgan, K.A.; McIntosh, R.; Bainbridge, S.; Dickinson, P.; Jamhawi, L.; Sansaridis, M.; Huggins, M.L.; Tighe, P.J. Mutant forms of tumour necrosis factor receptor I that occur in TNF-receptor-associated periodic syndrome retain signalling functions but show abnormal behaviour. *Immunology* **2004**, *113*, 65–79. [CrossRef] [PubMed]

54. Turk, V.; Bode, W. The cystatins: Protein inhibitors of cysteine proteinases. *FEBS Lett.* **1991**, *285*, 213–219. [CrossRef]
55. Van Wyk, S.G.; Kunert, K.J.; Cullis, C.A.; Pillay, P.; Makgopa, M.E.; Schlüter, U.; Vorster, B.J. The future of cystatin engineering. *Plant Sci.* **2016**, *246*, 119–127. [CrossRef] [PubMed]
56. Martínez, M.; Abraham, Z.; Carbonero, P.; Díaz, I. Comparative phylogenetic analysis of cystatin gene families from arabidopsis, rice and barley. *Mol. Genet. Genom.* **2005**, *273*, 423–432. [CrossRef] [PubMed]
57. Belenghi, B.; Acconcia, F.; Trovato, M.; Perazzolli, M.; Bocedi, A.; Polticelli, F.; Ascenzi, P.; Delledonne, M. *AtCYS1*, a cystatin from *Arabidopsis thaliana*, suppresses hypersensitive cell death. *Eur. J. Biochem.* **2003**, *270*, 2593–2604. [CrossRef] [PubMed]
58. Lei, J.J.; Yang, W.J.; Yuan, S.H.; Ying, F.Y.; Qiong, L.C. Study on transformation of cysteine proteinase inhibitor gene into cabbage (*Brassica oleracea* var. *capitata* L.). In Proceedings of the IV International Symposium on Brassicas and XIV Crucifer Genetics Workshop 706, Daejeon, Korea, 1 April 2006. [CrossRef]
59. Bouchard, E.; Michaud, D.; Cloutier, C. Molecular interactions between an insect predator and its herbivore prey on transgenic potato expressing a cysteine proteinase inhibitor from rice. *Mol. Ecol.* **2003**, *12*, 2429–2437. [CrossRef]
60. Gholizadeh, A. The possible involvement of d-amino acids or their metabolites in Arabidopsis cysteine proteinase/cystatin-dependent proteolytic pathway. *Cytol. Genet.* **2015**, *49*, 73–79. [CrossRef]
61. Irie, K.; Hosoyama, H.; Takeuchi, T.; Iwabuchi, K.; Watanabe, H.; Abe, M.; Abe, K.; Arai, S. Transgenic rice established to express corn cystatin exhibits strong inhibitory activity against insect gut proteinases. *Plant Mol. Biol.* **1996**, *30*, 149–157. [CrossRef]
62. Rassam, M.; Laing, W.A. Purification and characterization of phytocystatins from kiwifruit cortex and seeds. *Phytochemistry* **2004**, *65*, 19–30. [CrossRef] [PubMed]
63. Tan, Y.; Wang, S.; Liang, D.; Li, M.; Ma, F. Genome-wide identification and expression profiling of the cystatin gene family in apple (*Malus* × *domestica* Borkh.). *Plant Physiol. Biochem.* **2014**, *79*, 88–97. [CrossRef]
64. Zhao, P.; Zhou, X.M.; Zou, J.; Wang, W.; Wang, L.; Peng, X.B.; Sun, M.X. Comprehensive analysis of cystatin family genes suggests their putative functions in sexual reproduction, embryogenesis, and seed formation. *J. Exp. Bot.* **2014**, *65*, 5093–5107. [CrossRef] [PubMed]
65. Hwang, J.E.; Hong, J.K.; Je, J.H.; Lee, K.O.; Kim, D.Y.; Lee, S.Y.; Lim, C.O. Regulation of seed germination and seedling growth by an Arabidopsis phytocystatin isoform, *AtCYS6*. *Plant Cell Rep.* **2009**, *28*, 1623–1632. [CrossRef] [PubMed]
66. Pernas, M.; Sánchez-Monge, R.; Salcedo, G. Biotic and abiotic stress can induce cystatin expression in chestnut. *FEBS Lett.* **2000**, *467*, 206–210. [CrossRef]
67. Van der Vyver, C.; Schneidereit, J.; Driscoll, S.; Turner, J.; Kunert, K.; Foyer, C.H. Oryzacystatin I expression in transformed tobacco produces a conditional growth phenotype and enhances chilling tolerance. *Plant Biotechnol. J.* **2003**, *1*, 101–112. [CrossRef] [PubMed]
68. Papolu, P.K.; Dutta, T.K.; Tyagi, N.; Urwin, P.E.; Lilley, C.J.; Rao, U. Expression of a cystatin transgene in eggplant provides resistance to root-knot nematode, Meloidogyne incognita. *Front. Plant Sci.* **2016**, *7*, 1122. [CrossRef]
69. Wang, W.; Zhao, P.; Zhou, X.; Xiong, H.; Sun, M. Genome-wide identification and characterization of cystatin family genes in rice (*Oryza sativa* L.). *Plant Cell Rep.* **2015**, *34*, 1579–1592. [CrossRef] [PubMed]
70. Jiang, M.; Li, S.; Chen, M.; Cai, Y.; Xie, Y.; Li, B.; Sun, Q.; Jiang, H.; Pan, Z.; Gao, Y.L. Molecular cloning and expression of cDNA encoding the cysteine proteinase inhibitor from upland cotton. *J. Plant Biol.* **2009**, *52*, 426–432. [CrossRef]
71. Miyaji, T.; Murayama, S.; Kouzuma, Y.; Kimura, N.; Kanost, M.R.; Kramer, K.J.; Yonekura, M. Molecular cloning of a multidomain cysteine protease and protease inhibitor precursor gene from the tobacco hornworm (*Manduca sexta*) and functional expression of the cathepsin F-like cysteine protease domain. *Insect Biochem. Mol. Biol.* **2010**, *40*, 835–846. [CrossRef]
72. Zhang, X.Q.; Hou, P.; Zhu, H.T.; Li, G.D.; Liu, X.G.; Xie, X.M. Knockout of the *VPS22* component of the ESCRT-II complex in rice (*Oryza sativa* L.) causes chalky endosperm and early seedling lethality. *Mol. Biol. Rep.* **2013**, *40*, 3475–3481. [CrossRef]
73. Ariizumi, T.; Toriyama, K. Genetic regulation of sporopollenin synthesis and pollen exine development. *Ann. Rev. Plant Biol.* **2011**, *62*, 437–460. [CrossRef] [PubMed]
74. Shi, J.; Cui, M.; Yang, L.; Kim, Y.J.; Zhang, D. Genetic and biochemical mechanisms of pollen wall development. *Trends Plant Sci.* **2015**, *20*, 741–753. [CrossRef] [PubMed]
75. Shang, B.; Xu, C.; Zhang, X.; Cao, H.; Xin, W.; Hu, Y. Very-long-chain fatty acids restrict regeneration capacity by confining pericycle competence for callus formation in Arabidopsis. *Proc. Natl. Acad. Sci. USA* **2016**, *113*, 5101–5106. [CrossRef] [PubMed]
76. Acosta, I.F.; Laparra, H.; Romero, S.P.; Schmelz, E.; Hamberg, M.; Mottinger, J.P.; Moreno, M.A.; Dellaporta, S.L. tasselseed1 is a lipoxygenase affecting jasmonic acid signaling in sex determination of maize. *Science* **2009**, *323*, 262–265. [CrossRef]
77. Cai, Q.; Yuan, Z.; Chen, M.; Yin, C.; Luo, Z.; Zhao, X.; Liang, W.; Hu, J.; Zhang, D. Jasmonic acid regulates spikelet development in rice. *Nat. Comm.* **2014**, *5*, 1–13. [CrossRef]
78. Wang, F.; Yuan, Z.; Zhao, Z.; Li, C.; Zhang, X.; Liang, H.; Liu, Y.; Xu, Q.; Liu, H. Tasselseed5 encodes a cytochrome C oxidase that functions in sex determination by affecting jasmonate catabolism in maize. *J. Integr. Plant Biol.* **2020**, *62*, 247–255. [CrossRef]
79. Yan, Y.; Christensen, S.; Isakeit, T.; Engelberth, J.; Meeley, R.; Hayward, A.; Emery, R.N.; Kolomiets, M.V. Disruption of *OPR7* and *OPR8* reveals the versatile functions of jasmonic acid in maize development and defense. *Plant Cell* **2012**, *24*, 1420–1436. [CrossRef]
80. Zhang, J.; Li, J.; Garcia-Ruiz, H.; Bates, P.D.; Mirkov, T.E.; Wang, X. A stearoyl-acyl carrier protein desaturase, N b SACPD-C, is critical for ovule development in N icotiana benthamiana. *Plant J.* **2014**, *80*, 489–502. [CrossRef]

81. Kim, H.U.; Huang, A.H. Plastid lysophosphatidyl acyltransferase is essential for embryo development in Arabidopsis. *Plant Physiol.* **2004**, *134*, 1206–1216. [CrossRef]
82. Chen, Q.F.; Xiao, S.; Qi, W.; Mishra, G.; Ma, J.; Wang, M.; Chye, M.L. The Arabidopsis acbp1acbp2 double mutant lacking acyl-CoA-binding proteins ACBP1 and ACBP2 is embryo lethal. *New Phytol.* **2010**, *186*, 843–855. [CrossRef]
83. Ishiguro, S.; Kawai-Oda, A.; Ueda, J.; Nishida, I.; Okada, K. The *Defective in Anther Dehiscence1* gene encodes a novel phospholipase A1 catalyzing the initial step of jasmonic acid biosynthesis, which synchronizes pollen maturation, anther dehiscence, and flower opening in Arabidopsis. *Plant Cell* **2001**, *13*, 2191–2209. [CrossRef] [PubMed]
84. McConn, M. The critical requirement for linolenic acid is pollen development, not photosynthesis, in an Arabidopsis mutant. *Plant Cell* **1996**, *8*, 403–416. [CrossRef] [PubMed]
85. Stintzi, A. The Arabidopsis male-sterile mutant, opr3, lacks the 12-oxophytodienoic acid reductase required for jasmonate synthesis. *Proc. Natl. Acad. Sci. USA* **2000**, *97*, 10625–10630. [CrossRef]
86. Park, J.H.; Halitschke, R.; Kim, H.B.; Baldwin, I.T.; Feldmann, K.A.; Feyereisen, R. A knock-out mutation in allene oxide synthase results in male sterility and defective wound signal transduction in Arabidopsis due to a block in jasmonic acid biosynthesis. *Plant J.* **2002**, *31*, 1–12. [CrossRef] [PubMed]
87. Xiao, Y.; Chen, Y.; Charnikhova, T.; Mulder, P.P.; Heijmans, J.; Hoogenboom, A.; Agalou, A.; Michel, C.; Morel, J.B.; Dreni, L. *OsJAR1* is required for JA-regulated floret opening and anther dehiscence in rice. *Plant Mol. Biol.* **2014**, *86*, 19–33. [CrossRef]
88. Riemann, M.; Haga, K.; Shimizu, T.; Okada, K.; Ando, S.; Mochizuki, S.; Nishizawa, Y.; Yamanouchi, U.; Nick, P.; Yano, M. Identification of rice Allene Oxide Cyclase mutants and the function of jasmonate for defence against Magnaporthe oryzae. *Plant J.* **2013**, *74*, 226–238. [CrossRef]
89. Kim, E.H.; Kim, Y.S.; Park, S.H.; Koo, Y.J.; Do Choi, Y.; Chung, Y.Y.; Lee, I.J.; Kim, J.K. Methyl jasmonate reduces grain yield by mediating stress signals to alter spikelet development in rice. *Plant Physiol.* **2009**, *149*, 1751–1760. [CrossRef]
90. Li, X.; Wang, Y.; Duan, E.; Qi, Q.; Zhou, K.; Lin, Q.; Wang, D.; Wang, Y.; Long, W.; Zhao, Z. *OPEN GLUME1*: A key enzyme reducing the precursor of JA, participates in carbohydrate transport of lodicules during anthesis in rice. *Plant Cell Rep.* **2018**, *37*, 329–346. [CrossRef] [PubMed]
91. National Center for Biotechnology Information. 1988. Available online: http://blast.ncbi.nlm.nih.gov/Blast.cgi (accessed on 14 August 2018).
92. Xie, X.; Ma, X.; Zhu, Q.; Zeng, D.; Li, G.; Liu, Y.G. CRISPR-GE: A convenient software toolkit for CRISPR-based genome editing. *Mol. Plant* **2017**, *10*, 1246–1249. [CrossRef]
93. Liu, H.; Ding, Y.; Zhou, Y.; Jin, W.; Xie, K.; Chen, L.L. CRISPR-P 2.0: An improved CRISPR-Cas9 tool for genome editing in plants. *Mol. Plant* **2017**, *10*, 530–532. [CrossRef]
94. Ma, X.; Zhang, Q.; Zhu, Q.; Liu, W.; Chen, Y.; Qiu, R.; Wang, B.; Yang, Z.; Li, H.; Lin, Y. A robust CRISPR/Cas9 system for convenient, high-efficiency multiplex genome editing in monocot and dicot plants. *Mol. Plant* **2015**, *8*, 1274–1284. [CrossRef] [PubMed]
95. Hiei, Y.; Ohta, S.; Komari, T.; Kumashiro, T. Efficient transformation of rice (*Oryza sativa* L.) mediated by Agrobacterium and sequence analysis of the boundaries of the T-DNA. *Plant J.* **1994**, *6*, 271–282. [CrossRef] [PubMed]
96. Liu, W.; Xie, X.; Ma, X.; Li, J.; Chen, J.; Liu, Y.G. DSDecode: A web-based tool for decoding of sequencing chromatograms for genotyping of targeted mutations. *Mol. Plant* **2015**, *8*, 1431–1433. [CrossRef] [PubMed]
97. Wang, Z.Q.; Xu, X.Y.; Gong, Q.Q.; Xie, C.; Fan, W.; Yang, J.L.; Lin, Q.S.; Zheng, S.J. Root proteome of rice studied by iTRAQ provides integrated insight into aluminum stress tolerance mechanisms in plants. *J. Proteom.* **2014**, *98*, 189–205. [CrossRef]
98. Livak, K.J.; Schmittgen, T.D. Analysis of relative gene expression data using real-time quantitative PCR and the $2^{-\Delta\Delta CT}$ method. *Methods* **2001**, *25*, 402–408. [CrossRef] [PubMed]

International Journal of *Molecular Sciences*

MDPI

Article

OsCRP1, a Ribonucleoprotein Gene, Regulates Chloroplast mRNA Stability That Confers Drought and Cold Tolerance

Seung Woon Bang [1], Ho Suk Lee [1], Su-Hyun Park [1,2], Dong-Keun Lee [1,3], Jun Sung Seo [1], Youn Shic Kim [1,4], Soo-Chul Park [1] and Ju-Kon Kim [1,*]

1 Crop Biotechnology Institute, GreenBio Science and Technology, Seoul National University, Pyeongchang 25354, Korea; seungwoon93@gmail.com (S.W.B.); viphosuk17@snu.ac.kr (H.S.L.); suhyun@tll.org.sg (S.-H.P.); eastrootut@gmail.com (D.-K.L.); xfiles96@snu.ac.kr (J.S.S.); younshic@gmail.com (Y.S.K.); scpark1@snu.ac.kr (S.-C.P.)

2 Temasek Life Sciences Laboratory, 1 Research Link, National University of Singapore, Singapore 117604, Singapore

3 E GREEN GLOBAL, Gunpo 15843, Korea

4 Department of Agriculture and Life Industry, Kangwon National University, Chuncheon 24341, Korea

* Correspondence: jukon@snu.ac.kr; Tel.: +82-33-339-5826

Abstract: Chloroplast ribonucleoproteins (cpRNPs) are nuclear-encoded and highly abundant proteins that are proposed to function in chloroplast RNA metabolism. However, the molecular mechanisms underlying the regulation of chloroplast RNAs involved in stress tolerance are poorly understood. Here, we demonstrate that *CHLOROPLAST RNA-BINDING PROTEIN 1* (*OsCRP1*), a rice (*Oryza sativa*) cpRNP gene, is essential for stabilization of RNAs from the NAD(P)H dehydrogenase (NDH) complex, which in turn enhances drought and cold stress tolerance. An RNA-immunoprecipitation assay revealed that OsCRP1 is associated with a set of chloroplast RNAs. Transcript profiling indicated that the mRNA levels of genes from the NDH complex significantly increased in the *OsCRP1* overexpressing compared to non-transgenic plants, whereas the pattern in *OsCRP1* RNAi plants were opposite. Importantly, the *OsCRP1* overexpressing plants showed a higher cyclic electron transport (CET) activity, which is essential for elevated levels of ATP for photosynthesis. Additionally, overexpression of *OsCRP1* resulted in significantly enhanced drought and cold stress tolerance with higher ATP levels compared to wild type. Thus, our findings suggest that overexpression of *OsCRP1* stabilizes a set of mRNAs from genes of the NDH complex involved in increasing CET activity and production of ATP, which consequently confers enhanced drought and cold tolerance.

Keywords: drought tolerance; cold tolerance; *Oryza sativa*; *OsCRP1*; chloroplast ribonucleoproteins; NAD(P)H dehydrogenase (NDH) complex

Citation: Bang, S.W.; Lee, H.S.; Park, S.-H.; Lee, D.-K.; Seo, J.S.; Kim, Y.S.; Park, S.-C.; Kim, J.-K. *OsCRP1*, a Ribonucleoprotein Gene, Regulates Chloroplast mRNA Stability That Confers Drought and Cold Tolerance. *Int. J. Mol. Sci.* **2021**, *22*, 1673. https://doi.org/10.3390/ijms22041673

Academic Editors: Kiyosumi Hori and Matthew Shenton

Received: 21 January 2021
Accepted: 5 February 2021
Published: 7 February 2021

1. Introduction

Members of the green plant lineage have chloroplasts with their own organellar genomes that have their evolutionary origins in endosymbiotic cyanobacteria. Proteins encoded by chloroplast genes play crucial roles in photosynthesis and in the expression of photosynthesis-related nuclear genes. Expression of chloroplast mRNAs is regulated at both the transcriptional and posttranscriptional levels [1–3], and during post-transcriptional regulation, numerous nucleus-encoded RNA-binding proteins (RBPs) act as a regulator of cleavage, splicing, editing, or stabilization of chloroplast RNAs [3,4]. For example, pentatricopeptide repeat (PPR) proteins, the most abundant protein family in plants, are well-characterized RBPs that mediate RNA editing through interaction with specific chloroplast RNA sequences [5–8].

Chloroplast ribonucleoproteins (cpRNPs) comprise a small family of RBPs consisting of two RNA recognition motifs (RRMs) and participate in chloroplast RNA processing [9–11].

The consensus RNP structure of five tobacco (*Nicotiana sylvestris*) cpRNPs (cp28, cp29A, cp29B, cp31, and cp33) has been solved and their binding affinities to RNA homopolymers, such as poly (G) and poly (U), have been determined. These studies suggest that cpRNPs have a key function in chloroplast RNA metabolism [10,11]. As previously reported for spinach (*Spinacia oleracea*) 28RNP, a tobacco cp28 and cp31 ortholog, cpRNPs confer correct 3′-end processing of chloroplast mRNAs such as *psbA*, *rbcL*, *petD*, and *rps14* [12]. In *Arabidopsis thaliana*, in silico analysis indicated that the cpRNP protein family is composed of 10 members [13,14]. An *A. thaliana* null mutant, *CP31A*, was shown to have defects in RNA editing and to have a number of destabilized transcripts under normal growth conditions [15]. It has also been demonstrated that cpRNPs are required for activity of the NADH dehydrogenase-like (NDH) complex through stabilization of *ndhF* mRNA and editing of *ndhF*, *ndhB*, and *ndhD* mRNAs [15]. Interestingly, Kupsch et al. [16] found that CP31A and CP29A in *A. thaliana* are essential for cold stress tolerance through their stabilization of numerous chloroplast mRNAs. Moreover, it has been demonstrated that cpRNPs are highly regulated proteins that respond to various external and internal signals, including light and temperature, which affect both their expression levels and post-translational modification [16–18]. While a number experimental systems have been used to elucidate the molecular mechanisms and functions of cpRNPs, there are still several important crop species for which such information is absent, notably rice (*Oryza sativa*).

In later diverging land plants, the light reactions of photosynthesis involve at least two routes through which light energy is converted into NADPH and ATP. Through the first route, ATP and NADPH are generated by electrons released from water to photosystem II (PSII) and photosystem I (PSI) via linear electron transport (LET) [19]. However, while LET generally produces sufficient amounts of NADPH, this is not the case for ATP [20–22]. In the second route, an electron can be recycled from either reduced ferredoxin or NADPH to plastoquinone and subsequently to the Cyt b_6f complex. This cyclic electron transport (CET) requires only PSI photochemical reactions to produce ATP and does not involve the production of chloroplastic NADPH [19,23].

CET consists of two pathways: the PROTON GRADIENT REGULATION 5 (PGR5)/PGR-LIKE PHOTOSYNTHETIC PHENOTYPE 1 (PGRL1)-dependent pathway and the NDH complex-dependent pathway [24–27]. The former represents the major pathway under normal growth conditions, whereas many studies have shown that NDH-dependent CET is involved in protective or adaptive mechanisms in response to abiotic stresses, such as heat, high light, or drought. In rice, a *crr6* mutant, which has a defect in the *ndhK* gene, shows growth defects under low temperature, low light and fluctuating-light stress conditions [22,28,29], and a tobacco *ndhB* mutant that is deficient in NDH-dependent CET has decreased relative leaf water content and net CO_2 assimilation under water stress conditions [30]. The salt-tolerant soybean (*Glycine max*) variety S111-9, which under normal conditions has high expression levels of *ndhB* and *ndhH*, shows higher CET activity and ATP accumulation than the salt-sensitive variety Melrose, suggesting a correlation between salt tolerance and NDH-dependent CET [31]. These studies are congruent with the idea that NDH-dependent CET is important for the adaptation of plants to abiotic stress conditions.

In this current study, we investigated the significance of *OsCRP1*, a rice chloroplast ribonucleoprotein, in drought and cold stress tolerance. The OsCRP1 protein was found to have a broad range of binding affinities to chloroplast RNAs, and specifically to regulate NDH complex gene expression. Overexpression of *OsCRP1* in rice resulted in increased CET activity and accumulation of ATP, whereas knock-down lines had lower activity under stress conditions. We also found that *OsCRP1* overexpressing plants had enhanced drought and cold stress tolerance compared to non-transgenic (NT) plants, whereas the knock-down lines remained susceptible. Overall, these results suggest that overexpression of *OsCRP1* confers improved drought and cold tolerance through modulation of NDH-dependent CET.

2. Results

2.1. *OsCRP1 Is a Rice Nuclear-Encoded and Chloroplast Targeting Ribonucleoprotein*

Nuclear-encoded chloroplast ribonucleoproteins (cpRNPs) consist of a transit peptide (TP) and two RNA recognition motifs (RRM) that are involved in the interaction with RNA molecules (Figure S1). The *A. thaliana* genome is predicted to encode 10 cpRNPs (Figure 1a), and we identified 8 cpRNP proteins encoded by the rice genome based on the conserved RRM protein sequence of AtCP31A, using SmartBLAST (http://blast.ncbi.nlm.nih.gov). In order to name the rice cpRNPs in accordance with published classification, a phylogenetic tree was generated using full length protein sequences from the 10 *A. thaliana* and 8 rice cpRNPs (Figure 1a). cpRNP protein sequences show a high degree of sequence conservation between dicots and monocots (Figure S1), and we named the rice cpRNPs OsCRP1 (Os09g0565200), OsCRP2 (Os08g557100), OsCRP3 (Os07g0158300), OsCRP4 (Os03g0376600), OsCRP5 (Os07g0631900), OsCRP6 (Os02g0815200), OsCRP7 (Os09g279500) and OsCRP8 (Os08g0117100) (Figure 1a).

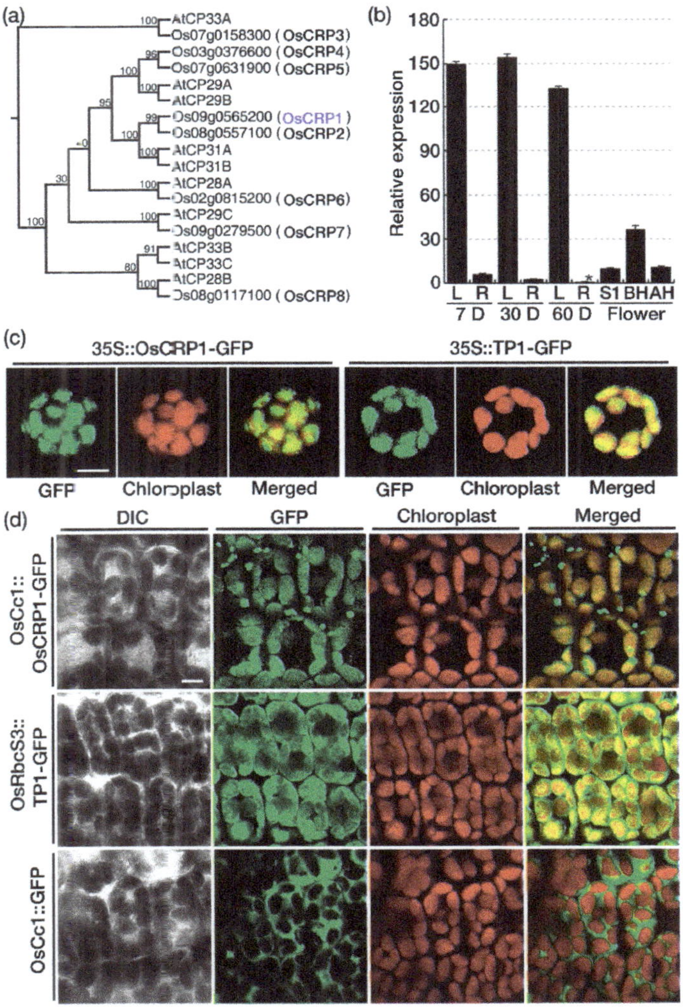

Figure 1. Expression patterns of *OsCRP1* and subcellular localization of the corresponding protein.

(**a**) Phylogenetic tree created using the neighbor-joining method in CLC sequence viewer using full-length amino acid sequences of the rice and *Arabidopsis thaliana* chloroplast ribonucleoproteins (cpRNPs). Bootstrap support (100 repetitions) is shown for each node. (**b**) Quantitative RT-PCR of *OsCRP1* in various tissues and at different growth stages. (D, day after germination; L, leaf; R, root; S1, <1 cm in panicle length; BH, before heading; AH, after heading). *OsUbi1* (AK121590) expression was used as an internal control, and were plotted relative to the level of mRNA in the lowest-expressing stages (indicated by the asterisk). Data bars represent the mean ± SD of two biological replicates, each of which with three technical replicates (* $p < 0.05$). (**c**) Subcellular localization of OsCRP1 in rice protoplasts. Rice leaf protoplasts were transformed with two different constructs and observed using a confocal microscope. (**d**) Localization of OsCRP1 was confirmed by the observation of GFP fluorescence in leaves of one-week old *OsCc1::OsCRP1-GFP* transgenic rice plants. *OsRbcS::TP1-GFP* and *OsCc1::GFP* were controls for localization in chloroplasts and the cytoplasm, respectively. Scale bar, 10 μm.

OsCRP1 was chosen for functional characterization since its transcripts were detected in all tissues from the various developmental stages. The *OsCRP1* expression was particulalry abundant in green tissues, including leaves and green flowers, while it remained low in roots at all developmental stages (Figure 1b).

To determine the subcellular localization of *OsCRP1*, we expressed the whole protein or the transit peptide in rice protoplasts as a fusion with green fluorescent protein (OsCRP1-GFP or TP1-GFP) under the control of the 35S promoter (Figure S2). The GFP fluorescence of OsCRP1-GFP and TP1-GFP, resulting from transformation of the protoplasts with vectors *pro35S::OsCRP1-GFP* or *pro35S::TP1-GFP*, respectively, overlapped with the red chloroplast autofluorescence (Figure 1c). To confirm OsCRP1 localization, we also generated transgenic rice plants expressing *OsCRP1-GFP*, *GFP* and *TP1-GFP* under the control of the *OsCc1* (rice *CYTOCHROME C1*) and *RbcS* (rice *small subunit of ribulose bisphosphate carboxylase/oxygenase*) promoter, respectively (Figure S2). The *OsCRP1-GFP* under the control of the *OsCc1* promoter (*OsCc1::OsCRP1-GFP*) showed uniform yet aggregated patterns of GFP fluorescence in chloroplasts (Figure 1d). This unique patterns of GFP fluorescence was different from those of two control constructs *RbcS::TP1-GFP* or *OsCc1::GFP* that showed either GFP fluorescence evenly distributed in all chloroplasts or no GFP fluorescence within chloroplasts, respectively (Figure 1d). These data suggest that the OsCRP1-GFP is targeted to a sub-structure of chloroplasts, presummably stroma.

2.2. OsCRP1 Is Required for the Accumulation of Chloroplast mRNAs

To determine whether the OsCRP1 protein was associated with chloroplast mRNA accumulation, we performed a RNA immunoprecipitation (RIP) assay. OsCRP1 levels in the leaves of transgenic rice lines transformed with *OsCc1::OsCRP1-GFP* were verified by Western blot analysis using a α-GFP antibody (Figure S3a). RNA-protein complexes in *OsCc1::OsCRP1-GFP* leaf extracts were precipitated using α-GFP antibodies, and *OsCc1::GFP* leaf extracts were used as a negative control. We then analyzed the RNA quantity of 23 plastid-encoded genes corresponding to four major chloroplast protein classes: ATP synthase (*atp*), photosystem I (*psa*), photosystem II (*psb*), and the NADH dehydrogenase complex (*ndh*), by quantitative real time (qRT)-PCR analysis. Most of the chloroplast mRNAs were enriched > 5-fold in extracts from the *OsCc1::OsCRP1-GFP* lines compared to the control (Figure 2a), indicating that OsCRP1 can bind to a broad range of chloroplast RNAs. To further elucidate the functions of *OsCRP1*, we generated overexpression (*OsCc1::OsCRP1^{OX}*) and RNAi (*OsCc1::OsCRP1^{RNAi}*) transgenic rice plants using the *OsCc1* promoter, which is constitutively active throughout the plant [32] (Figure S2). Thirty independent transgenic lines were generated, and those that grew normally were selected for further analysis to eliminate the effects of soma clonal variation. Based on the expression levels of *OsCRP1* in the transgenic plants, we chose three independent single-copy homozygous lines from each transgene type (*OsCc1::OsCRP1^{OX}*; #7, 9, 12 and *OsCc1::OsCRP1^{RNAi}* ; #4, 6, 8) for further study (Figure S3b).

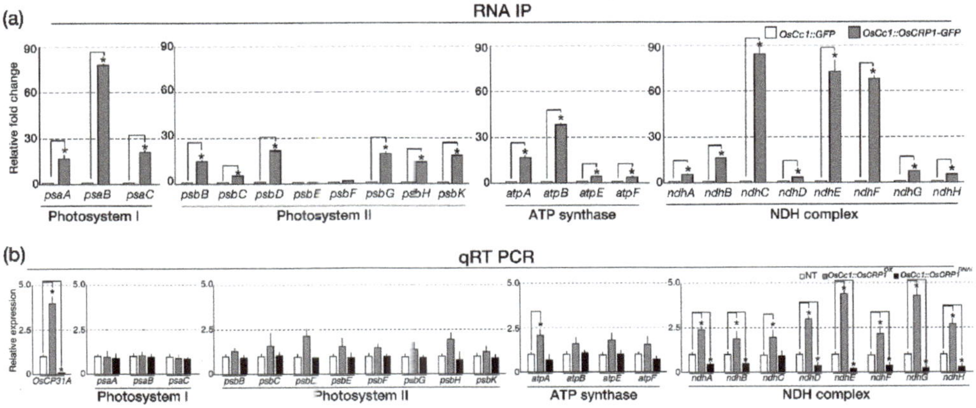

Figure 2. Identification of OsCRP1 target cpRNAs. The GFP-tagged transgenic rice plants were generated using the *OsCc1::OsCRP1-GFP* vector. Soluble protein and total RNA extractions and RNA-immunoprecipitation (RIP) assays were conducted with 2-week-old *OsCc1::OsCRP1-GFP* transgenic leaves. (**a**) Identification of OsCRP1 target chloroplast RNAs (cpRNAs) by RIP. cDNAs were synthesized using the immunoprecipitated RNAs and α-GFP antibodies, prior to quantitative RT-PCR. All the values were normalized based on total input RNA per sample, and bars represent the mean ± SD of four repeats. (**b**) Relative expression levels of cpRNAs in total RNA samples from *OsCc1::OsCRP1*^{OX}, non-transgenic (NT) and *OsCc1::OsCRP1*^{RNAi} plants. qRT-PCR with cDNA from NT and transgenic leaves was performed using 23 chloroplast gene-specific primer sets. All the values were normalized to the internal *OsUbi1* control gene, and data bars represent the mean ± SD of two biological replicates, each of which had three technical replicates. Significant differences from the control are indicated by asterisks (Student's *t*-test, * $p < 0.05$).

To verify the *OsCRP1* target genes, we performed an RNA-seq analysis with leaves from *OsCc1::OsCRP1*^{OX} (#7, 9, 12), non-transgenic control (NT), and *OsCc1::OsCRP1*^{RNAi} (#4, 6, 8) plants grown under normal conditions (Table S2). When we analyzed the mRNA levels of 23 chloroplast genes in these plants, we observed differences for almost all NDH complex genes in the transgenic plants compared to the control, with higher expression in *OsCc1::OsCRP1*^{OX} plants and lower expression in *OsCc1::OsCRP1*^{RNAi} plants. Their increased and decreased levels of expression in the *OsCRP1*^{OX} and the *OsCRP1*^{RNAi} leaves, respectively, were validated by qRT-PCR (Figure 2b). In summary, our results suggested that OsCRP1 directly binds to a set of cpRNAs, causing an increase in the mRNA stability of NDH complex genes.

2.3. Down-Regulation of OsCRP1 Results in Chlorosis under Light Stress Conditions

The NDH complex is known to catalyze electron transfer from the stromal pool of reductants to plastoquinone (PQ), which activate the cyclic electron transport (CET) under abiotic stress [30,33–36]. We observed an increase in chlorophyll fluorescence after the offset of actinic light, which is caused by the NDH complex catalyzing a reduction of the PQ pool [25]. Moderate heat stress (e.g., 35–42 °C) can affect photosynthesis and cause a significant increase in CET [37], and so we exposed plants in a dark chamber to different temperatures (22 °C, 28 °C and 35 °C) before taking measurements (Figure 3a). Under normal conditions (22 °C and 28 °C), a similar increase in chlorophyll fluorescence was observed in all plants following illumination. After heat stress, the responsiveness of NDH-dependent CET under dark conditions was enhanced in *OsCc1::OsCRP1*^{OX} plants compared to NT plants. In contrast, *OsCc1::OsCRP1*^{RNAi} plants did not exhibit this characteristic rise in post-illumination fluorescence.

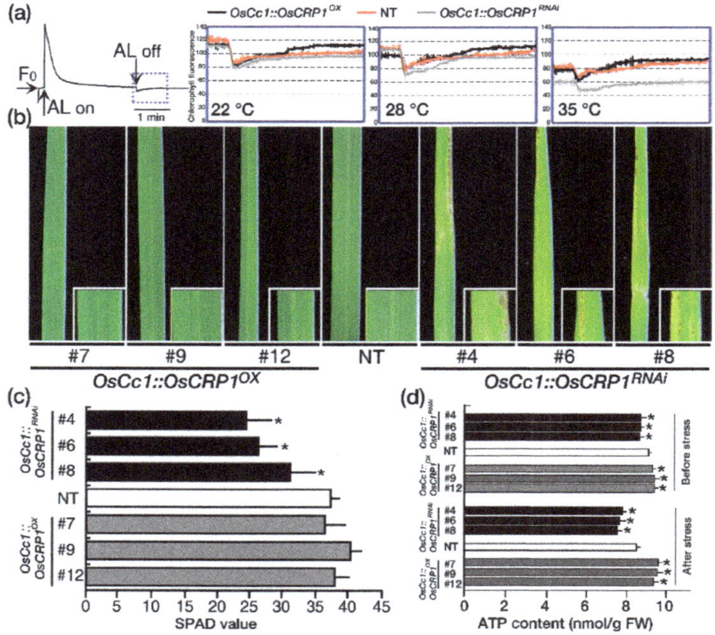

Figure 3. Monitoring of NDH-dependent CET activity by chlorophyll a fluorescence. (**a**) Chlorophyll a fluorescence in *OsCc1::OsCRP1*^{OX}, NT and *OsCc1::OsCRP1*^{RNAi}. 5-week-old plants grown under chamber conditions. The middle portions of leaves were used for measurements. The post-illumination chlorophyll fluorescence curve, which represents NDH-dependent CET, was therefore magnified from the blue box area for analysis. Values are means (±SD) of three independent measurements. (**b**) *OsCc1::OsCRP1*^{OX}, NT, and *OsCc1::OsCRP1*^{RNAi} plants were grown for 2 weeks under chamber conditions and light intensities of 170~180 μmol m^{-2} s^{-1} prior to stress treatment. Plants were then transferred to light intensities of 240~250 μmol m^{-2} s^{-1} and phenotyped. All light measurements were made with a LI-250A Light Meter (LI-COR, Lincoln, NE, USA), and photos were obtained 2 weeks after treatment. The analysis was carried out for three biologicals with three technical replicates each. (**c**) SPAD values for *OsCc1::OsCRP1*^{OX}, NT and *OsCc1::OsCRP1*^{RNAi} leaves, representing the amount of chlorophyll per leaf. The values were measured for 10 leaves of three representative transgenic lines and NT plants using a Chlorophyll Meter SPAD-502Plus. Data bars represent the mean ± SD of two biological replicates, each of which had three technical replicates. Asterisks indicate significant differences compared with NT (* $p < 0.05$, One-way ANOVA). (**d**) ATP contents in leaves of the *OsCc1::OsCRP1*^{OX}, NT and *OsCc1::OsCRP1*^{RNAi} plants under before and after light stress conditions. Ten plants were used for each line, and the middle portion of the second leaf from the top was taken for analysis. Data bars represent the mean ± SD of three biological replicates, each of which had two technical replicates. Asterisks indicate significant differences compared with NT (* $p < 0.05$, One-way ANOVA).

It has been reported that strong light can cause severe irreversible photodamage, as evidenced by chlorosis in NDH-defective plants [33]. To confirm this phenomenon, *OsCc1::OsCRP1*^{OX}, NT, and *OsCc1::OsCRP1*^{RNAi} plants were grown for 2 weeks under chamber conditions of moderate light (170 ~ 180 μmol m^{-2} s^{-1}) and then exposed to light stress conditions (240 ~ 250μmol m^{-2} s^{-1}). *OsCc1::OsCRP1*^{RNAi} plants showed chlorosis after 2 weeks of light stress treatments, while no visual symptoms were observed for *OsCc1::OsCRP1*^{OX} and NT plants (Figure 3b). This phenotype was confirmed by measuring the leaf chlorophyll using a Soil Plant Analysis Development (SPAD) chlorophyll meter. As shown in Figure 3c, chlorophyll content was similar between *OsCc1::OsCRP1*^{OX} and

NT plants, while significantly lower in *OsCc1::OsCRP1*^{RNAi} plants. These results suggest a correlation between *OsCRP1* expression and NDH-dependent CET activity under stress conditions. It has also been shown that the NDH-dependent CET activity is involved in a mechanism by which plants protect against drought, light, and high temperature stresses through increased production of ATP [33,38]. We set out to analyze the ATP contents of the *OsCc1::OsCRP1*^{OX}, NT and *OsCc1::OsCRP1*^{RNAi} plants before and after exposure to high light stress. Before light stress treatments, the ATP contents of *OsCc1::OsCRP1*^{OX} leaves were higher than NT leaves by 3.4%, whereas those of the *OsCc1::OsCRP1*^{RNAi} leaves were lower than NT leaves by 4.2% without difference in chlorosis. However, after light stress treatments, the ATP contents of the *OsCc1::OsCRP1*^{OX} leaves were higher than NT leaves by 10.8%, whereas those of the *OsCc1::OsCRP1*^{RNAi} leaves were lower than NT leaves by 10.2% (Figure 3d). The ATP levels were higher in *OsCc1::OsCRP1*^{OX} plants and lower in *OsCc1::OsCRP1*^{RNAi} plants compared to NT plants, indicating that *OsCRP1* is involved in the increased production of ATP through elevated NDH-dependent CET activity under high light stress conditions.

2.4. Overexpression of OsCRP1 Confers Cold Stress Tolerance

Chloroplast RNPs have been shown to confer cold stress tolerance to *A. thaliana* by influencing multiple chloroplast RNA processing steps [16]. We found that the expression level of *OsCRP1* also increased under cold stress conditions (Figure 4a). These observations led us to examine the cold stress tolerance of 2-week-old *OsCc1::OsCRP1* plants that had been treated with 4 °C for three days and then allowed to recover for seven days (Figure 4b). Most of the *OsCc1::OsCRP1* plants survived (~85% survival rate), whereas only ~50% of the NT and ~30% of the *OsCc1::OsCRP1*^{RNAi} plants survived (Figure 4c), suggesting that overexpression of *OsCRP1* significantly enhanced cold tolerance. Since cold stress has been reported to reduce the efficiency of photosystem II [39], we measured *Fv/Fm* values, an indicator of the photochemical efficiency of photosystem II, in plants after exposure to cold stress (Figure 4d). The *Fv/Fm* values of the *OsCc1::OsCRP1*^{OX} plants were higher than those of the NT and *OsCc1::OsCRP1*^{RNAi} plants during cold stress, indicating that the photochemical efficiency of photosystem II in the *OsCc1::OsCRP1*^{OX} plants was less damaged by the cold stress treatments than in NT and *OsCc1::OsCRP1*^{RNAi} plants. The NDH complex drives CET around photosystem I and enhances the production of ATP for photosynthesis and increases abiotic stress tolerance [31]. Thus, we set out to analyze the ATP contents of the *OsCc1::OsCRP1*^{OX}, NT and *OsCc1::OsCRP1*^{RNAi} plants before and after exposure to cold stress. After cold stress treatments, the ATP contents of the *OsCc1::OsCRP1*^{OX} leaves were higher than NT leaves by 5.7%, whereas those of the *OsCc1::OsCRP1*^{RNAi} leaves were lower than NT leaves by 18.4%. ATP levels in *OsCc1::OsCRP1*^{OX} and *OsCc1::OsCRP1*^{RNAi} plants were higher and lower, respectively, than those in NT plants, (Figure 4e), indicating that overexpression of *OsCRP1* confers cold tolerance via enhancement of NDH-dependent CET under cold stress conditions.

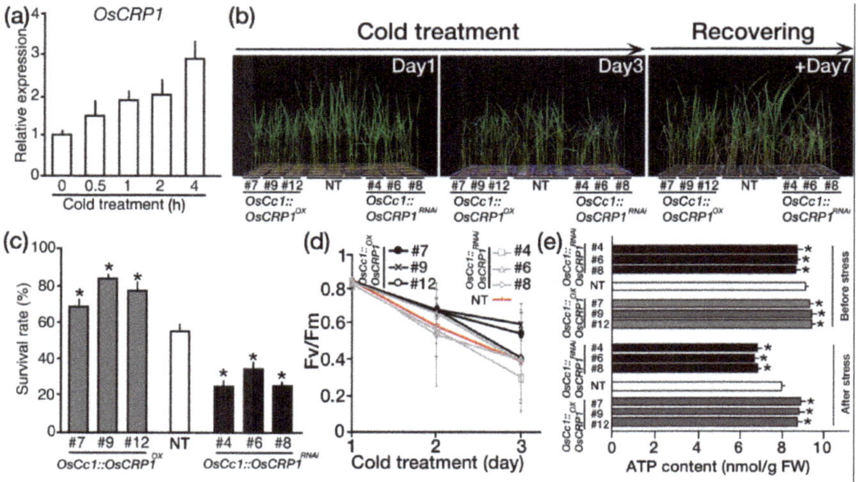

Figure 4. *OsCRP1* overexpression in rice confers cold tolerance. (**a**) Relative expression of *OsCRP1* in response to cold stress. Two-week-old seedlings were exposed to at 4 °C (low temperature) for the indicated times. *OsUbi1* expression was used as an internal control. Values are the means ± SD of three independent experiments. (**b**) Phenotypes of *OsCc1::OsCRP1*^{OX} and *OsCc1::OsCRP1*^{RNAi} transgenic rice plants under cold stress at the vegetative stage. Three independent homozygous *OsCc1::OsCRP1*^{OX} and *OsCc1::OsCRP1*^{RNAi} lines and NT control plants were grown in soil for 2 weeks and exposed to cold stress for 3 days, followed by recovery. (**c**) Survival rate scored 7 days after recovery. Values represent means ± SD of three repeated tests. Asterisks indicate significant differences compared with NT (* $p < 0.05$, One-way ANOVA). (**d**) Chlorophyll fluorescence (Fv/Fm) of *OsCc1::OsCRP1*^{OX}, NT and *OsCc1::OsCRP1*^{RNAi} plants during a 3-day cold treatment. Fv/Fm values were measured in the dark to ensure sufficient dark adaptation. Data are shown as the mean ± SD (n = 30). (**e**) ATP contents in leaves of the *OsCc1::OsCRP1*^{OX}, NT and *OsCc1::OsCRP1*^{RNAi} plants under before and after cold stress conditions. Ten two-week-old plants were used for each line, and the middle portion of the second leaf from the top was taken for analysis. Data bars represent the mean ± SD of three biological replicates, each of which had two technical replicates. Asterisks indicate significant differences compared with NT (* $p < 0.05$, One-way ANOVA).

2.5. Overexpression of OsCRP1 Confers Drought Stress Tolerance

We exposed the *OsCc1::OsCRP1*^{OX} and *OsCc1::OsCRP1*^{RNAi} plants to drought stress by withholding water for 3 consecutive days, during which drought-induced visual symptoms were observed (Figure 5a). Soil moisture content decreased similarly in all the pots, indicating that the drought stress was uniformly applied (Figure 5b). The *OsCc1::OsCRP1*^{OX} plants showed delayed visual symptoms of drought-induced damage, such as leaf rolling and wilting, compared to NT and *OsCc1::OsCRP1*^{RNAi} plants. After rehydration, the *OsCc1::OsCRP1*^{OX} plants rapidly recovered, whereas the NT and *OsCc1::OsCRP1*^{RNAi} plants did not recover well (Figure 5a). The *OsCc1::OsCRP1*^{RNAi} plants showed similar sensitivity to NT plants in their response to the drought stress. Collectively these results suggest that *OsCRP1* overexpression enhanced drought stress tolerance. To verify the performance of the plants under drought stress conditions, *Fv/Fm* values were measured. In *OsCc1::OsCRP1*^{RNAi} and NT plants the values decreased one day after exposure to drought stress, while only a slightly decrease was observed in *OsCc1::OsCRP1*^{OX} plants on day 2 (Figure 5c). Before drought stress treatments, ATP levels were similarly high in *OsCc1::OsCRP1*^{OX} and NT plants, but significantly lower in *OsCc1::OsCRP1*^{RNAi} plants. ATP levels in NT and *OsCc1::OsCRP1*^{RNAi} plants rapidly declined after exposure to drought stress conditions, whereas *OsCc1::OsCRP1*^{OX} plants showed a slow decrease (Figure 5d). Taken together, our results indicate that in rice plants, *OsCRP1* modulates CET activity via changing mRNA stability of NDH complex genes, which consequently confers drought stress tolerance.

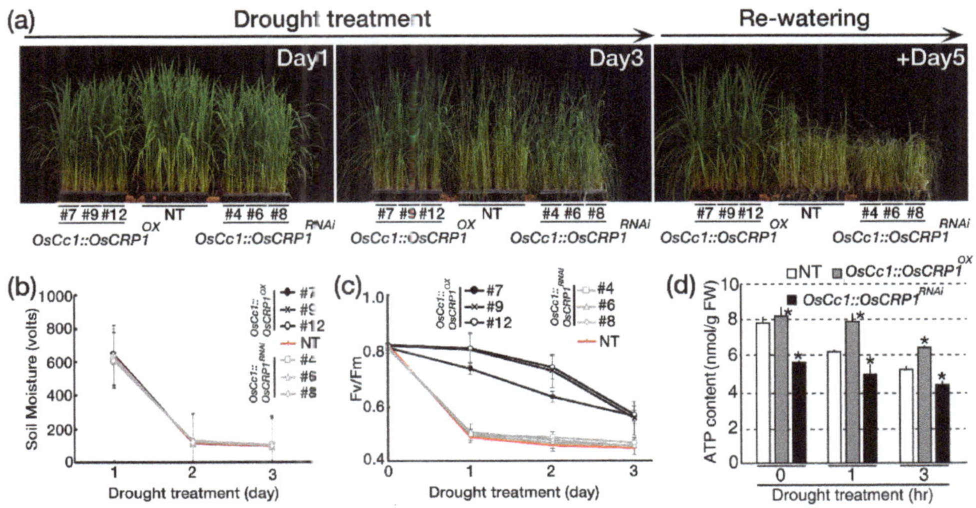

Figure 5. *OsCRP1* overexpression in rice confers drought tolerance. (**a**) Phenotypes of *OsCc1::OsCRP1*^{OX} and *OsCc1::OsCRP1*^{RNAi} transgenic rice plants under drought stress at the vegetative stage. Three independent homozygous *OsCc1::OsCRP1*^{OX} and *OsCc1::OsCRP1*^{RNAi} lines and NT control plants were grown in soil for 5 weeks and exposed to drought for 3 days, followed by re-watering. (**b**) Soil moisture in the pots exposed to drought treatment at the indicated time points. Values are the means ± SD (n = 10). (**c**) Chlorophyll fluorescence (Fv/Fm) of *OsCc1::OsCRP1*^{OX} and *OsCc1::OsCRP1*^{RNAi} transgenic rice plants and NT plants during a 3-day drought treatment. Fv/Fm values were measured in the dark to ensure sufficient dark adaptation. Data are shown as the mean ± SD (n = 30). (**d**) ATP contents in leaves of the *OsCc1::OsCRP1*^{OX}, NT and *OsCc1::OsCRP1*^{RNAi} plants under drought stress conditions. Five-week-old plants were transferred to water for 2 days prior to drought treatments. Air-dried plants were taken at 1 and 3 h of treatment to compare ATP contents. Ten plants were used for each line. Data bars represent the mean ± SD of three biological replicates, each of which had two technical replicates. Asterisks indicate significant differences compared with NT (* $p < 0.05$, One-way ANOVA).

3. Discussion

Chloroplast RNA metabolism is affected by various environmental changes, including light and temperature, and chloroplast RNA-binding proteins (cpRNPs) are known to play a central role in their post-transcriptional processing, such as splicing, editing, and stabilization [7]. It has been reported that in the model dicotyledon, *A. thaliana*, several cpRNPs enhance abiotic stress tolerance through their function as RNA chaperones [16,40]. However, the underlying molecular mechanisms of their abiotic stress effect have not been well studied in the monocotyledon, rice.

Several reports have shown that cpRNPs are also involved in editing and 3′-end processing of chloroplast mRNAs [12,16,41], and that regulation of chloroplast mRNA stability by cpRNPs is important for development and abiotic stress responses [16,42]. For example, RIP analyses have demonstrated that *A. thaliana* CP33A is associated with the stability of multiple chloroplast mRNAs. Moreover, loss of *CP33A* results in an albino plants that also show aberrant leaf development [42]. *A. thaliana* CP31A and CP29A are known to interact with and stabilize multiple chloroplast mRNAs that are associated with limiting the effects of cold stress on chloroplast development [16]. Our analysis of the chloroplast-localized cpRNP, OsCRP1, revealed that it has a broad range of target chloroplast RNAs (Figure 2a). Notably, mRNA level of most *ndh* genes was decreased in NT plant after drought treatment and it suggested that drought treatment could reduce the transcription or stability of those mRNA (Figure S4a). However, transcript level of most *ndh* genes was significantly higher in *OsCRP1* overexpressing plants compared to NT plants under both normal and drought condition (Figure 2b and Figure S4b). These results suggest that OsCRP1 directly interacts with a set of cpRNAs, improving drought tolerance by enhancing the mRNA stability of NDH complex genes. We propose that the RNA stabilizing mechanism of OsCRP1 involves protecting the target RNA against 3′-exonucleolytic activity; analogous to the mechanism exhibited by *A. thaliana* CP31A [16].

The chloroplast NDH complex is a ferredoxin (Fd)-dependent PQ reductase that associates with the CET around PSI to catalyze electron transfer [43,44], which in turn leads to a transient increase in chlorophyll a fluorescence after the offset of actinic light [23]. NDH activity was not detectable in *A. thaliana CP31* deficient mutants where fluorescence phenotypes were identical with *ndhB*, *ndhD* or *ndhF* mutant lines, suggesting that cpRNPs are critical for chloroplastic NDH enzyme activity [15,24,25,45,46]. We also observed increases in fluorescence in *OsCc1::OsCRP1*^{OX} but not in *OsCc1::OsCRP1*^{RNAi} plants, under heat stress conditions (Figure 3a). These independent lines of evidence support the idea that OsCRP1 modulates NDH complex activity.

It was previously reported that NDH-defective mutants exhibited leaf chlorosis under high light stress, especially in ΔndhB [33]. Moreover, constitutively high CET elevation in the *hcef1* mutant does not occur in the *hcef1 crr2-2* (NDH-defective) double mutant, suggesting that NDH modulates CET activity [47]. Here, we also found that leaves of *OsCc1::OsCRP1*^{RNAi} plants exhibited chlorosis under light stress conditions (Figure 3b). Furthermore, a decreased SPAD value in the knock-down plants indicated a reduction in CET activity, which correlated with low ATP accumulation in *OsCc1::OsCRP1*^{RNAi} plants (Figure 3c,d).

When plants are exposed to abiotic stress conditions, such as high light, drought, high salt and cold, large amounts of cellular ATP are needed to support adaptive responses [24,31,48]. We observed improved tolerance of *OsCc1::OsCRP1*^{OX} plants to both drought and cold stress, whereas *OsCc1::OsCRP1*^{RNAi} and NT plants remained sensitive to drought and cold stress (Figures 4a and 5a). The stress-tolerant phenotype of the overexpressing plants can be explained by enhanced accumulation of ATP (Figures 4e and 5d), and a previous study proposed that increased ATP production by NDH-dependent CET involves vacuolar proton ATPases driving proton import. A study with soybean showed that an outward proton gradient across the tonoplast generated by a proton ATPase enhanced the vacuolar sequestration of Na^+, resulting in enhanced salt tolerance [31]. Similarly, the *OsCc1::OsCRP1*^{OX} plants generated in this current study accumulated higher levels of ATP than NT plants under cold (Figure 4e) and drought (Figure 5d) stress conditions, whereas the *OsCc1::OsCRP1*^{RNAi} plants had lower levels of ATP than NT plants under the same stress conditions. These observations suggest that *OsCRP1* increases ATP generation by enhancing NDH-dependent CET under cold and drought stress conditions, leading to increased abiotic stress tolerance.

In summary, we hypothesize that OsCRP1-mediated mRNA stabilization of NDH complex genes results in increased ATP production under stress conditions via enhancement of NDH-dependent CET. During activation of NDH-dependent CET, protons from the stroma are transferred into the thylakoid lumen, causing acidification. Increased proton levels inside the thylakoid lumen drive ATP synthesis and help maintain an ideal NADPH/ATP ratio, enhancing higher stress tolerance in *OsCc1::OsCRP1*^{OX} plants (Figure 6). Decline of photosynthesis activity is one of the key features of plant abiotic-stress response and directly related to crop productivity. Our study provided an additional evidence that cpRNPs could be a promising target locus to develop the abiotic-stress tolerant crops preparing for climate change and sustainable agriculture.

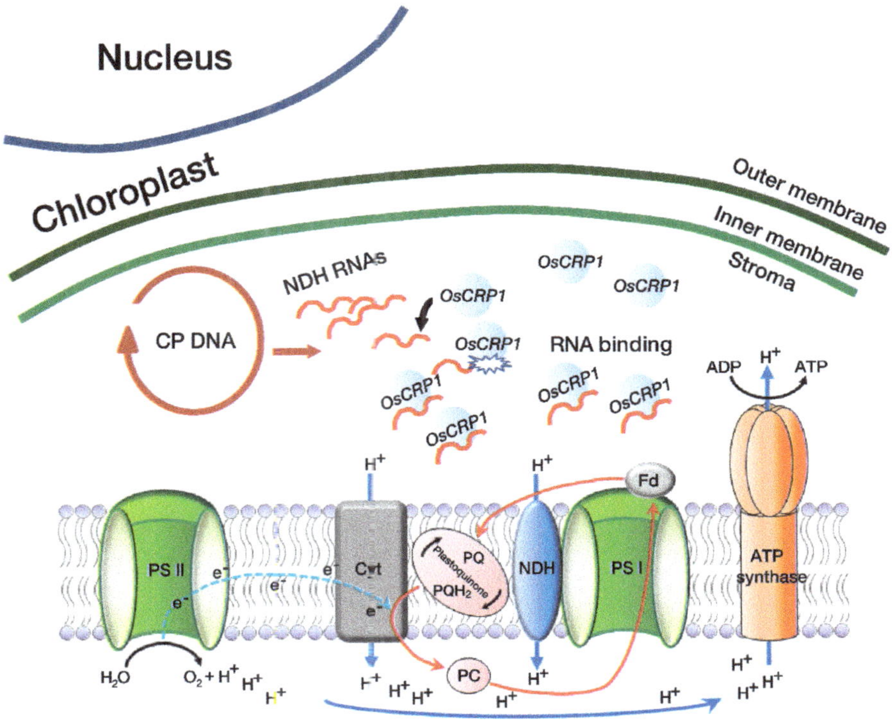

Figure 6. Schematic representation of NDH-dependent CET in *OsCRP1* overexpressing plants under abiotic stress conditions. In *OsCRP1* overexpressing plants, binding affinity of OsCRP1 to NDH complex RNAs was increased under stress conditions, leading to stabilization of transcripts from NDH complex genes. Hence, the activity of NDH-dependent CET was increased, and protons from the stroma were transferred into the thylakoid lumen, resulting in acidification. The protons drive ATP synthesis, maintaining an optimal NADPH/ATP ratio. Drought and cold stress both induce an increase in ATP demand that may be fulfilled by NDH-dependent CET around photosystem I (PSI).

4. Materials and Methods

4.1. Plasmid Construction and Agrobacterium-Mediated Rice Transformation

To generate *OsCRP1* (Os09g0565200) overexpression lines, the 969 base pair coding sequence (CDS) was isolated from rice (*Oryza sativa* cv. Dongjin) cDNA and cloned into the pSB11 vector using the Gateway™ cloning system (Invitrogen, USA). The rice *OsCc1* promoter was used as a constitutive promoter [32], and the potato-derived (*Solanum tuberosum*) 3′pinII as a terminator (*OsCc1::OsCRP1*^{OX}). The *OsCRP1* CDS without the stop codon was isolated from rice (*O.sativa* cv. Nakdong) cDNA and fused to GFP (*OsCc1::OsCRP1-GFP*) under control of the *OsCc1* promoter with the 3′pinII as a terminator, as before. The bar gene controlled by the *CaMV 35S* promoter and the 3′nos terminator were used for herbicide resistance selection. For the knockdown construct (*OsCc1::OsCRP1*^{RNAi}), the CDS was isolated from rice (*O. sativa* cv Dongjin) cDNA and cloned into the pGOS2-RNAi vector [49] containing the bar selection marker using the Gateway™ cloning system. Primers used for vector construction are listed in Table S1. All transgenic plants were produced by *Agrobacterium tumefaciens* (LBA4404)-mediated transformation and tissue culture as previously described [50]. Three representative T_5 homozygote transgenic lines were selected for further studies based on gene expression levels.

4.2. Subcellular Localization of OsCP31A

The detailed method for rice protoplast preparation and transient protoplast transformation has been previously described [51]. The plasmid *OsCc1::OsCRP1:GFP* DNA transformed into the protoplasts using the polyethylene glycol-mediated method with approximately 10^6 cells per reaction. The transformed protoplasts were incubated for 16 h at 28 °C under dark conditions, and the GFP fluorescence of the transfected protoplasts was observed using a confocal laser scanning microscope (Leica TCS SP8 STED, Wetzlar, Germany) as in Park et al. [51].

4.3. qRT-PCR Analysis

The cDNAs of total and/or immunoprecipitated RNAs were synthesized using the RevertAid™ First Strand cDNA Synthesis kit with an oligo(dT) primer (Thermo Scientific, Waltham, MA, USA). Based on RNA amount, 20 ng of cDNA was used as a template for qRT-PCR analysis. The PCR enzymes and fluorescent dye was used with the 2× Real-time PCR Pre-mix with Evagreen (SolGent, Seoul, Korea), and the q-RT-PCR experiments were performed with an MX3005p qPCR system (Agilent Technologies, CA, USA). The thermocycling conditions were 95 °C for 10 min followed by 40 cycles of 95 °C for 30 s, 60 °C for 1 min. The gene-specific primer pairs are listed in Supplemental Table S1 and were checked by melting curve analysis (55–95 °C at a heating rate of 0.1 °C s^{-1}). The qRT-PCR values of cDNAs synthesized from total RNAs were normalized to the *OsUbi1* (Os06g0681400) gene, whereas the total input per experiment was used for the cDNA from immunoprecipitated RNAs. Total RNA samples were extracted from the leaves of *OsCc1::OsCRP1*^{OX}, NT and *OsCc1::OsCRP1*^{RNAi} plants using the Hybrid-R kit (GeneALL, Lisbon, Portugal). Each sample was treated with 70μL of DNase reaction buffer (DRB) containing 2 μL of DNase I (GeneALL, Lisbon, Portugal) for 10 min to avoid DNA contamination. To synthesize cDNA, 1 μL of RNA was used with oligo dT primers and 1 μL of RevertAidTM reverse transcriptase (Thermo Fischer Scientific, Waltham, MA, USA). Reverse transcription was performed at 42 °C for 90 min and terminated by incubating the reaction mixture for 5 min at 70 °C. qRT-PCR was carried out on a Mx3000p real-time PCR machine with the Mx3000p software and in a 20 μL reaction mixture containing 1 μL of cDNA template, 2 μL of primer, 0.04 μL of ROX reference dye (Invitrogen, Carlsbad, CA, USA), 1 μL of 20X Evagreen (SolGent, Daejeon, Korea), 10 μL of 2× premix, and dH2O. Cycling conditions were 1 cycle at 95 °C for 10 min and 55 cycles at 95 °C for 30 s, 58 °C for 30 s and at 72 °C for 30 s. The analysis was carried out with three biological and three technical replicates. *OsUbi1* (Os06g0681400) was used as an internal control in all experiments. Primers used for qRT-PCR are listed in Table S1.

4.4. Stress Treatments and Tolerance Evaluation

OsCRP1 transgenic and non-transgenic plants (*O. Sativa* cv. Dongjin) were sown on MS (Murashige and Skoog) media and incubated in a dark growth chamber for 4 days at 28 °C. Seedlings were then transferred to a growth chamber with a light/dark cycle of 16 h light/8 h dark and grown for 1 additional day before transplanting to soil. For cold stress treatments, fifteen plants from each line were transplanted into five soil pots (4 cm × 4 cm × 6 cm: three plants per pot) within a container (59 cm × 38.5 cm × 15 cm) and grown for 2 additional weeks in a growth chamber (16h light/8 h dark cycle) at 30 °C. Cold stress was imposed by exposing the plants to 4 °C for 3 days and the plants were then left recover for 7 days of 30 °C. For drought stress treatments, thirty plants from each line were transplanted into ten soil pots (4 cm × 4 cm × 6 cm: three plants per pot) within a container (59 cm × 38.5 cm × 15 cm) and grown for an additional 5 weeks in a greenhouse (16 h light/8 h dark cycle) at 30 °C. Drought stress was imposed by withholding water for 3 days and re-watering for 5 days. Stress-induced symptoms were monitored by imaging transgenic and NT plants at the indicated time points using a NEX-5N camera (Sony, Tokyo, Japan). The soil moisture contents were measured at the indicated time points using a SM150 Soil Moisture Sensor (Delta-T Devices, Cambridge, UK). Transient chlorophyll a fluorescence was measured using the Handy-PEA fluorimeter (Hansatech Instruments, Norfolk, UK) as previously described [52]. Chlorophyll a fluorescence was measured for the longest leaves of each plant after 1 h of dark adaptation to ensure sufficient opening of the reaction center.

4.5. RNA-Immunoprecipitation (RIP) Analysis

RIP experiments were performed as previously described [53,54] at 4 °C, with minor modifications. Leaf tissue from 14-day-old rice seedlings was powdered in liquid nitrogen and the powder incubated with polysome lysis buffer consisting of 100 mM KCl, 5 mM $MgCl_2$, 10 mM HEPES, pH 7.0, 0.5% Nonidet P-40, 1 mM DTT, RNase Out RNase inhibitor, 100 units mL^{-1} (Invitrogen, Carlsbad, CA, USA), 2 mM vanadyl ribonucleoside complexes solution (Sigma-Aldrich, St. Louis, MO, USA), and protease inhibitor cocktail tablets (Roche, Mannheim, Germany) for 20 min with shaking. The supernatant was separated from the crude extract by centrifuging at 16,000× *g* for 20 min, and after quantification of the soluble proteins using the Bradford method, lysate containing 1 mg protein was used for the next step. To confirm the quality of the OsCRP1-GFP protein, a preliminary immuno-blotting experiment was carried out. Before the immunoprecipitation step, the lysate was clarified by rotating at 4 °C for 2 h with 50% slurry containing protein A-agarose beads equilibrated in lysis buffer containing 1 mg mL^{-1} bovine serum albumin (BSA). After incubation of the lysate with specific antibodies, the protein-RNA complexes were pulled-down using protein A-agarose beads. The beads were washed four times with polysome lysis buffer without RNase and proteinase inhibitors and an additional four times with the same buffer containing 1 M urea. Finally, the RNA was eluted from the beads with the polysome lysis buffer containing 0.1% SDS and 30 μg proteinase K. The RNA was purified and enriched using Trizol reagent (Invitrogen Life Technologies) and 20 μg glycogen was added during the ethanol precipitation step.

4.6. RNA-Seq

Total RNA was prepared from leaf tissue of two-week-old transgenic and NT plants using the RNeasy plant mini kit (Qiagen, Valencia, Spain), according to the manufacturer's instruction. RNA quality and purity was assessed with a Thermo Scientific Nanodrop 2000 and an Agilent Bioanalyzer 2100. RNA-seq libraries were prepared using the TruSeq RNA Library Prep Kit (Illumina, San Diego, CA, USA) according to the manufacturer's instructions and sequenced (MACROGEN Inc., Seoul, Korea) using the Illumina HiSeq2000 (Illumina, San Diego, USA). Single-end sequences were generated and raw sequence reads were trimmed to remove adaptor sequences, and those with a quality lower than Q30 were removed using the clc quality trim software (CLCBIO). All reads were assembled with the clc_ref_assemble 6 (version 4.06; Aarhus, Denmark) program, using annotated gene and sequences from the rapdb (http://rapdb.dna.affrc.go.jp; 2 February 2019; Chloroplast_GCF_001433935.1_IRGSP-1.0).

4.7. Analysis of NDH-Dependent CET

NDH-dependent CET was determined by monitoring chlorophyll a fluorescence with a mini-PAM (Waltz, Germany) as previously described [25]. Plants were adapted in growth chambers (22 °C, 28 °C and 35 °C dark) for at least 30 min prior to measurements. Leaves were exposed to actinic light (AL: 200 μmol photons m^{-2} s^{-1}) for 5 min after the light was turned on (Fo level: minimum yield of Chlorophyll a fluorescence) to drive electron transport between photosystem II and photosystem I. Maximum fluorescence (Fm) and steady-state fluorescence (Fs) were determined under these conditions. The transient increase in chlorophyll a fluorescence was monitored after actinic light was turned off.

4.8. Determination of ATP Content

ATP measurements were performed as described in the ENLITEN® ATP Assay Kit (Promega, USA) protocol. Leaf samples (0.05 g) were transferred to 2 mL tubes containing

1ml of Tris-HCl (pH 7.8) and the tubes were then heated in a water bath at 100 °C for 10 min and cooled to room temperature. To determine the ATP content, 10 µL of the cooled samples were added to wells containing 100 µL of rL/L reagent each. The ATP standard curve was obtained using ATP standard samples provided with the kit. Luminescence was measured with a Infinite M200 System (Tecan, Seestrasse, Mannedorf, Switzerland) using the ATP standard curve.

4.9. Accession Numbers

Genes from this article can be found in the National Center for Biotechnology Information (http://www.ncbi.nlm.nih.gov; 2 February 2019) with the following accession numbers: PRJNA631899 (RNA-seq), OsCRP1 (Os09g0565200), OsCRP2 (Os08g557100), OsCRP3 (Os07g0158300), OsCRP4 (Os03g0376600), OsCRP5 (Os07g0631900), OsCRP6 (Os02g0815200), OsCRP7 (Os09g279500) and OsCRP8 (Os08g0117100).

Supplementary Materials: The following are available online at https://www.mdpi.com/1422-0067/22/4/1673/s1.

Author Contributions: S.W.B., S.-H.P. and D.-K.L. designed experiments; S.W.B., H.S.L. and S.-H.P. performed experiments; Y.S.K. helped in collection of plant materials; S.W.B., J.S.S., S.-C.P. and J.-K.K. wrote the manuscript and prepare the figures. All authors have read and agreed to the published version of the manuscript.

Funding: This work was supported by the New Breeding Technologies Development Program (Project No. PJ01477201 to J.-K.K.), Rural Development Administration, Republic of Korea and by the National Research Foundation of Korea funded by the Korean Government (grant no. NRF-2020R1C1C1009494 to S.W.B.).

Institutional Review Board Statement: Not applicable.

Informed Consent Statement: Not applicable.

Data Availability Statement: Not applicable.

Acknowledgments: We thank the Rural Development Administration and Kyungpook National University for providing the rice paddy fields.

Conflicts of Interest: The authors declare no conflict of interest.

References

1. Schmitz-Linneweber, C.; Barkan, A. RNA splicing and RNA editing in chloroplasts. In *Cell and Molecular Biology of Plastids*; Bock, R., Ed.; Springer: Berlin/Heidelberg, Germany, 2007; pp. 213–248.
2. Kroeger, T.S.; Watkins, K.P.; Friso, G.; van Wijk, K.J.; Barkan, A. A plant-specific RNA-binding domain revealed through analysis of chloroplast group II intron splicing. *Proc. Natl. Acad. Sci. USA* **2009**, *106*, 4537–4542. [CrossRef] [PubMed]
3. Stern, D.B.; Goldschmidt-Clermont, M.; Hanson, M.R. Chloroplast RNA metabolism. *Annu. Rev. Plant Biol.* **2010**, *61*, 125–155. [CrossRef] [PubMed]
4. Barkan, A. Expression of plastid genes: Organelle-specific elaborations on a prokaryotic scaffold. *Plant Physiol.* **2011**, *155*, 1520–1532. [CrossRef] [PubMed]
5. Kotera, E.; Tasaka, M.; Shikanai, T. A pentatricopeptide repeat protein is essential for RNA editing in chloroplasts. *Nature* **2005**, *433*, 326–330. [CrossRef] [PubMed]
6. Okuda, K.; Myouga, F.; Motohashi, R.; Shinozaki, K.; Shikanai, T. Conserved domain structure of pentatricopeptide repeat proteins involved in chloroplast RNA editing. *Proc. Natl. Acad. Sci. USA* **2007**, *104*, 8178–8183. [CrossRef]
7. Jacobs, J.; Kuck, U. Function of chloroplast RNA-binding proteins. *Cell Mol. Life Sci.* **2011**, *68*, 735–748. [CrossRef]
8. Germain, A.; Hotto, A.M.; Barkan, A.; Stern, D.B. RNA processing and decay in plastids. *Wiley Interdiscip. Rev. RNA* **2013**, *4*, 295–316. [CrossRef]
9. Li, Y.; Sugiura, M. Nucleic acid-binding specificities of tobacco chloroplast ribonucleoproteins. *Nucleic Acids Res.* **1991**, *19*, 2893–2896. [CrossRef]
10. Li, Y.; Ye, L.; Sugita, M.; Sugiura, M. Tobacco nuclear gene for the 31 kd chloroplast ribonucleoprotein: Genomic organization, sequence analysis and expression. *Nucleic Acids Res.* **1991**, *19*, 2987–2991. [CrossRef]
11. Ye, L.; Sugiura, M. Domains required for nucleic acid binding activities in chloroplast ribonucleoproteins. *Nucleic Acids Res.* **1992**, *20*, 6275–6279. [CrossRef]

12. Schuster, G.; Gruissem, W. Chloroplast mRNA 3 end processing requires a nuclear-encoded RNA-binding protein. *EMBO J.* **1991**, *10*, 1493–1502. [CrossRef]
13. Lorkovic, Z.J.; Barta, A. Genome analysis: RNA recognition motif (RRM) and K homology (KH) domain RNA-binding proteins from the flowering plant *Arab. Thaliana. Nucleic Acids Res.* **2002**, *30*, 623–635. [CrossRef]
14. Ruwe, H.; Kupsch, C.; Teubner, M.; Schmitz-Linneweber, C. The RNA-recognition motif in chloroplasts. *J. Plant Physiol.* **2011**, *168*, 1361–1371. [CrossRef]
15. Tillich, M.; Hardel, S.L.; Kupsch, C.; Armbruster, U.; Delannoy, E.; Gualberto, J.M.; Lehwark, P.; Leister, D.; Small, I.D.; Schmitz-Linneweber, C. Chloroplast ribonucleoprotein *CP31A* is required for editing and stability of specific chloroplast mRNAs. *Proc. Natl. Acad. Sci. USA* **2009**, *106*, 6002–6007. [CrossRef]
16. Kupsch, C.; Ruwe, H.; Gusewski, S.; Tillich, M.; Small, I.; Schmitz-Linnewebera, C. *Arabidopsis* chloroplast RNA binding proteins *CP31A* and *CP29A* associate with large transcript pools and confer cold stress tolerance by influencing multiple chloroplast RNA processing steps. *Plant Cell* **2012**, *24*, 4266–4280. [CrossRef] [PubMed]
17. Churin, Y.; Hess, W.; Boerner, T. Cloning and characterization of three cDNAs encoding chloroplast RNA binding proteins from barley: Differential regulation of expression by light and plastid development. *Curr. Genet.* **1999**, *36*, 173–181. [CrossRef] [PubMed]
18. Kleffmann, T.; von Zychlinski, A.; Russenberger, D.; Hirsch-Hoffmann, M.; Gehrig, P.; Gruissem, W.; Baginsky, S. Proteome dynamicsduring plastid differentiation in rice. *Plant Physiol.* **2007**, *143*, 912–923. [CrossRef] [PubMed]
19. Yamori, W.; Shikanai, T. Physiological functions of cyclic electron transport around Photosystem I in sustaining photosynthesis and plant growth. *Annu. Rev. Plant Biol.* **2016**, *67*, 81–106. [CrossRef] [PubMed]
20. Seelert, H.; Poetsch, A.; Dencher, N.A.; Engel, A.; Stahlberg, H.; Müller, D.J. Structural biology: Proton-powered turbine of a plant motor. *Nature* **2000**, *405*, 418–419. [CrossRef]
21. Wang, C.; Yamamoto, H.; Shikanai, T. Role of cyclic electron transport around photosystem I in regulating proton motive force. *Biochim. Biophys. Acta.* **2015**, *1847*, 931–938. [CrossRef]
22. Yamori, W.; Makino, A.; Shikanai, T. A physiological role of cyclic electron transport around photosystem I in sustaining photosynthesis under fluctuating light in rice. *Sci. Rep.* **2016**, *6*, 20147. [CrossRef]
23. Munekage, Y.; Hashimoto, M.; Miyake, C.; Tomizawa, K.I.; Endo, T.; Tasaka, M.; Shikanai, T. Cyclic electron flow around photosystem I is essential for photosynthesis. *Nature* **2004**, *429*, 579–582. [CrossRef] [PubMed]
24. Burrows, P.A.; Sazanov, L.A.; Svab, Z.; Maliga, P.; Nixon, P.J. Identification of a functional respiratory complex in chloroplasts through analysis of tobacco mutants containing disrupted plastid *ndh* genes. *EMBO J.* **1998**, *17*, 868–876. [CrossRef]
25. Shikanai, T.; Endo, T.; Hashimoto, T.; Yamada, Y.; Asada, K.; Yokota, A. Directed disruption of the tobacco *ndhB* gene impairs cyclic electron flow around photosystem I. *Proc. Natl. Acad. Sci. USA* **1998**, *95*, 9705–9709. [CrossRef]
26. DalCorso, G.; Pesaresi, P.; Masiero, S.; Aseeva, E.; Schünemann, D.; Finazzi, G.; Joliot, P.; Barbato, R.; Leister, D. A complex containing *PGRL1* and *PGR5* is involved in the switch between linear and cyclic electron flow in *Arabidopsis*. *Cell* **2008**, *132*, 273–285. [CrossRef] [PubMed]
27. Munekage, Y.; Hojo, M.; Meurer, J.; Endo, T.; Tasaka M.; Shikanai, T. *PGR5* is involved in cyclic electron flow around photosystem I and is essential for photoprotection in *Arabidopsis*. *Cell* **2002**, *110*, 361–371. [CrossRef]
28. Yamori, W.; Sakata, N.; Suzuki, Y.; Shikanai, T.; Makino, A. Cyclic electron flow around photosystem I via chloroplast NAD (P) H dehydrogenase (NDH) complex performs a significant physiological role during photosynthesis and plant growth at low temperature in rice. *Plant J.* **2011**, *68*, 966–976. [CrossRef]
29. Yamori, W.; Shikanai, T.; Makino, A. Photosystem I cyclic electron flow via chloroplast NADH dehydrogenase-like complex performs a physiological role for photosynthesis at low light. *Sci. Rep.* **2015**, *5*, 13908. [CrossRef] [PubMed]
30. Munne-Bosch, S.; Shikanai, T.; Asada, K. Enhanced ferredoxin dependent cyclic electron flow around photosystem I and α-tocopherol quinone accumulation in water-stressed *ndhB* inactived tobacco mutants. *Planta* **2005**, *222*, 502–511. [CrossRef]
31. He, Y.; Fu, J.; Yu, C.; Wang, X.; Jiang, Q.; Hong, J.; Lu, K.; Xue, G.; Yan, C.; James, A. Increasing cyclic electron flow is related to Na^+ sequestration into vacuoles for salt tolerance in soybean. *J. Exp. Bot.* **2015**, *66*, 6877–6889. [CrossRef]
32. Jang, I.C.; Choi, W.B.; Lee, K.H.; Song, S.I.; Nahm, B.H.; Kim, J.K. High-level and ubiquitous expression of the rice cytochrome c gene *OsCc1* and its promoter activity in transgenic plants provides a useful promoter for transgenesis of monocots. *Plant Physiol.* **2002**, *129*, 1473–1481. [CrossRef] [PubMed]
33. Endo, T.; Shikanai, T.; Takabayashi, A.; Asada, K.; Sato, F. The role of chloroplastic NAD(P)H dehydrogenase in photoprotection. *FEBS Lett.* **1999**, *457*, 5–8. [CrossRef]
34. Horváth, E.M.; Peter, S.O.; Joët, T.; Rumeau, D.; Cournac, L.; Horváth, G.V.; Kavanagh, T.A.; Schäfer, C.; Peltier, G.; Medgyesy, P. Targeted inactivation of the plastid ndhB gene in tobacco results in an enhanced sensitivity of photosynthesis to moderate stomatal closure. *Plant Physiol.* **2000**, *123*, 1337–1350. [CrossRef]
35. Li, X.-G.; Duan, W.; Meng, Q.-W.; Zou, Q.; Zhao, S.-J. The function of chloroplastic NAD (P) H dehydrogenase in tobacco during chilling stress under low irradiance. *Plant Cell Physiol.* **2004**, *45*, 103–108. [CrossRef] [PubMed]
36. Wang, P.; Duan, W.; Takabayashi, A.; Endo, T.; Shikanai, T.; Ye, J.Y.; Mi, H. Chloroplastic NAD(P)H dehydrogenase in tobacco leaves functions in alleviation of oxidative damage caused by temperature stress. *Plant Physiol.* **2006**, *141*, 465–474. [CrossRef]
37. Zhang, R.; Sharkey, T.D. Photosynthetic electron transport and proton flux under moderate heat stress. *Photosynth. Res.* **2009**, *100*, 29–43. [CrossRef]

38. Rumeau, D.; Peltier, G.; Cournac, L. Chlororespiration and cyclic electron flow around PSI during photosynthesis and plant stress response. *Plant Cell Environ.* **2007**, *30*, 1041–1051. [CrossRef] [PubMed]
39. Hajihashemi, S.; Noedoost, F.; Geuns, J.M.; Djalovic, I.; Siddique, K.H. Effect of cold stress on photosynthetic traits, carbohydrates, morphology, and anatomy in nine cultivars of stevia rebaudiana. *Front. Plant Sci.* **2018**, *9*, 1430. [CrossRef]
40. Wang, S.; Bai, G.; Wang, S.; Yang, L.; Yang, F.; Wang, Y.; Zhu, J.-K.; Hua, J. Chloroplast RNA-binding protein RBD1 promotes chilling tolerance through 23S rRNA processing in Arabidopsis. *PLoS Genet.* **2016**, *12*, e1006027. [CrossRef]
41. Hirose, T.; Sugiura, M. Involvement of a site-specific trans-acting factor and a common RNA-binding protein in the editing of chloroplast mRNAs: Development of a chloroplast in vitro RNA editing system. *EMBO J.* **2001**, *20*, 1144–1152. [CrossRef]
42. Teubner, M.; Fuß, J.; Kühn, K.; Krause, K.; Schmitz-Linneweber, C. The RNA recognition motif protein *CP33A* is a global ligand of chloroplast mRNAs and is essential for plastid biogenesis and plant development. *Plant J.* **2017**, *89*, 472–485. [CrossRef] [PubMed]
43. Shikanai, T. Central role of cyclic electron transport around photosystem I in the regulation of photosynthesis. *Curr. Opin. Biotechnol.* **2014**, *26*, 25–30. [CrossRef] [PubMed]
44. Shikanai, T. Chloroplast NDH: A different enzyme with a structure similar to that of respiratory NADH dehydrogenase. *Biochim. Biophys. Acta.* **2016**, *7*, 1015–1022. [CrossRef]
45. Kofer, W.; Koop, H.U.; Wanner, G.; Steinmuller, K. Mutagenesis of the genes encoding subunits A, C, H, I, J and K of the plastid NAD(P)H-plastoquinone-oxidoreductase in tobacco by polyethylene glycol-mediated plastome transformation. *Mol. Gen. Genet.* **1998**, *258*, 166–173. [CrossRef] [PubMed]
46. Rumeau, D.; Becuwe-Linkaa, N.; Beylya, A.; Louwagieb, M.; Garinb, J.; Peltiera, G. New subunits NDH-M, -N, and -O, encoded by nuclear genes, are essential for plastid Ndh complex functioning in higher plants. *Plant Cell* **2005**, *17*, 219–232. [CrossRef] [PubMed]
47. Livingston, A.K.; Cruz, J.A.; Kohzuma, K.; Dhingra, A.; Kramer, D.M. An *Arabidopsis* mutant with high cyclic electron flow around photosystem I (hcef) involving the NADPH dehydrogenase complex. *Plant Cell* **2010**, *22*, 221–233. [CrossRef]
48. Hou, Q.Z.; Sun, K.; Zhang, H.; Su, X.; Fan, B.Q.; Feng, H.Q. The responses of photosystem II and intracellularATP production of Arabidopsis leaves to salt stress are affected by extracellular ATP. *J. Plant Res.* **2018**, *131*, 331–339. [CrossRef]
49. Lee, D.-K.; Jung, H.; Jang, G.; Jeong, J.S.; Kim, Y.S.; Ha, S.-H.; Choi, Y.D.; Kim, J.-K. Overexpression of the *OsERF71* transcription factor alters rice root structure and drought resistance. *Plant Physiol.* **2016**, *172*, 575–588. [CrossRef]
50. Hiei, Y.; Ohta, S.; Komari, T.; Kumashiro, T. Efficient transformation of rice (*Oryza sativa* L.) mediated by *Agrobacterium* and sequence analysis of the boundaries of the T-DNA. *Plant J.* **1994**, *6*, 271–282. [CrossRef]
51. Park, S.H.; Jeong, J.S.; Choi, Y.D.; Kim, J.K. Characterization of the rice *RbcS3* promoter and its transitpeptide for use in chloroplast-targeted expression. *Plant Biotechnol. Rep.* **2015**, *9*, 395–403. [CrossRef]
52. Bang, S.W.; Lee, D.-K.; Jung, H.; Chung, P.J.; Kim, Y.S.; Choi, Y.D.; Suh, J.-W.; Kim, J.-K. Overexpression of *OsTF1L*, a rice HD-Zip transcription factor, promotes lignin biosynthesis and stomatal closure that improves drought tolerance. *Plant Biotechnol. J.* **2019**, *17*, 118–131. [CrossRef] [PubMed]
53. Keene, J.D.; Komisarow, J.M.; Friedersdorf, M.B. RIP-Chip: The isolation and identification of mRNAs, microRNAs and protein components of ribonucleoprotein complexes from cell extracts. *Nat. Protoc.* **2006**, *1*, 302–307. [CrossRef] [PubMed]
54. Peritz, T.; Zeng, F.; Kannanayakal, T.J.; Kilk, K.; Eiriksdottir, E.; Langel, U.; Eberwine, J. Immunoprecipitation of mRNA-protein complexes. *Nat. Protoc.* **2006**, *1*, 577–580. [CrossRef] [PubMed]

International Journal of
Molecular Sciences

MDPI

Article

Rice Transcription Factor OsWRKY55 Is Involved in the Drought Response and Regulation of Plant Growth

Kai Huang †, Tao Wu †, Ziming Ma, Zhao Li, Haoyuan Chen, Mingxing Zhang, Mingdi Bian, Huijiao Bai, Wenzhu Jiang * and Xinlin Du *

Jilin Province Engineering Laboratory of Plant Genetic Improvement, College of Plant Science, Jilin University, Changchun 130062, China; 15584159829@163.com (K.H.); jidawt@jlu.edu.cn (T.W.); mzm1040722406@163.com (Z.M.); zklizhao@jlu.edu.cn (Z.L.); hychen18@mails.jlu.edu.cn (H.C.); mingxing19@jlu.edu.cn (M.Z.); bianmd@jlu.edu.cn (M.B.); bhj20212021@163.com (H.B.)

* Correspondence: jwz1975@jlu.edu.cn (W.J.); duxinglin2004@163.com (X.D.); Tel.: +86-25-87835716 (W.J.); +86-25-87836251 (X.D.)

† These authors have contributed equally to this work.

Abstract: WRKY transcription factors (TFs) have been reported to respond to biotic and abiotic stresses and regulate plant growth and development. However, the molecular mechanisms of WRKY TFs involved in drought stress and regulating plant height in rice remain largely unknown. In this study, we found that transgenic rice lines overexpressing *OsWRKY55* (*OsWRKY55*-OE) exhibited reduced drought resistance. The *OsWRKY55*-OE lines showed faster water loss and greater accumulation of hydrogen peroxide (H_2O_2) and superoxide radical ($O_2^{\cdot-}$) compared to wild-type (WT) plants under drought conditions. *OsWRKY55* was expressed in various tissues and was induced by drought and abscisic acid (ABA) treatments. Through yeast two-hybrid assays, we found that OsWRKY55 interacted with four mitogen-activated protein kinases (MAPKs) that could be induced by drought, including OsMPK7, OsMPK9, OsMPK20-1, and OsMPK20-4. The activation effects of the four OsMPKs on OsWRKY55 transcriptional activity were demonstrated by a GAL4-dependent chimeric transactivation assay in rice protoplasts. Furthermore, OsWRKY55 was able to reduce plant height under normal conditions by decreasing the cell size. In addition, based on a dual luciferase reporter assay, OsWRKY55 was shown to bind to the promoter of *OsAP2-39* through a yeast one-hybrid assay and positively regulate *OsAP2-39* expression. These results suggest that OsWRKY55 plays a critical role in responses to drought stress and the regulation of plant height in rice, further providing valuable information for crop improvement.

Keywords: rice; transcription factor; OsWRKY55; drought response; plant growth; *OsAP2-39*

Citation: Huang, K.; Wu, T.; Ma, Z.; Li, Z.; Chen, H.; Zhang, M.; Bian, M.; Bai, H.; Jiang, W.; Du, X. Rice Transcription Factor OsWRKY55 Is Involved in the Drought Response and Regulation of Plant Growth. *Int. J. Mol. Sci.* **2021**, *22*, 4337. https://doi.org/10.3390/ijms22094337

Academic Editor: Kiyosumi Hori

Received: 18 March 2021
Accepted: 20 April 2021
Published: 21 April 2021

1. Introduction

As an adverse environmental factor, drought stress greatly constrains plant growth and productivity [1]. To withstand this stress, plants have evolved a complex series of actions that manipulate the expression levels of certain sets of genes. These actions include signal perception and transmission. Transcription factors (TFs) are involved and play a vital role in these signaling cascades [2,3]. In rice, which is often used as a monocot plant model, many TFs have been characterized using forward and/or reverse genetic approaches, including MYB, WRKY, Apetala2 (AP2)/ethylene-responsive factor (ERF), basic region-leucine zipper (bZIP), NAC (NAM, ATAF1/2, CUC1/2), and basic helix-loop-helix (bHLH) [4–8].

Plant-specific WRKY TFs, one of the largest families of transcriptional regulators, are characterized by possessing a WRKY-DNA-binding domain [9]. Investigations of the overexpression or knockdown lines have shown that most WRKY TFs are involved in the plant immune response [10–13]. WRKY TFs also play crucial roles in various abiotic stress responses. The heterologous overexpression of either *ZmWRKY106* or *ZmWRKY40* in *Arabidopsis* was found to confer drought tolerance [14,15], and the overexpression of *GmWRKY12* in *Arabidopsis* was found to enhance tolerance to drought and salt. Similarly, *GmWRKY54* was found to confer drought tolerance in soybean [16,17], while the overexpression of *OsWRKY76* improved tolerance to cold stress in rice [18]. However, the mechanisms by which WRKY TFs participate in drought stress responses in rice have not been fully elucidated. Plant mitogen-activated protein kinase (MAPK) cascades play important roles in different physiological responses, including biotic and abiotic stresses [19]. A few MAPK cascade genes have been identified to function in response to biotic and abiotic stresses through WRKY TFs [20,21]. OsWRKY30 is phosphorylated by MAP kinases, thereby enhancing drought tolerance in rice [22], and OsBWMK1 mediates defense responses by activating *OsWRKY33* to regulate the expression of several pathogenesis-related genes [23]. OsWRKY53 interacts with OsMPK3/OsMPK6 and inhibits their activity [24].

The roles of several WRKY genes in plant growth and developmental processes, such as seed dormancy, germination, leaf senescence, and plant height, were also identified in previous studies. For example, in rice, the *Oswrky29* mutant displayed enhanced seed dormancy [25]. In *Arabidopsis*, a lack of *WRKY41* reduced primary seed dormancy [26]. The heterologous overexpression of *OsWRKY23* in *Arabidopsis* accelerated leaf senescence in darkness [27]. WRKY53 was found to act as a positive regulator of senescence, while WRKY70 negatively regulated leaf senescence in *Arabidopsis* [28,29]. Moreover, the enhanced expression of *OsWRKY11.2* resulted in a semi-dwarf phenotype [30]. However, the functions of a number of WRKY TFs in rice remain to be elucidated.

Previously, overexpression of the WRKY family gene *OsWRKY55* (LOC_Os03g20550, named Os*WRKY55* by Xie et al. [31] and *OsWRKY31* by Zhang et al. [32]) in rice was shown to enhance resistance to *Magnaporthe grisea* infection and induce the expression of defense-related genes, as well as affect root growth and auxin responses [32]. In this study, we characterized the other underlying roles of *OsWRKY55* in rice. OsWRKY55 not only modulates rice height but is also involved in drought stress. The overexpression of *OsWRKY55* in rice leads to a reduction in drought tolerance and plant height. Our findings reveal that OsWRKY55 plays crucial roles in the responses to drought stress and in the determination of plant height.

2. Results

2.1. Overexpression of OsWRKY55 Reduces Drought Tolerance in Rice

Previously, we constructed transgenic rice plants overexpressing different TFs to explore the roles of TFs in response to environmental stresses [33]. We identified two transgenic lines overexpressing *OsWRKY55* by surveying the phenotypes of transgenic lines overexpressing the *WRKY* gene under drought stress treatment. Using quantitative reverse transcription PCR (qRT-PCR), we verified that *OsWRKY55* was significantly upregulated in two independent transgenic lines, OE-1 and OE-2, and the expression level of the endogenous *OsWRKY55* gene in transgenic rice plants was similar to that in wild type (WT) (Figure 1A,B and Figure S1).

The seedlings overexpressing *OsWRKY55* (*OsWRKY55*-OE) and wild-type (WT) plants were phenotypically indistinguishable under normal conditions. However, leaf wilt was much more severe in the transgenic seedlings than in the WT seedlings under drought stress (Figure 1C). After 7 days of recovery (re-watering), the survival rates of the WT seedlings (≈65–80%) were significantly greater than those of the transgenic *OsWRKY55*-OE seedlings (≈45–53%; Figure 1D), suggesting that *OsWRKY55* negatively regulates drought resistance in rice. This hypothesis was tested by a detached leaf water loss assay. The transgenic *OsWRKY55*-OE lost water faster than the WT (Figure 1E). Since abiotic stress may lead to the accumulation of excessive reactive oxygen species (ROS) in plants, resulting in cell damage, we measured ROS accumulation using histochemical assays. Ultimately, 3, 3′-diaminobenzidine (DAB) staining indicated that under drought conditions, more hydrogen peroxide (H_2O_2) was accumulated in the leaves of the *OsWRKY55*-OE plants than in those of the WT plants, as evidenced by the abundance of brown spots in the

OsWRKY55-OE leaves; there were no visible differences between *OsWRKY55*-OE and WT leaves under normal conditions (Figure 1F). Although both WT and *OsWRKY55*-OE plants showed evidence of superoxide radical ($O_2^{-\cdot}$) accumulation under drought conditions (as indicated by blue dots on the leaves after nitro-blue tetrazolium (NBT) staining), fewer, lighter dots were also observed on the WT leaves; there were no visible differences between the *OsWRKY55*-OE and WT leaves under normal conditions (Figure 1G). These results indicate that the overexpression of *OsWRKY55* may increase ROS production, thereby decreasing drought resistance.

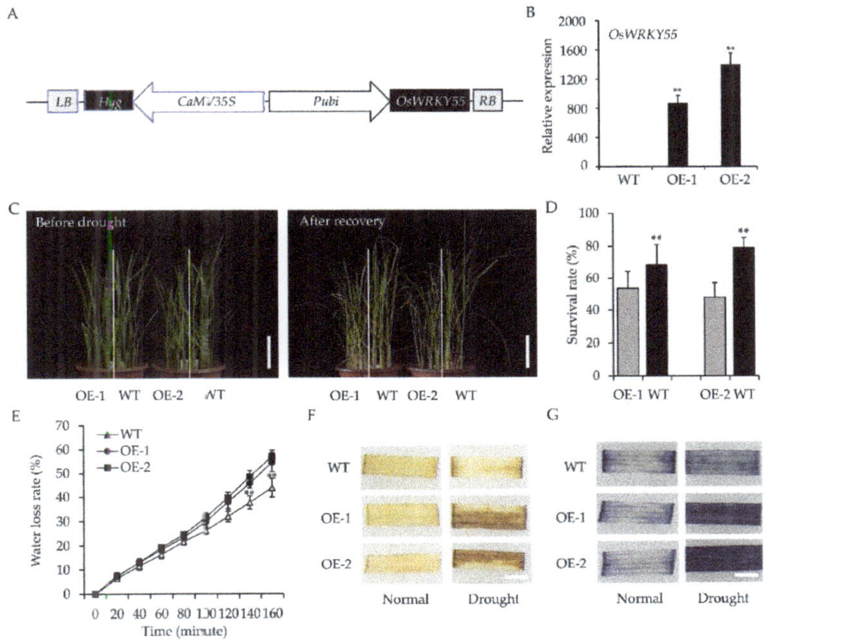

Figure 1. Overexpression of *OsWRKY55* increases the sensitivity of rice seedlings to drought stress. (**A**) Diagrams of the overexpression plasmid of *OsWRKY55*. (**B**) The *OsWRKY55* mRNA level in each genotype was analyzed by quantitative reverse transcription PCR (qRT-PCR). The rice *OsActin* gene was used as the internal control. (**C**) Appearance of wild-type (WT) and transgenic seedlings (OE-1 and OE-2) before drought stress and after 7 days of recovery (rewatering). Scale bars = 5 cm. (**D**) Survival rates of WT and transgenic seedlings after recovery for 7 days. (**E**) Water loss assay of WT, OE-1, and OE-2. (**F**) 3, 3′-diaminobenzidine (DAB) staining showing the level of hydrogen peroxide (H_2O_2) in the leaves of WT and transgenic plants under normal and drought-stress conditions. Scale bars = 0.5 cm. (**G**) Nitro-blue tetrazolium (NBT) staining showing the level of superoxide radical ($O_2^{-\cdot}$) in the leaves of WT and transgenic plants under normal and drought-stress conditions. Scale bars = 0.5 cm. In all graphs, values shown are means ± SD (n = 3), ** $p < 0.01$ (Student's *t*-test).

2.2. *Expression Pattern of OsWRKY55*

To further explore the function of *OsWRKY55* in rice, we first detected the expression profile of *OsWRKY55* under drought stress and abscisic acid (ABA) treatment. *OsWRKY55* was strongly up-regulated after drought stress. *OsWRKY55* expression peaked in the leaves after 6 h of drought (Figure 2A) and in the roots after 0.5 h of drought (Figure 2B). QRT-PCR analyses showed that the induced expression patterns of *OsWRKY55* in leaves and roots under ABA treatment were similar to those after drought stress (Figure 2C,D). In addition, we also assessed the tempo-spatial expression profile of this gene via qRT-PCR. *OsWRKY55* was constitutively expressed in all tissues analyzed, with relatively higher expression levels in roots and flag leaves and relatively lower expression levels in young leaves and stems (Figure 2E). As the preliminary results suggested that *OsWRKY55* is associated with drought response, we next analyzed the promoter sequence of the *OsWRKY55* gene using the PlantCARE database (http://sphinx.rug.ac.be:8080/PlantCARE/, accessed on 15 September, 2020). Analysis of the *OsWRKY55* gene promoter sequence identified many putative stress response-related cis-elements, including the MYB recognition site (5 hits), ABA-responsive element (ABRE) (5 hits), W-box (3 hits), and MYC recognition site (2 hits) (Figure 2F).

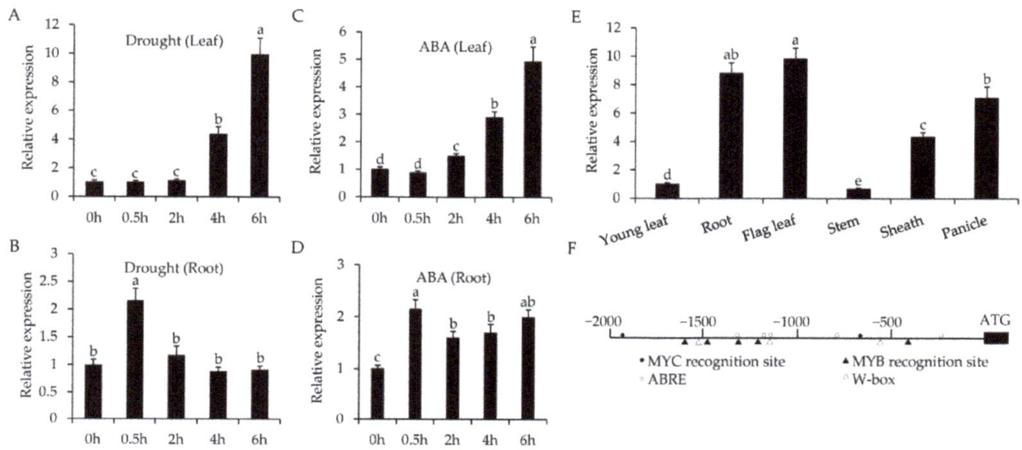

Figure 2. Expression pattern analysis of *OsWRKY55*. (**A**,**B**) Relative expression of *OsWRKY55* in leaves (**A**) and roots (**B**) after drought treatment (dehydration). (**C**,**D**) Relative expression of *OsWRKY55* in leaves (**C**) and roots (**D**) after treatment with 100 μM of the phytohormone abscisic acid (ABA). (**E**) Expression profile of *OsWRKY55* in various rice tissues of the wild-type (WT). (**F**) Distribution of stress-response-related cis-elements in the 2 kb promoter region of *OsWRKY55*. In all graphs, values shown are means ± SD (n = 3). The rice *OsActin* gene was used as the internal control. Different letters above the bars indicate significant differences ($p < 0.05$).

2.3. *OsWRKY55 Interacts with Several OsMPKs*

To determine whether OsWRKY55 interacts with OsMPKs, the full-length coding sequences (CDSs) of 15 *OsMPKs* [34] were fused to separate pGADT7 as prey, and pGBKT7-OsWRKY55 was used as bait in the yeast two-hybrid assays. The results showed that all transformants grew well on the SD/-Trp-Leu-His plates containing 0 mM 3-amino-1, 2, 4-triazole (3-AT) (Figure 3A and Figure S2) due to the auto-transcriptional activity of OsWRKY55 in yeast [32]. Only three transformants (carrying BD-OsWRKY55/AD-OsMPK20-1, BD-OsWRKY55/AD-OsMPK20-4, or BD-OsWRKY55/AD-OsMPK9) grew well on the SD/-Trp-Leu-His plates containing 0.2 and 0.4 mM 3-AT (Figure 3A and Figure S2). However, the growth of cells carrying BD-OsWRKY55/AD-OsMPK7 was reduced on the 0.2 mM 3-AT plate and inhibited on the 0.4 mM 3-AT plate (Figure 3A). Negative control yeast cells carrying BD-OsWRKY55/AD did not grow well on SD/-Trp-Leu-His plates containing 0.2 and 0.4 mM 3-AT. These results indicate that, in yeast, OsWRKY55 interacts strongly with OsMPK20-1, OsMPK20-4, and OsMPK9 but weakly with OsMPK7 (Figure 3A).

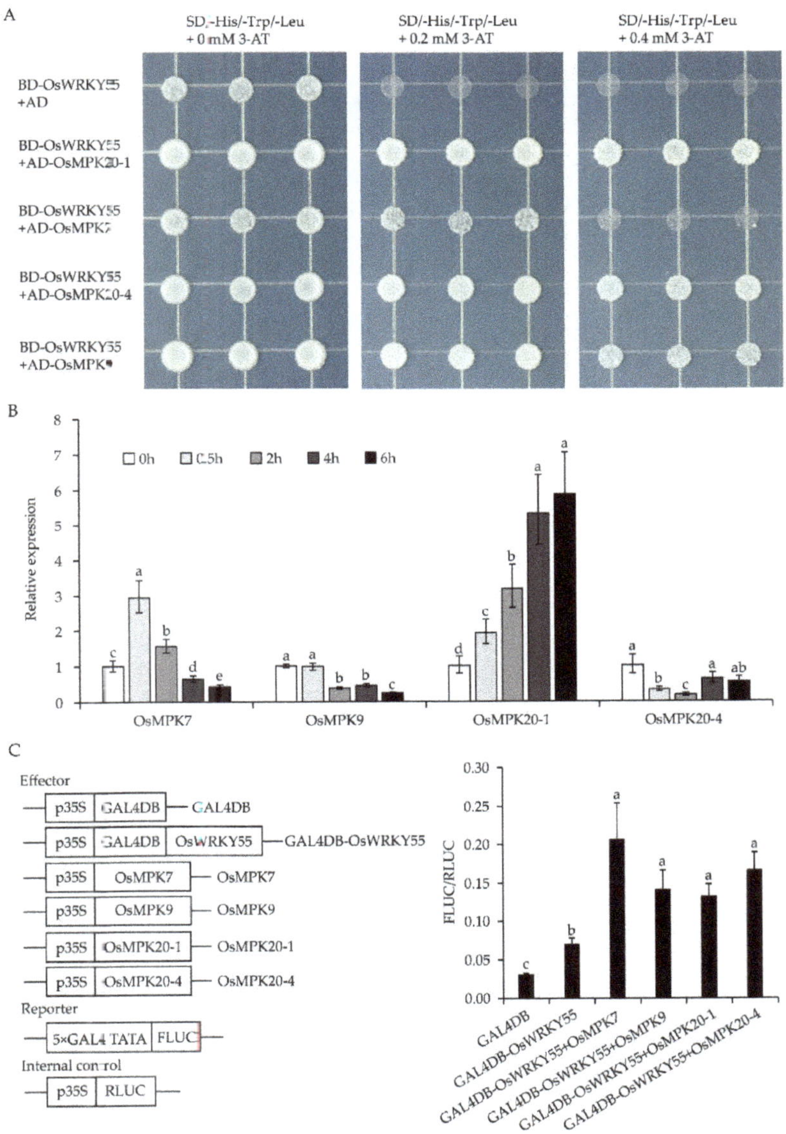

Figure 3. OsMPKs interact with OsWRKY55 and increase its transcription activity. (**A**) The interactions between OsWRKY55 and four OsMPKs in the yeast two-hybrid assays. Growth phenotypes of yeast cells on selective media plates (SD/-Trp-Leu-His) containing 0, 0.2, or 0.4 mM 3-amino-1, 2, 4-triazole (3-AT). (**B**) Relative expressions of *OsMPK7*, *OsMPK9*, *OsMPK20-1*, and *OsMPK20-4* after drought treatment (dehydration). (**C**) GAL4-dependent chimeric transactivation assay of OsWRKY55 in rice protoplasts. A schematic illustration of the effector and reporter vectors is shown on the left. FLUC, firefly luciferase; RLUC, renilla luciferase. The right part is the transcription activities by co-transformation of different effector vector(s) with the reporter vector and internal control. Values shown are means ± SD (n = 3). The rice *OsActin* gene was used as the internal control. Different letters above the bars indicate significant differences ($p < 0.05$).

To examine the possible involvement of these four *OsMPKs* in the drought stress response, we assessed the expression patterns of *OsMPKs* under drought stress via qRT-PCR. We found that the expression level of *OsMPK7* was increased within 0.5 h and that *OsMPK9* expression was down-regulated rapidly after 2 h of drought treatment. Under drought stress, the expression of *OsMPK20-1* was significantly induced, and *OsMPK20-4* was suppressed after 0.5 h, but the expression resumed slightly after 4 h (Figure 3B).

To further explore whether the four OsMPKs affect the activity of OsWRKY55, a GAL4-dependent chimeric transactivation assay in rice protoplasts was performed. The results illustrated the activation effects of the four OsMPKs on OsWRKY55 transcriptional activity (Figure 3C), which also suggested that the interactions between OsWRKY55 and OsMPKs occurred in vivo.

2.4. Overexpression of OsWRKY55 Reduces Plant Height

OsWRKY55 was not only associated with the drought stress response but was also observed to affect plant growth. The transgenic *OsWRKY55*-OE lines grew more slowly than the WT plants (Figure 4A). At maturity, the two *OsWRKY55*-OE lines were significantly shorter than those of the WT plants—OE1 and OE2 were 25.4% and 27.8% shorter than the WT, respectively (Figure 4B). All four internodes of each *OsWRKY55*-OE line were significantly shorter than the corresponding WT internodes (Figure 4C,D). A microscopic examination of the longitudinal sections of the uppermost internodes of the main culms of the WT and *OsWRKY55*-OE (OE-1) plants showed that there were noticeable differences in cell shape between the WT and transgenic plants. Indeed, there were significant differences in both cell length and cell width between the WT and *OsWRKY55*-OE (OE-1) plants (Figure 5A,B). Thus, our results indicate that *OsWRKY55* affects plant height by negatively regulating cell expansion.

Figure 4. Overexpression of *OsWRKY55* reduces plant height. (**A**) Phenotypes of wild-type (WT) plants and transgenic lines overexpressing *OsWRKY55* (OE-1 and OE-2). Scale bar = 10 cm. (**B**) Heights of mature WT and OE plants. (**C**) Panicles (P) and internodes (I-IV) of mature WT and *OsWRKY55*-OE plants. Scale bar = 5 cm. (**D**) Lengths of panicles (P) and internodes (I-IV) in mature WT and *OsWRKY55*-OE plants. In all graphs, values shown are means ± SD (n ≥ 15), ** $p < 0.01$ (Student's *t*-test).

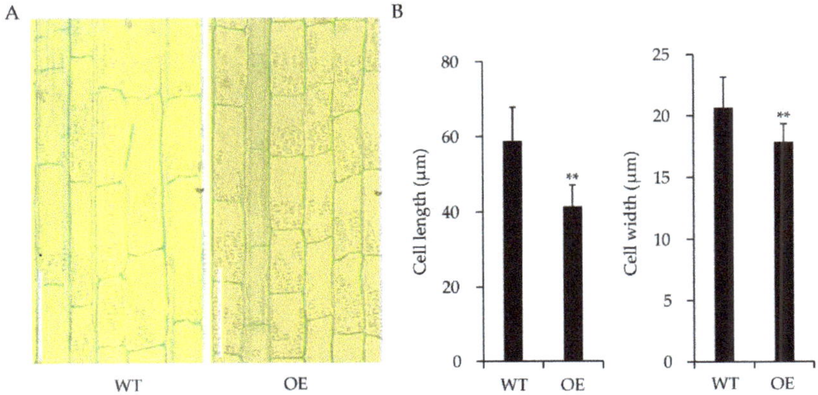

Figure 5. Overexpression of *OsWRKY55* decreases cell size. (**A**) Longitudinal sections of the uppermost internodes of the main culms of the wild-type (WT) plants and transgenic lines overexpressing *OsWRKY55* (*OsWRKY55*-OE-1). Scale bars = 50 μm. (**B**) Cell size comparison of the uppermost internodes of the main culms in WT and *OsWRKY55*-OE (OE-1) lines. Values shown are means ± SD (n = 3), ** $p < 0.01$ (Student's *t*-test).

2.5. *OsAP2-39 Is Directly Regulated by OsWRKY55*

To elucidate the molecular function of *OsWRKY55*, we identified the target genes of OsWRKY55 through the genome-wide expression profile changes in WT and *OsWRKY55*-OE lines via an RNA-sequencing (RNA-seq) assay under normal growth conditions at the four-leaf stage. The result revealed that about 162 genes in overexpressing *OsWRKY55* rice experienced two-fold changes compared to the genes in the WT rice. Among these genes, 98 were up-regulated in the OE lines. Although the predicted functions of the up-regulated genes were extremely diverse, there were many stress-related genes, such as *OsAP2-39* (LOC_Os04g52090), *OsERF48/OsDRAP1* (LOC_Os08g31580), *OsFbx352* (LOC_Os10g03850), *OsRAV2* (LOC_Os01g04800), *OsMYB48* (LOC_Os01g74410), and Zinc finger protein (LOC_Os03g55540, LOC_Os03g60570, LOC_Os01g61420, LOC_Os02g45710) (Table S2). Therefore, these genes might be direct or indirect targets of *OsWRKY55*. As *OsAP2-39*, the APETALA-2-like transcription factor is known to negatively regulate plant growth and drought tolerance [35], we further investigated an experimentally verified gene, *OsAP2-39* (Figure 6A). The up-regulation of *OsAP2-39* in transgenic lines overexpressing *OsWRKY55* was verified by qRT-PCR (Figure 6B).

To further identify whether *OsAP2-39* is the target gene of OsWRKY55, a yeast one-hybrid assay and a dual luciferase reporter assay were performed. The β-galactosidase activity level in the yeast cells co-transformed with pLacZi-Pro and pGAD424-OsWRKY55 was significantly greater than that of the control (a ≈10-fold increase in activity; Figure 6C), indicating that OsWRKY55 binds to the *OsAP2-39* promoter in yeast. Furthermore, the dual luciferase reporter assay showed that the FLUC/RLUC ratio, which reflects transcriptional activity, was significantly up-regulated in rice protoplasts harboring the effector vector pAN580-OsWRKY55 compared to that in protoplasts carrying the vector pAN580 (Figure 6D). These results suggest that *OsAP2-39* is a target gene of OsWRKY55.

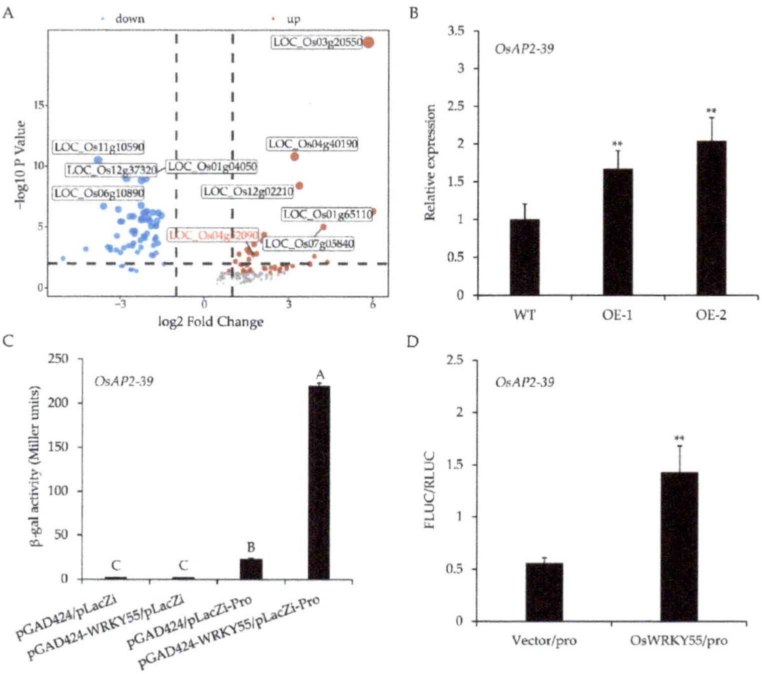

Figure 6. OsWRKY55 positively regulates the expression of *OsAP2-39*. (**A**) Identification of differentially expressed genes (DEGs) in transgenic lines overexpressing *OsWRKY55* compared to the wild type (WT) under normal conditions via RNA-sequencing (RNA-seq) analysis. The blue and red spots represent down-regulated and up-regulated genes, respectively. The sizes of the spots indicate the -$\log_{10}$FDR. (**B**) Relative expression of the *OsAP2-39* gene in WT plants and transgenic lines overexpressing *OsWRKY55* (OE-1 and OE-2). The rice *OsActin* gene was used as the internal control. Values shown are means ± SD (n = 3), ** $p < 0.01$ (Student's *t*-test). (**C**) β-galactosidase activity levels in co-transformed yeast cells. Values shown are means ± SD (n = 3). Different uppercase letters above the bars indicate significant differences (least significant differences, $p < 0.01$). (**D**) Transcriptional activity of *OsAP2-39* in rice protoplasts harboring the effector vector pAN580-OsWRKY55 compared to that in protoplasts carrying the empty vector. The transient expression of FLUC driven by the promoter of *OsAP2-39* was normalized to the internal control reporter (RLUC). Values shown are means ± SD (n = 3), ** $p < 0.01$ (Student's *t*-test).

3. Discussion

Rice is negatively affected by various abiotic and biotic stresses, resulting in serious yield losses worldwide. Previous studies have identified that many WRKY TFs participate in the rice response to biotic stresses [36]. However, to date, our understanding of the molecular mechanisms by which WRKY TFs in rice are involved in abiotic stresses, especially drought stress, remains largely unknown. Furthermore, the roles of WRKY TFs in the regulation of growth and development in rice have not yet been studied extensively. A prior study showed that *OsWRKY55* enhanced resistance against infection with *M. grisea* and affected root growth and auxin response [32]. In the present work, we focused on the response of *OsWRKY55* to drought stress and the regulation of plant height. Under drought treatment, by withholding the water supply, *OsWRKY55*-OE plants became more sensitive to drought stress than WT plants (Figure 1C,D), which is likely because the *OsWRKY55*-OE plants lost water at a faster rate (Figure 1E) and had fewer, shorter lateral roots than the WT plants [32]. Furthermore, the *OsWRKY55*-OE plants accumulated more H_2O_2 and O_2^- under drought conditions (Figure 1F,G). The associated increase in oxidative damage might also explain the increased susceptibility of the *OsWRKY55*-OE transgenic lines to drought. We also found that the overexpression of *OsWRKY55* reduced plant growth by decreasing cell size under normal conditions: *OsWRKY55*-OE plants were shorter than WT plants, with smaller, thinner cells in the internodes (Figures 4 and 5). A previous study revealed that *OsWRKY55* is an auxin-inducible gene and that the overexpression of *OsWRKY55* enhances auxin-related phenotypes in rice [32]. Auxin regulates a variety of processes during plant development [37]. However, it remains unclear whether the involvement of *OsWRKY55* in auxin signaling leads to the dwarfism observed in the *OsWRKY55*-OE plants.

MAPK cascades in plants are involved in various physiological functions. WRKY TFs and MAPK cascades have been shown to modulate plant stress response [19,38–42]. Under biotic and abiotic stress, many WRKY TFs are activated by MAPKs. For example, OsMPK6 phosphorylates and activates OsWRKY45. However, under cold and salinity stress, a tyrosine protein phosphatase dephosphorylates and inactivates OsMPK6, which reduces OsWRKY45 defense [43]. OsMPK7and OsMPK20-4 interact with OsWRKY30, and OsMPK7 further phosphorylates OsWRKY30, which are both crucial processes for OsWRKY30 to confer drought tolerance in rice [22]. In this study, we demonstrated that OsWRKY55 interacts with four different MAPKs in yeast, including OsMPK7, OsMPK9, OsMPK20-1, and OsMPK20-4 (Figure 3A), and that the four OsMPKs can activate the transcriptional activity of OsWRKY55. The qRT-PCR results showed that these four *OsMPKs* may be related to drought stress response (Figure 3B). These results suggest that the decreased drought tolerance associated with *OsWRKY55* may, to a large extent, be due to the interactions between OsWRKY55 and these four OsMPKs. OsWRKY30 and OsWRKY55 can be activated by OsMPK7, but they confer different drought tolerance levels in rice, which indicates that the regulatory network of responses to drought stress in rice is complex. Overexpression of the *OsWRKY55* gene was found to enhance resistance against infection with *M. grisea* [32]. It was suggested that OsWRKY55 plays an opposite role in biotic and abiotic stress, which may be due to the regulation of different downstream genes by OsWRKY55.

AP2/ERF proteins play important roles in plant responses to biotic and abiotic stress, as well as in the growth and development in rice. For example, *OsAP2-39*, a member of the APETALA2 (AP2) family, affects dehydration tolerance and plant height in rice by controlling the ABA/gibberellin (GA) balance [35]. However, overexpressing *OsAP2-39* lines display different response results depending on the different ways in which they are grown under drought stress. When grown in the same pot, transgenic *OsAP2-39* plants exhibited lower dehydration tolerance than WT plants, which is likely due to smaller root systems of overexpressing *OsAP2-39* lines, which are similar to those of the drought-sensitive phenotype of *OsWRKY55*-OE plants. When grown in separate pots, the responses were reversed, which was likely because the transpiration rates of the WT plants were greater. The definitive conclusion, that overexpressing *OsAP2-39* lines had lower dehydration tolerance than the WT, was reached via an excised leaf water loss assay. It was reported that an ERF protein, OsDERF1, activates *OsERF3* and *OsAP2-39* to negatively modulate ethylene synthesis and drought tolerance in rice [44]. The present study revealed that OsWRKY55 can bind to the promoter of *OsAP2-39* to up-regulate the expression of *OsAP2-39* (Figure 6), suggesting that the OsWRKY55 and OsAP2-39 transcriptional cascade negatively modulates drought response. OsDERF1 and OsWRKY55 both regulate drought tolerance by activating *OsAP2-39* in rice, but whether they function in the same pathway remains to be elucidated.

In conclusion, the overexpression of *OsWRKY55* could reduce drought resistance and plant height in rice. Four OsMPKs interacted with OsWRKY55 and activated OsWRKY55. Moreover, OsWRKY55 was able to directly regulate *OsAP2-39*. This work demonstrates the novel function of *OsWRKY55* in rice.

4. Materials and Methods

4.1. Plant Materials and Growth Conditions

Wild-type rice seedlings (WT, *Oryza sativa japonica* cv Kitaake) and two homozygous T3 transgenic rice lines overexpressing *OsWRKY55* (OE-1 and OE-2) were used in these experiments. To generate *OsWRKY55*-OE overexpression lines, the full-length coding sequence (CDS) of *OsWRKY55* was amplified from rice seedlings of Kitaake and cloned into the vector pCUbi1390 under the control of the maize ubiquitin (Ubi) promoter. The constructed vector *pUbi:OsWRKY55* was introduced into rice (Kitaake) using *Agrobacterium tumefaciens* (EHA105)-mediated transformation. The homozygous lines were selected via hygromycin resistance evaluation. Rice overexpressing *OsWRKY55* is denoted as OE, and the different transgenic lines are indicated with numbers.

Seeds were submerged in water at 28 °C for 3 days. Then, the uniformly germinated seeds were transplanted into containers filled with Yoshida's culture solution and boxes filled with soil from a paddy field in a growth chamber (28 °C, 80% relative humidity and a 14/10 h day/night photoperiod), respectively.

4.2. Stress Treatments

Phenotypic analysis of drought stress was conducted at the seedling stage. When *OsWRKY55*-OE lines and WT plants growing in the soil reached the four-leaf stage, irrigation was withheld for 7 days, and then the seedlings were allowed to recover for 7 days. Seedlings that did not grow were considered not to have survived. The survival rates were calculated as the percentage of seedlings that survived among the total treated seedlings. WT and transgenic plants were grown to maturity in the paddy field under normal conditions. At the booting stage, we measured lengths of the panicles and internodes and plant heights.

To determine the *OsWRKY55* gene expression profile under stress conditions, four-leaf-stage WT rice seedlings growing in the hydroponic culture medium were subjected to drought stress (by removing the water supply) and phytohormone ABA treatment (100 μM ABA added to the culture medium) before being sampled at 0, 0.5, 2, 4, and 6 h. To characterize the spatiotemporal expression of *OsWRKY55*, young leaves and roots were collected from the WT seedlings at the four-leaf stage in the soil under normal conditions, and samples of the flag leaves, stems, sheaths, and panicles were collected from WT plants at the booting stage described above.

4.3. Water Loss Assay

The *OsWRKY55*-OE and WT plants were grown in pots under normal growth conditions for 4 weeks. The leaves were detached from the plants and weighed immediately as the initial weight. The samples were exposed to air at room temperature and then weighed at the same time intervals (20 min) after being cut down. The rate of water loss was calculated based on the initial weights of the samples. Three replicates were made for each line.

4.4. Quantification of Gene Expression

Total RNA was extracted from each sample using the RNAprep Pure Plant Kit (Tiangen Biotech). First-strand cDNA was synthesized with 2 μg of total RNA using a RevertAid First Strand cDNA Synthesis Kit (Thermo Scientific Fermentas). Quantitative RT-PCR reaction was performed using a Mx3005P instrument (Stratagene) with the SYBR Green Real-time PCR Master Mix (Toyobo) according to the following conditions: 1 cycle of 2 min at 95 °C, 40 cycles of 15 s at 95 °C, and 30 s at 60 °C, followed by 1 cycle of 60 s at 95 °C, 30 s at 55 °C, and 30 s at 95 °C for the melting curve analysis. The rice *OsActin* gene and *Ubiquitin5* gene (Supplemental Figure S3) were used as the internal controls against which to calculate relative expression. Relative expression levels were calculated using the $2^{-\Delta\Delta Ct}$ method [45]. The experiment was conducted using three biological replicates. The primers used are listed in Table S1.

4.5. RNA Sequencing

Four-leaf-stage WT and the *OsWRKY55*-OE plants under normal conditions were sampled for RNA sequencing (RNA-seq). Total RNA of the leaves was extracted using the RNAprep Pure Plant Kit (Tiangen Biotech). cDNA library construction and next-generation sequencing were performed by Biomarker Bio-Tech Company (Beijing, China) on an Illumina Hiseq2500 platform (2 × 101 bp). Three biological replicates were performed. The raw reads were filtered using the FASTQ_Quality_Filter tool from the FASTX-Toolkit (http://hannonlab.cshl.edu/fastx_toolkit, accessed on 11 May, 2020). Next, the clean reads were mapped to the rice reference genome (MSU, Rice Genome Annotation Release 7) using Hisat2, a spliced read mapper for RNA-seq (http://ccb.jhu.edu/software/hisat2/index.shtml, accessed on 11 May, 2020). Gene expression profiles were analyzed using Stringtie (Transcript assembly for RNA-seq), and transcript abundance was measured using the Fragment Per Kilobase of transcript sequence per Millions base pairs Sequenced (FPKM) method. Differentially expressed genes (DEGs) were identified using the DESeq package in R. The volcano map of DEGs was made using R package ggplot2.

4.6. Yeast Two-Hybrid Assay

To identify the interactions of OsWRKY55 and OsMPKs in yeast, a yeast two-hybrid assay was carried out according to the manufacturer's instructions (Clontech). The full-length coding DNA sequence (CDS) of *OsWRKY55* was fused to the pGBKT7 vector (BD-OsWRKY55) as bait, and the full-length CDSs of the *OsMPKs* were cloned into separate pGADT7 vectors as prey. The bait vector was co-transformed with each prey vector into the Y2H Gold yeast strain and cultured on SD/-Trp-Leu plates at 30 °C for 2–3 days. Positive transformants were then cultivated on SD/-Trp-Leu-His plates containing 0, 0.2, or 0.4 mM 3-amino-1, 2, 4-triazole (3-AT) at 30 °C for 3 days.

4.7. Yeast One-Hybrid Assay

For the yeast one-hybrid assay, the promoter (≈1500 bp) of *OsAP2-39* was cloned into the reporter vector pLacZi (pLacZi-Pro). The full-length CDS of *OsWRKY55* was amplified and fused to the vector pGAD424 (Clontech), which encodes the activation domain of the yeast GAL4 transcriptional activator. Then, the reporter vector pLacZi-Pro was digested using a restriction enzyme (either *NcoI* or *ApaI*) and co-transformed with the fusion protein into the yeast strain YM4271 (Clontech). The transformants were cultured on SD/-Leu-Ura plates at 30 °C for 3 days. Putative positive clones were identified using DNA sequencing. Finally, β-galactosidase liquid assays were performed to quantify the DNA–protein interactions, following the instructions of the manufacturer (Clontech).

4.8. GAL4-Dependent Chimeric Transactivation Assay

The assay was performed in rice protoplasts, as described previously [46]. The *OsWRKY55* coding region was cloned into the effector vector with the GAL4 DNA-binding domain, while the full-length coding regions of OsMPK7, OsMPK9, OsMPK20-1, and OsMPK20-4 were cloned into the effector plasmid without the GAL4 DNA-binding domain. GAL4DB was used as a negative effector. The reporter was a plasmid containing the firefly luciferase (FLUC) gene. A plasmid harboring the Renilla LUC (RLUC) gene under control of the *CaMV* 35S promoter was used as an internal control. The effector plasmids were co-transformed with the reporter plasmid and internal control into rice protoplasts. After incubation at 28 °C for 12–16 h, the protoplasts were collected, and the results were reported as the ratios between the activity of FLUC and RLUC using a luciferase activity assay (Promega).

4.9. Dual Luciferase Reporter Assay

For the dual luciferase reporter assay, the *OsAP2-39* promoter (~1500 bp) was inserted into the vector pGreen II 0800 (which contained a firefly luciferase (FLUC) reporter gene and an internal control Renilla luciferase (RLUC) driven by the *CaMV* 35S promoter) to generate the reporter vector. The CDS of *OsWRKY55* was inserted into the pAN580 vector to generate the effector vector. Then, the reporter vector was co-transformed into rice protoplasts via PEG-mediated transformation [47] with either the empty pAN580 vector (as the negative control) or the effector vector (pAN580-OsWRKY55). After 12–16 h of culture at 28 °C, the FLUC and RLUC activity levels in the rice protoplasts were measured using a Dual-Luciferase Reporter Assay System (E1910, Promega). The transient expression of LUC, driven by the promoter of *OsAP2-39*, was normalized to the expression of the internal control reporter (RLUC).

4.10. Histochemical Analysis

For 3, 3′-diaminobenzidine (DAB) staining, rice leaves were immersed in 1 mg mL^{-1} DAB in 50 mM Tris-acetate buffer (pH 5.0) for 12 h at room temperature in the dark. To detect superoxide radical ($O_2^{-\cdot}$), leaves were incubated in 1 mg mL^{-1} nitro-blue tetrazolium (NBT) in a 10 mM potassium phosphate solution (pH 7.8) at 25 °C for 12 h in the dark. Then, the leaves were de-stained with 95% ethanol for 10 min at 100 °C. De-staining was repeated several times until the chlorophyll was completely removed.

WT and *OsWRKY55*-OE (OE-1) internodes were immobilized in paraffin and sectioned as described previously [48]. In brief, the uppermost internodes of the main culms of the WT and *OsWRKY55*-OE (OE-1) plants were longitudinally sectioned, fixed in formaldehyde/glacial acetic acid/70% ethanol (1:1:18, *v*/*v*/*v*) for 48 h, and softened with 15% hydrofluoric acid for 2 weeks. Then, the samples were dehydrated using a graded ethanol series. The dehydrated samples were embedded in paraffin (Sigma-Aldrich) and microsectioned to a thickness of 8 μm using a microtome. The sections were stained with toluidine blue and viewed under a light microscope after paraffin removal with xylene.

4.11. Accession Numbers

Sequence data from this article can be found in the MSU Rice Genome Annotation Project (http://rice.plantbiology.msu.edu/analyses_search_locus.shtml, accessed on 18 April, 2020) databases under the following accession numbers: *OsWRKY55/OsWRKY31* (LOC_Os03g20550), *OsMPK7* (LOC_Os05g49140), *OsMPK9* (LOC_Os05g50560), *OsMPK20-1* (LOC_Os01g43910), *OsMPK20-4* (LOC_Os01g47530), *OsMPK3* (LOC_Os03g17700), *OsMPK4-1* (LOC_Os10g38950), *OsMPK6* (LOC_Os06g06090), *OsMPK14* (LOC_Os02g05480), *OsMPK16* (LOC_Os11g17080), *OsMPK17-1* (LOC_Os06g49430), *OsMPK17-2* (LOC_Os02g04230), *OsMPK20-3* (LOC_Os06g26340), *OsMPK4-2* (LOC_Os06g48590), *OsMPK21-1* (LOC_Os05g50120), *OsMPK21-2* (LOC_Os01g45620), *OsActin* (LOC_Os03g50885), *Ubiquitin5* (LOC_Os01g22490), and *OsAP2-39* (LOC_Os04g52090).

Supplementary Materials: The following are available online at https://www.mdpi.com/article/10.3390/ijms22094337/s1.

Author Contributions: X.D. and W.J. designed the study; K.H., Z.M., H.C., M.Z. and H.B. executed the experiments; Z.L. analyzed the data; K.H. and T.W. wrote the manuscript; X.D., W.J. and M.B. modified the manuscript. All authors have read and agreed to the published version of the manuscript.

Funding: This research was supported by National Natural Science Foundation of China (31701396 and 31901434), Science and Technology Development Plan of Jilin Province, China (20160204007NY), and Jilin Provincial and School Co-construction project (No. SXGJXX2019-2).

Institutional Review Board Statement: Not applicable.

Informed Consent Statement: Not applicable.

Conflicts of Interest: The authors declare no conflict of interest.

References

1. Fang, Y.J.; Liao, K.F.; Du, H.; Xu, Y.; Song, H.Z.; Li, X.H.; Xiong, L.Z. A stress-responsive NAC transcription factor SNAC3 confers heat and drought tolerance through modulation of reactive oxygen species in rice. *J. Exp. Bot.* **2015**, *66*, 6803–6817. [CrossRef] [PubMed]
2. Xiong, L.M.; Schumaker, K.S.; Zhu, J.K. Cell signaling during cold, drought, and salt stress. *Plant Cell* **2002**, *14*, S165–S183. [CrossRef] [PubMed]
3. Yamaguchi-Shinozaki, K.; Shinozaki, K. Transcriptional regulatory networks in cellular responses and tolerance to dehydration and cold stresses. *Annu. Rev. Plant Biol.* **2006**, *57*, 781–803. [CrossRef]
4. Tang, Y.H.; Bao, X.X.; Zhi, Y.L.; Wu, Q.; Guo, Y.; Yin, X.H.; Zeng, L.Q.; Li, J.; Zhang, J.; He, W.L.; et al. Overexpression of a MYB family gene, *OsMYB6*, increases drought and salinity stress tolerance in transgenic rice. *Front. Plant Sci.* **2019**, *10*, 168. [CrossRef] [PubMed]
5. Seo, J.S.; Joo, J.; Kim, M.J.; Kim, Y.K.; Nahm, B.H.; Song, S.I.; Cheong, J.J.; Lee, J.S.; Kim, J.K.; Choi, Y.D. OsbHLH148, a basic helix-loop-helix protein, interacts with OsJAZ proteins in a jasmonate signaling pathway leading to drought tolerance in rice. *Plant J.* **2011**, *65*, 907–921. [CrossRef] [PubMed]
6. Oh, S.-J.; Kim, Y.S.; Kwon, C.-W.; Park, H.K.; Jeong, J.S.; Kim, J.-K. Overexpression of the transcription factor AP37 in rice improves grain yield under drought conditions. *Plant Physiol.* **2009**, *150*, 1368–1379. [CrossRef] [PubMed]
7. Yang, S.Q.; Xu, K.; Chen, S.J.; Li, T.F.; Xia, H.; Chen, L.; Liu, H.Y.; Luo, L.J. A stress-responsive bZIP transcription factor OsbZIP62 improves drought and oxidative tolerance in rice. *BMC Plant Biol.* **2019**, *19*, 260. [CrossRef]
8. Singh, D.; Laxmi, A. Transcriptional regulation of drought response: A tortuous network of transcriptional factors. *Front. Plant Sci.* **2015**, *6*, 895. [CrossRef]
9. Eulgem, T.; Rushton, P.J.; Robatzek, S.; Somssich, I.E. The WRKY superfamily of plant transcription factors. *Trends Plant Sci.* **2000**, *5*, 199–206. [CrossRef]
10. Chujo, T.; Takai, R.; Akimoto-Tomiyama, C.; Ando, S.; Minami, E.; Nagamura, Y.; Kaku, H.; Shibuya, N.; Yasuda, M.; Nakashita, H.; et al. Involvement of the elicitor-induced gene *OsWRKY53* in the expression of defense-related genes in rice. *Biochim. Biophys. Acta* **2007**, *1769*, 497–505. [CrossRef]
11. Liu, X.Q.; Bai, X.J.; Wang, X.J.; Chu, C.C. OsWRKY71, a rice transcription factor, is involved in rice defense response. *J. Plant Physiol.* **2007**, *164*, 969–979. [CrossRef] [PubMed]
12. Wang, H.H.; Hao, J.J.; Chen, X.J.; Hao, Z.N.; Wang, X.; Lou, Y.G.; Peng, Y.L.; Guo, Z.J. Overexpression of rice *WRKY89* enhances ultraviolet B tolerance and disease resistance in rice plants. *Plant Mol. Biol.* **2007**, *65*, 799–815. [CrossRef] [PubMed]
13. Peng, Y.; Bartley, L.E.; Chen, X.W.; Dardick, C.; Chern, M.; Ruan, R.; Canlas, P.E.; Ronald, P.C. OsWRKY62 is a negative regulator of basal and *Xa21*-mediated defense against *Xanthomonas oryzae* pv. *oryzae in rice. Mol. Plant* **2008**, *1*, 446–458. [CrossRef] [PubMed]
14. Wang, C.T.; Ru, J.N.; Liu, Y.W.; Li, M.; Zhao, D.; Yang, J.F.; Fu, J.D.; Xu, Z.S. Maize WRKY transcription factor ZmWRKY106 confers drought and heat tolerance in transgenic plants. *Int. J. Mol. Sci.* **2018**, *19*, 3046. [CrossRef]
15. Wang, C.T.; Ru, J.N.; Liu, Y.W.; Yang, J.F.; Li, M.; Xu, Z.S.; Fu, J.D. The Maize WRKY transcription factor ZmWRKY40 confers drought resistance in transgenic *Arabidopsis*. *Int. J. Mol. Sci.* **2018**, *19*, 2580. [CrossRef]
16. Shi, W.Y.; Du, Y.T.; Ma, J.; Min, D.H.; Jin, L.G.; Chen, J.; Chen, M.; Zhou, Y.B.; Ma, Y.Z.; Xu, Z.S.; et al. The WRKY transcription factor GmWRKY12 confers drought and salt tolerance in Soybean. *Int. J. Mol. Sci.* **2018**, *19*, 4087. [CrossRef]
17. Wei, W.; Liang, D.W.; Bian, X.H.; Shen, M.; Xiao, J.H.; Zhang, W.K.; Ma, B.; Lin, Q.; Lv, J.; Chen, X.; et al. GmWRKY54 improves drought tolerance through activating genes in abscisic acid and Ca^{2+} signaling pathways in transgenic soybean. *Plant J.* **2019**, *100*, 384–398. [CrossRef]
18. Yokotani, N.; Sato, Y.; Tanabe, S.; Chujo, T.; Shimizu, T.; Okada, K.; Yamane, H.; Shimono, M.; Sugano, S.; Takatsuji, H.; et al. WRKY76 is a rice transcriptional repressor playing opposite roles in blast disease resistance and cold stress tolerance. *J. Exp. Bot.* **2013**, *64*, 5085–5097. [CrossRef]
19. Ma, H.G.; Chen, J.; Zhang, Z.Z.; Ma, L.; Yang, Z.Y.; Zhang, Q.L.; Li, X.H.; Xiao, J.H.; Wang, S.P. MAPK kinase 10.2 promotes disease resistance and drought tolerance by activating different MAPKs in rice. *Plant J.* **2017**, *92*, 557–570. [CrossRef]
20. Asai, T.; Tena, G.; Plotnikova, J.; Willmann, M.R.; Chiu, W.L.; Gomez-Gomez, L.; Boller, T.; Ausubel, F.M.; Sheen, J. MAP kinase signalling cascade in *Arabidopsis* innate immunity. *Nature* **2002**, *415*, 977–983. [CrossRef]
21. Andreasson, E.; Jenkins, T.; Brodersen, P.; Thorgrimsen, S.; Petersen, N.H.; Zhu, S.; Qiu, J.L.; Micheelsen, P.; Rocher, A.; Petersen, M.; et al. The MAP kinase substrate MKS1 is a regulator of plant defense responses. *EMBO J.* **2005**, *24*, 2579–2589. [CrossRef]
22. Shen, H.S.; Liu, C.T.; Zhang, Y.; Meng, X.P.; Zhou, X.; Chu, C.C.; Wang, X.P. OsWRKY30 is activated by MAP kinases to confer drought tolerance in rice. *Plant Mol. Biol.* **2012**, *80*, 241–253. [CrossRef]
23. Koo, S.C.; Moon, B.C.; Kim, J.K.; Kim, C.Y.; Sung, S.J.; Kim, M.C.; Cho, M.J.; Cheong, Y.H. OsBWMK1 mediates SA-dependent defense responses by activating the transcription factor OsWRKY33. *Biochem. Biophys. Res. Commun.* **2009**, *387*, 365–370. [CrossRef]
24. Hu, L.F.; Ye, M.; Li, R.; Zhang, T.F.; Zhou, G.X.; Wang, Q.; Lu, J.; Lou, Y.G. The rice transcription factor WRKY53 suppresses herbivore-induced defenses by acting as a negative feedback modulator of mitogen-activated protein kinase activity. *Plant Physiol.* **2015**, *169*, 2907–2921. [CrossRef]
25. Zhou, C.L.; Lin, Q.B.; Lan, J.; Zhang, T.Y.; Liu, X.; Miao, R.; Mou, C.L.; Nguyen, T.; Wang, J.C.; Zhang, X.; et al. WRKY transcription factor OsWRKY29 represses seed dormancy in rice by weakening abscisic acid response. *Front. Plant Sci.* **2020**, *11*, 691. [CrossRef]

26. Ding, Z.J.; Yan, J.Y.; Li, G.X.; Wu, Z.C.; Zhang, S.Q.; Zheng, S.J. WRKY41 controls *Arabidopsis* seed dormancy via direct regulation of *ABI3* transcript levels not downstream of ABA. *Plant J.* **2014**, *79*, 810–823. [CrossRef]
27. Jing, S.J.; Zhou, X.; Song, Y.; Yu, D.Q. Heterologous expression of *OsWRKY23* gene enhances pathogen defense and dark-induced leaf senescence in *Arabidopsis*. *Plant Growth Regul.* **2009**, *58*, 181–190. [CrossRef]
28. Miao, Y.; Laun, T.; Zimmermann, P.; Zentgraf, U. Targets of the WRKY53 transcription factor and its role during leaf senescence in *Arabidopsis*. *Plant Mol. Biol.* **2004**, *55*, 853–867. [CrossRef]
29. Ulker, B.; Mukhtar, M.S.; Somssich, I.E. The WRKY70 transcription factor of *Arabidopsis* influences both the plant senescence and defense signaling pathways. *Planta* **2007**, *226*, 125–137. [CrossRef]
30. Cai, Y.H.; Chen, X.J.; Xie, K.; Xing, Q.K.; Wu, Y.W.; Li, J.; Du, C.H.; Sun, Z.X.; Guo, Z.J. Dlf1, a WRKY transcription factor, is involved in the control of flowering time and plant height in rice. *PLoS ONE* **2014**, *9*, e102529. [CrossRef]
31. Xie, Z.; Zhang, Z.L.; Zou, X.; Huang, J.; Ruas, P.; Thompson, D.; Shen, Q.J. Annotations and functional analyses of the rice WRKY gene superfamily reveal positive and negative regulators of abscisic acid signaling in aleurone cells. *Plant Physiol.* **2005**, *137*, 176–189. [CrossRef] [PubMed]
32. Zhang, J.; Peng, Y.L.; Guo, Z.J. Constitutive expression of pathogen-inducible *OsWRKY31* enhances disease resistance and affects root growth and auxin response in transgenic rice plants. *Cell Res.* **2008**, *18*, 508–521. [CrossRef] [PubMed]
33. Wu, T.; Zhang, M.X.; Zhang, H.J.; Huang, K.; Chen, M.J.; Chen, C.; Yang, X.; Li, Z.; Chen, H.Y.; Ma, Z.M.; et al. Identification and characterization of *EDT1* conferring drought tolerance in rice. *J. Plant Biol.* **2019**, *62*, 39–47. [CrossRef]
34. Hamel, L.P.; Nicole, M.C.; Sritubtim, S.; Morency, M.J.; Ellis, M.; Ehlting, J.; Beaudoin, N.; Barbazuk, B.; Klessig, D.; Lee, J.; et al. Ancient signals: Comparative genomics of plant MAPK and MAPKK gene families. *Trends Plant Sci.* **2006**, *11*, 192–198. [CrossRef]
35. Yaish, M.W.; El-Kereamy, A.; Zhu, T.; Beatty, P.H.; Good, A.G.; Bi, Y.M.; Rothstein, S.J. The APETALA-2-like transcription factor OsAP2-39 controls key interactions between abscisic acid and gibberellin in rice. *PLoS Genet.* **2010**, *6*, e1001098. [CrossRef]
36. Rushton, P.J.; Somssich, I.E.; Ringler, P.; Shen, Q.J. WRKY transcription factors. *Trends Plant Sci.* **2010**, *15*, 247–258. [CrossRef]
37. Xu, M.; Zhu, L.; Shou, H.X.; Wu, P. A *PIN1* family gene, *OsPIN1*, involved in auxin-dependent adventitious root emergence and tillering in rice. *Plant Cell Physiol.* **2005**, *46*, 1674–1681. [CrossRef]
38. Kim, C.-Y.; Vo, K.T.X.; Nguyen, C.D.; Jeong, D.-H.; Lee, S.-K.; Kumar, M.; Kim, S.-R.; Park, S.-H.; Kim, J.-K.; Jeon, J.-S. Functional analysis of a cold-responsive rice WRKY gene, *OsWRKY71*. *Plant Biotechnol. Rep.* **2016**, *10*, 13–23. [CrossRef]
39. Deeba, F.; Sultana, T.; Mahmood, T.; O'Shea, C.; Skriver, K.; Naqvi, S.M.S. Involvement of WRKY, MYB and DOF DNA-binding proteins in interaction with a rice germin-like protein gene promoter. *Acta Physiol. Plant.* **2017**, *39*. [CrossRef]
40. Zhang, Z.Y.; Li, J.H.; Li, F.; Liu, H.H.; Yang, W.S.; Chong, K.; Xu, Y.Y. OsMAPK3 phosphorylates OsbHLH002/OsICE1 and inhibits its ubiquitination to activate *OsTPP1* and enhances rice chilling tolerance. *Dev. Cell* **2017**, *43*, 731–743.e5. [CrossRef]
41. Zhang, Z.Y.; Liu, H.H.; Sun, C.; Ma, Q.B.; Bu, H.Y.; Chong, K.; Xu, Y.Y. A C2H2 zinc-finger protein OsZFP213 interacts with OsMAPK3 to enhance salt tolerance in rice. *J. Plant Physiol.* **2018**, *229*, 100–110. [CrossRef]
42. Moustafa, K.; AbuQamar, S.; Jarrar, M.; Al-Rajab, A.J.; Trémouillaux-Guiller, J. MAPK cascades and major abiotic stresses. *Plant Cell Rep.* **2014**, *33*, 1217–1225. [CrossRef]
43. Ueno, Y.; Yoshida, R.; Kishi-Kaboshi, M.; Matsushita, A.; Jiang, C.J.; Goto, S.; Takahashi, A.; Hirochika, H.; Takatsuji, H. Abiotic stresses antagonize the rice defence pathway through the tyrosine-dephosphorylation of OsMPK6. *PLoS Pathog.* **2015**, *11*, e1005231. [CrossRef]
44. Wan, L.Y.; Zhang, J.F.; Zhang, H.W.; Zhang, Z.J.; Quan, R.D.; Zhou, S.R.; Huang, R.F. Transcriptional activation of OsDERF1 in *OsERF3* and *OsAP2-39* negatively modulates ethylene synthesis and drought tolerance in rice. *PLoS ONE* **2011**, *6*, e25216. [CrossRef]
45. Livak, K.J.; Schmittgen, T.D. Analysis of relative gene expression data using real-time quantitative PCR. *Methods* **2002**, *25*, 402–408. [CrossRef]
46. Guo, X.L.; Hou, X.M.; Fang, J.; Wei, P.W.; Xu, B.; Chen, M.L.; Feng, Y.Q.; Chu, C.C. The rice GERMINATION DEFECTIVE 1, encoding a B3 domain transcriptional repressor, regulates seed germination and seedling development by integrating GA and carbohydrate metabolism. *Plant J.* **2013**, *75*, 403–416. [CrossRef]
47. Yoo, S.D.; Cho, Y.H.; Sheen, J. *Arabidopsis* mesophyll protoplasts: A versatile cell system for transient gene expression analysis. *Nat. Protoc.* **2007**, *2*, 1565–1572. [CrossRef]
48. Ma, Y.M.; Yang, C.; He, Y.; Tian, Z.H.; Li, J.X. Rice OVATE family protein 6 regulates plant development and confers resistance to drought and cold stresses. *J. Exp. Bot.* **2017**, *68*, 4885–4898. [CrossRef]

International Journal of
Molecular Sciences

MDPI

Article

Submergence Gene *Sub1A* Transfer into Drought-Tolerant *japonica* Rice DT3 Using Marker-Assisted Selection

Yong-Pei Wu [1], Shu-Mei Wang [2], Yu-Chi Chang [1], Chi Ho [1] and Yu-Chia Hsu [3,*]

1 Department of Agronomy, Chiayi Agricultural Experiment Station, Taiwan Agricultural Research Institute, Chiayi 60014, Taiwan; wuypei@tari.gov.tw (Y.-P.W.); ycc71719@gmail.com (Y.-C.C.); chief5411@gmail.com (C.H.)

2 Department of Bio-Industry Communication and Development, National Taiwan University, Taipei 10617, Taiwan; wangsm@ntu.edu.tw

3 Department of Agronomy, National Chiayi University, Chiayi 600355, Taiwan

* Correspondence: hsuychia@mail.ncyu.edu.tw; Tel.: +886-5-2717382; Fax: +886-5-2717386

Abstract: Flash flooding is a major environmental stressor affecting rice production worldwide. DT3 is a drought-tolerant, recurrent parent with a good yield, edible quality, and agronomic traits akin to those of an elite Taiwanese variety, Taiken9 (TK9). Progenies carrying *Sub1A* can enhance submergence stress tolerance and can be selected using the marker-assisted backcross (MAB) breeding method. For foreground selection, Sub1A and SubAB1 were utilized as markers on the BC_2F_1, BC_3F_1, and BC_3F_2 generations to select the submergence-tolerant gene, *Sub1A*. Background selection was performed in the *Sub1A*-BC_3F_2 genotypes, and the percentages of recurrent parent recovery within individuals ranged from 84.7–99.55%. BC_3F_3 genotypes (N = 100) were evaluated for agronomic traits, yield, and eating quality. Four of the eleven BC_3F_4 lines showed good yield, yield component, grain, and eating quality. Four BC_3F_4 lines, SU39, SU40, SU89, and SU92, exhibited desirable agronomic traits, including grain quality and palatability, consistent with those of DT3. These genotypes displayed a high survival rate between 92 and 96%, much better compared with DT3 with 64%, and demonstrated better drought tolerance compared to IR64 and IR96321-345-240. This study provides an efficient and precise MAB strategy for developing climate-resilient rice varieties with good grain quality for flood-prone regions.

Keywords: *japonica* DT3; drought tolerance; submergence tolerance; marker-assisted backcross; foreground selection; background selection

Citation: Wu, Y.-P.; Wang, S.-M.; Chang, Y.-C.; Ho, C.; Hsu, Y.-C. Submergence Gene *Sub1A* Transfer into Drought-Tolerant *japonica* Rice DT3 Using Marker-Assisted Selection. *Int. J. Mol. Sci.* **2021**, *22*, 13365. https://doi.org/10.3390/ijms222413365

Academic Editors: Kiyosumi Hori and Matthew Shenton

Received: 10 November 2021
Accepted: 9 December 2021
Published: 13 December 2021

1. Introduction

Rice (*Oryza sativa*) is one of the most important cereal crops grown worldwide and is widely cultivated in Asia. It provides approximately 50% of the calories for more than half of the world's population. By 2050, the global population will reach 9.1 to 9.5 billion, 34 percent higher than that of today [1], and global rice consumption is estimated to increase to 650 million tons over time [2]. Nearly 640 million tons are grown in Asia, meeting the global requirement; however, climate change and heavy rainfall have affected global rice production and threaten long-term food security [3]. In recent years, the frequency of abnormal floods in Taiwan has increased substantially, particularly in Southeast Asian countries, such as Bangladesh, India, Indonesia, Nepal, and the Philippines [4]. In Bangladesh, 1.6 million hectares of rice fields are periodically affected by floods, while in India, about 32.2% of 16.1 million hectares of rice-growing fields are occasionally affected by floods [5,6]. Floods can cause up to a 100% reduction in rice yield, depending on the environmental and floodwater conditions [7]. Thus, the impact of catastrophic floods on rice production must be addressed urgently.

The *Sub1* QTL was mapped on chromosome 9, contributing to 70% of the phenotypic variation conferring an enhanced survival rate to rice under submergence [8]. It has been fine-mapped and positional cloned, and the cluster genes, including *Sub1A*, *Sub1B*, and *Sub1C*, were obtained in *indica* rice FR13A derivative [9,10] IR40931-26. The submergence-intolerant *japonica* rice Nipponbare contains *Sub1B* and *Sub1C* in the genome but lacks *Sub1A*, which encodes a putative DNA-binding protein with a single ERF/APETELA2 domain [10]. *Sub1A* repressed GA signal transduction in the submergence-tolerant rice under submergence stress, which affected the metabolic pathway of carbohydrates and the growth of FR13A for energy conservation, and rice growth resumed after the flood receded [11,12]. Thus, *Sub1* QTLs can be transferred into several different submergence-intolerant rice varieties via a conventional breeding program embedded with marker-assisted selection (MAS). In 2003, the International Rice Research Institute initiated a program and successfully introduced the *Sub1* QTL into six mega-varieties, Swarna, Samba Mahsuri, BR11, IR64, GR1009, and TDK1, using MAS [7,13–16]. Two high-yield potentials with submergence-tolerant rice varieties, BRRI dhan51 and BRRI dhan52, were released by the Bangladesh Rice Research Institute (BRRI) in 2010 [17]. Rice production is expected to surge in the submergence-prone areas of Bangladesh because the *Sub1* QTL increased the survival of submerged rice plants for a short duration.

Typhoons and heavy rainfall are the major causes of agriculture-related disasters in Taiwan. According to statistics from the Bureau of Meteorology, a total of 370 typhoons have hit Taiwan between 1911 and 2020. In the past hundred years, the annual average hit frequency of typhoons was 3–4 times. The number of typhoons has increased in Taiwan due to climate change in recent decades. The typhoon-related record-breaking increase in rainfall is a manifestation of global warming [18]. The heavy rain in 2021 has caused agricultural losses of TWD 150 million based on the statistics from the Council of Agriculture. Seedlings of rice of the second crop season in Taiwan are usually transplanted in August, the most vulnerable month to typhoons, which affects their survival rate. Introgression of *Sub1* QTLs in Taiwan *japonica* rice varieties is expected to enhance the survival rate of seedlings under submergence.

DT3, named Tainung83 in 2020, is an elite *japonica* rice line with good grain quality and drought tolerance. This variety does not contain the *Sub1A* gene; hence, it is susceptible to submergence stress. The introduction of *Sub1* QTL in IR96321-315-240, an *indica* rice of Swarna genetic background, conferred strong tolerance under submergence. This study was performed to convert DT3 into a submergence-tolerant genotype by introgressing *Sub1A* from IR96321-315-240, using the MAB breeding method. The study aimed to (i) develop DT3-*Sub1A* lines using MAB, (ii) evaluate the survival rate and related effects of DT3-*Sub1A* lines after submergence, and (iii) select good performance on agronomic traits and grain quality of individuals from the resulting submergence-tolerant lines. The development of submergence-tolerant lines can be a good genetic resource and provide new resilient rice cultivars for the future in Taiwan.

2. Results

2.1. Development of BC_3F_6 Submergence-Tolerant Lines Using Marker-Assisted Breeding

To develop a submergence-tolerant *japonica* cultivar, DT3 was used as a recurrent parent to backcross with IR96321-315-240 for three generations and then self-crossed to produce a BC_3F_6 population. The polymorphism was detected between the donor parent, IR96321-315-240, and recurrent parent DT3 with the markers Sub1A1 and Sub1AB1 for the *Sub1A* gene. In addition, 100 molecular markers (Figure S1), showing polymorphism between IR96321-315-240 and DT3, were used for background selection. During the breeding procedure, 33 BC_1F_1 plants were crossed with the recurrent parent DT3 (Figure 1). A total of 41 of 395 BC_2F_1 plants containing *Sub1A* were confirmed and selected by MAS. Functional marker selection was performed in three generations, including BC_2F_1, BC_3F_1, and BC_3F_2 generation. A total of 45 of the 88 BC_3F_1 were subjected to foreground selection.

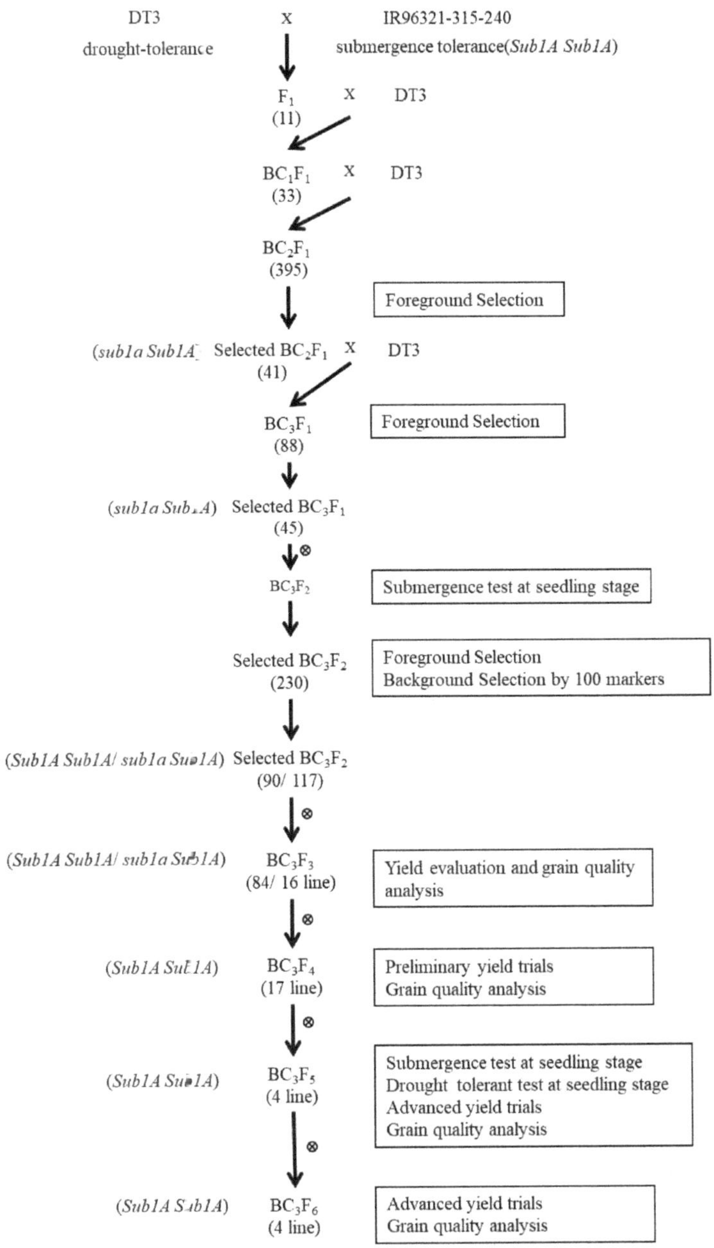

Figure 1. The schematic diagram for transferring *Sub1A* gene into Taiwanese *japonica* rice line, DT3, using marker-assisted backcrossing, detailing of submergence test and markers used for background selection. Numbers of plants selected in each generation are indicated in parentheses.

The 23 genotypes with "H" score are shown in the gel picture of foreground selection with marker Sub1AB1 (Figure 2). In the BC_3F_2, 800 plants were phenotyped at the seedling stage using submergence screening in the tank. A total of 230 plants were recovered after submergence treatment. In addition, surviving plants were selected for foreground and background selection. Out of 230 plants, 90 plants scored "A" (*Sub1A* homozygous), 117 plants scored "H" (*Sub1A/sub1a* heterozygous), and 23 plants scored "B" (*sub1a* homozygous) (Figure 1).

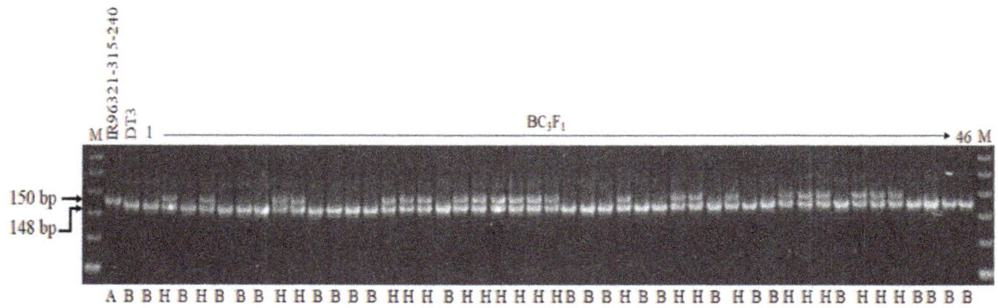

Figure 2. Partial view of the gel picture of the foreground selection with the marker Sub1AB1. M, 20 bp DNA ladder; 1–46, BC_3F_1 plants; A, homozygous donor allele; B, homozygous recurrent allele; H, heterozygous allele.

Background selection was conducted using 100 markers for 90 plants in the BC_3F_2 generation, which contained the *Sub1A* gene and possessed the recurrent genome content of DT3. The recovery of the recurrent parent genome ranged from 84.7 to 99.55%, with an average of 92.87% (Table S1). The highest percentage of recipient alleles was obtained in plant number 66 (99.55%), plant number 89 (98.5%), plant number 17 (98.4%), plant number 28 (98.1%), and plant number 77 (98%). Among them, 35 plants had a background recovery rate between 95 and 100%, 31 plants were 90–95%, 23 plants were 85–90%, and one plant was less than 85% (Figure 3). A total of 100 plants, 84 plants with homozygous *Sub1A* and 16 heterozygous *Sub1A* plants with good agronomic performance, were selected as BC_3F_3 lines (SU1-SU100) for further comparative yield tests.

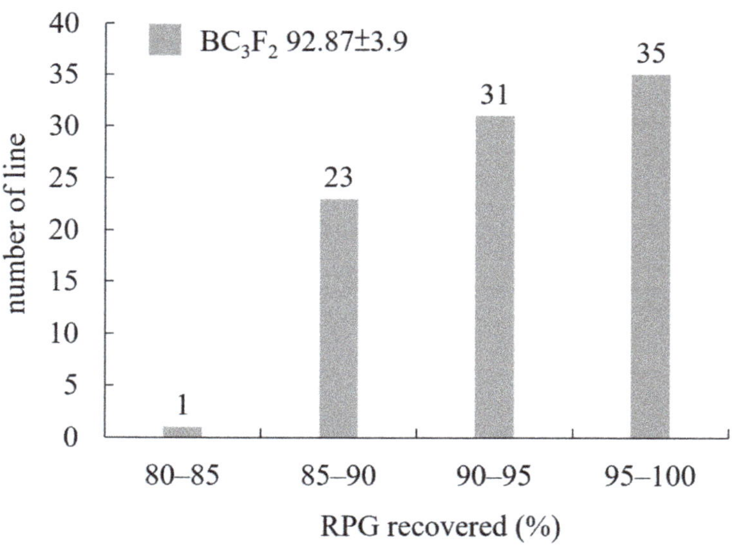

Figure 3. The frequency distribution of recurrent parent genome (RPG) recovered rate using marker-assisted backcrossing in BC_3F_2 population.

2.2. Phenotyping and Evaluation of the Agronomic Performance of Newly Developed Sub1A Lines

One hundred BC_3F_3 rice plants were evaluated for their agronomic performance in the first crop season of 2019 at the Taiwan Agricultural Research Institute, Taiwan. The

28 lines were selected based on their agronomic traits, such as plant height, productive tiller number, heading uniformity, panicle type, and panicle weight. The yields among BC_3F_3 lines varied between 3667.4 kg ha^{-1} (SU33) and 9350 kg ha^{-1} (SU1) (Table S2). The rice palatability value for the BC_3F_3 lines varied from 59 to 75, with an average of 68.5; the highest and the lowest palatability were for SU48 and SU98, respectively. Seventeen BC_3F_3 lines were selected for preliminary yield trials based on the following threshold conditions: yield of more than 4000 kg ha^{-1}, palatability value higher than 60, and percentage of chalky rice less than 40%.

Seventeen BC_3F_4 lines and their parents were evaluated for agronomic performance in the second crop season of 2019. In the range of agronomic traits of the BC_3F_4 lines, plant height varied between 93.4 and 120 cm, panicle number from 8 to 13, and grain yield from 4557.1 to 7857.1 kg ha^{-1} (Table S3). In addition, we classified the disease resistance reaction into five categories, resistance (R) to high sensitivity (HS), and evaluated the blast resistance of all the lines in the test nursery. The 17 BC_3F_4 lines displayed resistance (R) and moderate resistance (MR) to blast disease.

To evaluate the agronomic performance of these lines, 11 BC_3F_4 lines were selected for further investigation of the yield component and grain quality. The range of panicle weight varied from 3.56 to 4.96 g, spikelets per panicle from 148.56 to 185.1, percentage of seed set from 85.76 to 95.57%, and 1000-grain weight from 19 to 29.8 g (Table S4). In the grain quality analysis, the recurrent parent, DT3, had a recorded palatability value of 83.5, while the donor parent, IR96321-315-240, was 44. The rice palatability value for the BC_3F_4 lines differed from 72.5 to 87.5% (Table S5). Among them, the palatability values of SU39, SU40, and SU92 were higher than those of their parent line DT3 (83.5), which were 86, 87.5, and 84, respectively. After a comprehensive analysis of agronomic traits, yield, yield components, grain appearance, and eating quality, four lines (SU39, SU40, SU89, and SU92) met the following criteria: yield index higher than 90% (6647 kg ha^{-1}) of the recurrent parent, percentage of chalky rice lower than 10%, or palatability value higher than 80, and were selected for the BC_3F_5 generation for advanced yield trials.

Four lines at the BC_3F_5 generation, and recurrent and donor parents, were evaluated in the first crop season of 2020. Significant differences were observed between the *Sub1A* lines and parental rice varieties for grain yield, panicle weight, and 1000-grain weight (Table 1). Table 1 shows that the selected line, SU92, has the potential for a higher yielding ability than those of IR96321-315-240 and DT3. In addition, three lines, SU39, SU89, and SU92, had significantly higher performance on 1000-grain weight compared to the parental lines. Grain quality is one of the goals of Taiwan's breeding program. The mature rice grains of DT3 showed an 80.2% chalky rice rate, while the BC_3F_5 lines demonstrated a varied range of 33.3–86.5% with an average of 55.6%. There was no significant difference between SU39 and DT3 in the chalky rice rate, which was significantly lower than that of DT3 in three lines (SU40, SU89, and SU92), and SU89 was the lowest at 33.3% (Table 1).

Table 1. The grain yield evaluation and yield components of IR96321-315-240, DT3 and 4 BC_3F_5 backcross lines in advanced yield trial at the first cropping season in 2020.

Line or Variety	Grain Yield Evaluation		Yield Components					
	Grain Yield (kg ha^{-1})	Yield Index ^{z} (%)	Panicle Number (No.)	Panicle Length (cm)	Panicle Weight (g)	Spikelets per Panicle (No.)	Seed Set (%)	1000-Grain Weight (g)
IR96321-315-240	4617 ^{b}	73.7	19.8	19.0	2.08 ^{c}	117.0	84.7	19.0 ^{c}
DT3	6309 ^{ab}	100.0	12.8	19.8	2.72 ^{b}	138.3	79.6	22.3 ^{b}
SU39	5853 ^{ab}	95.9	14.2	18.9	3.39 ^{a}	150.4	80.4	25.9 ^{a}
SU40	6168 ^{ab}	101.8	14.4	20.4	2.88 ^{ab}	154.5	84.5	20.9 ^{b}
SU89	6101 ^{ab}	97.8	14.1	19.5	3.14 ^{ab}	143.2	84.2	24.7 ^{a}
SU92	7345 ^{a}	114.6	15.1	19.7	3.01 ^{ab}	133.6	85.0	25.3 ^{a}

Means followed by the different letters are significantly different at the 5% level by least significant difference (LSD) test. ^{z} The yield indices are compared with that of DT3.

The grain length of DT3 was 4.28 mm, those of BC_3F_5 lines were between 4.35 and 4.51 mm, which were significantly longer than that of DT3. The grain width of DT3 was

2.69 mm, and the BC_3F_5 lines were between 2.36 and 2.81 mm. Among them, SU39 and SU40 were significantly different from DT3 in grain length, while the other two lines, SU89 and 92, showed no difference.

We evaluated six physicochemical properties related to rice eating and cooking quality, including palatability and Rapid Visco Analyzer (RVA) parameters, in two parents and four BC_3F_5 lines planted in the first crop season. The palatability values of DT3 and BC_3F_5 lines were 58.6 and between 60.5 and 62.1, respectively. Except for SU89, the other three lines were not significantly different from the parental line DT3 (Table 2). In the analysis of the viscosity characteristics of rice grain, the peak viscosity (PKV) of SU92 was found to be 3551.25 cP, which was significantly higher than the 3159.5 cP of DT3. The PKV of three BC_3F_5 lines was between 3159.50 and 3404.25 cP and showed no significant difference compared to DT3 (Table 3). In addition, the breakdown viscosity (BDV) of SU92 was 2221 cP, which was significantly higher than that 1723.75 cP of DT3. The other three lines were between 1691.00–1882.75 cP, which was not significantly different from DT3.

Table 2. The seed appearance and palatability of IR96321-315-240, DT3 and 4 BC3F5 backcross lines in advanced yield trial at the first cropping season in 2020.

Line or Variety	Seed Appearance					Palatability
	Chalky Rice (%)	Seed Length (mm)	Seed Width (mm)	Seed Length-Width Ratio	Seed Thickness (mm)	
IR96321-315-240	53.1 ^{bc}	4.69 ^{a}	2.11 ^{d}	2.23 ^{a}	1.81 ^{d}	35.5 ^{c}
DT3	80.2 ^{a}	4.28 ^{d}	2.69 ^{b}	1.59 ^{d}	2.04 ^{a}	58.6 ^{b}
SU39	86.5 ^{a}	4.38 ^{c}	2.81 ^{a}	1.56 ^{e}	2.03 ^{a}	61.4 ^{ab}
SU40	59.2 ^{b}	4.35 ^{c}	2.36 ^{c}	1.85 ^{b}	1.90 ^{c}	61.9 ^{ab}
SU89	33.3 ^{d}	4.51 ^{b}	2.70 ^{b}	1.67 ^{c}	2.03 ^{a}	62.1 ^{a}
SU92	43.3 ^{c}	4.51 ^{b}	2.69 ^{b}	1.67 ^{c}	2.00 ^{b}	60.5 ^{ab}

Means followed by the different letters are significantly different at 5% level by least significant difference (LSD) test.

Table 3. Viscograph pasting of IR96321-315-240, DT3 and 4 BC_3F_5 backcross lines in advanced yield trial at the first cropping season in 2020.

Line or Variety	Peak Viscosity (PKV)	Hot Paste Viscosity (HPV)	Breakdown Viscosity (BDV)	Cool Paste Viscosity (CPV)	Setback Viscosity (CSV)
IR96321-315-240	2350.33 ^{c}	1463.33 ^{b}	887.00 ^{c}	3470.33 ^{a}	2007.00 ^{a}
DT3	3159.50 ^{b}	1435.75 ^{b}	1723.75 ^{b}	2302.25 ^{b}	866.50 ^{b}
SU39	3197.75 ^{b}	1506.75 ^{a}	1691.00 ^{b}	2414.25 ^{b}	907.50 ^{b}
SU40	3229.25 ^{b}	1393.75 ^{b}	1835.50 ^{b}	2228.50 ^{b}	834.75 ^{b}
SU89	3404.25 ^{ab}	1521.50 ^{a}	1882.75 ^{b}	2424.00 ^{b}	902.50 ^{b}
SU92	3551.25 ^{a}	1330.25 ^{b}	2221.00 ^{a}	2200.50 ^{b}	870.25 ^{b}

Means followed by the different letters are significantly different at 5% level by least significant difference (LSD) test.

2.3. *Performance of the Sub1A Lines (BC_3F_5) under Submergence and Drought Conditions*

The 21-day-old seedlings of *Sub1A* lines were grown in plastic trays along with submergence-tolerant control IR96321-315-240 and susceptible control DT3 (Figure 4a). After 14 days of complete submergence, all lines showed leaves hanging down due to elongation once the water level dropped (Figure 4b). After one day of recovery from submergence treatment, some leaf tips appeared dry and withered (Figure 4c). However, after being submerged and recovering for 14 day, most of the plants grew fresh green leaves and recovered normal growth and development due to the presence of *Sub1A*-acquired submergence tolerance (Figure 4d). Seedlings of the submergence-tolerant parent IR96321-315–240 had an average elongation of 4.9 cm in plant height under 14 days of submergence and the total stem elongation rate (TSE%) of 130% (Figure 4e,f). The average elongation of DT3 was 14.8 cm in plant height, and the TSE% was 181%. In addition, the average elongation and TSE% of four *Sub1A* lines varied between 10.6 and 15.5 cm, and 164% and 181%, respectively (Figure 4e,f). Except for the lower survival rate (an average of 64%) of susceptible parent DT3, the other *Sub1A* lines had a similar higher survival rate to IR96321-315-240 (Figure 4g).

Figure 4. Characterization of BC_3F_5 lines and its parental lines, IR96321-315-240 and DT3, at rice seedling stage under submergences. (**a**) The phenotype of 21-day-old plants. (**b**) Phenotype of rice seedlings dewatering after 14 day submergence. (**c**) The phenotype of plants after 1 day of recovery from 14 day of submergence. (**d**) The phenotype of rice seedlings submerged for 14 day and then allowed to recover for 14 day. (**e**) Changes in plant height after submergence. (**f**) Total stem elongation rate (TSE%) of seedlings after 14 day submergence followed by 14 day recovery. (**g**) Survival rates of seedlings after 14 day submergence followed by 14 day recovery. Data are presented as the mean of three replications, and the bars represent the standard error of the mean ($n = 3$). Values with the different letters are significantly different ($p < 0.05$ by LSD test).

The recurrent parent DT3 was used in this study selected from a crossing combination between Taikeng9 and the drought-tolerant cultivar Hang-yu15 [19]. Four *Sub1A* lines and the parental lines were used to evaluate the drought tolerance under 28% PEG-6000 simulated drought stress. Five days after osmotic stress in the seedling stage, two control cultivars IR96321-315-240 and IR64, were grouped on scale 7 of drought sensitivity, but all the *Sub1A* lines had a varied scale between 4 and 5.25 similar to that of recurrent parent DT3 (Table 4). According to the daily scale trend chart of drought sensitivity, for all the rice varieties (lines) under 28% PEG-6000 drought stress, the performance of the four *Sub1A* lines was significantly different from that of the control varieties under stress (Figure 5, Table 4).

Table 4. The scale of drought tolerance among IR96321-315-240, DT3 and 4 BC_3F_5 backcross lines under the treatment of 28% PEG 6000 solution at seedling stage.

Line or Variety	Days after Treatment					
	1	2	3	4	5	6
IR96321-315-240	0.00 ^{b}	1.00 ^{a}	3 ^{a}	4.00 ^{a}	7.00 ^{a}	9.00 ^{a}
DT3	0.00 ^{b}	0.00 ^{b}	1 ^{b}	2.00 ^{b}	5.50 ^{b}	7.75 ^{ab}
IR64 ^{z}	0.25 ^{a}	1.25 ^{a}	3 ^{a}	5.00 ^{a}	7.00 ^{a}	9.00 ^{a}
SU39	0.00 ^{b}	0.00 ^{b}	1 ^{b}	2.25 ^{b}	5.00 ^{b}	8.50 ^{ab}
SU40	0.00 ^{b}	0.00 ^{b}	1 ^{b}	1.00 ^{b}	4.00 ^{c}	7.25 ^{b}
SU89	0.00 ^{b}	0.00 ^{b}	1 ^{b}	2.00 ^{b}	4.75 ^{bc}	8.25 ^{ab}
SU92	0.00 ^{b}	0.00 ^{b}	1 ^{b}	1.75 ^{b}	5.25 ^{b}	8.25 ^{ab}

Means followed by different letters are significantly different at the 5% level as calculated by the least significant difference (LSD) test. ^{z} The sensitive control of drought experiment.

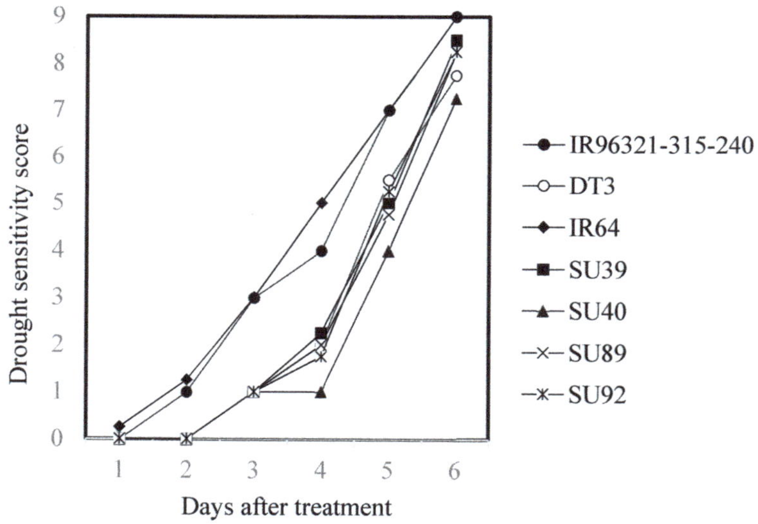

Figure 5. The drought sensitivity score of drought-tolerant donor parent IR96321-315-240, drought-susceptible variety IR64, recurrent parent DT3 and the 4 BC_3F_5 backcrossing line under the 28% PEG-6000 osmotic solution.

3. Discussion

In addition to the second crop season of rice planting in Taiwan being affected by typhoons, the first crop season faces water shortage crises as well, which affect rice production. Taiwan has experienced a serious water shortage since late 2020. It was also the first time in the last 56 years, and no typhoons passed through Taiwan. According to the Council of Agriculture statistics, a total of 24 percent of planted areas were affected by drought in Taiwan, which evaluated the irrigation of 74,000 hectares of first crop rice in 2021.

Therefore, the development of submergence and drought-tolerant varieties can overcome the Taiwan rice production crisis in the future. DT3 is described as a *japonica* rice line with a high yield potential (6–7 t/ha), excellent grain quality, and good drought tolerance [19]. As an extremely valuable yet submergence-susceptible line, DT3 was selected as the focus of this study to increase the submergence tolerance through the introgression of *Sub1A*.

Typical flash flooding results in rapidly rising water levels with submergence for 1–2 weeks. Complete submergence results in accelerated energy consumption in rice seedlings and affects plant growth. It is prone to lodging after the water level recedes [20,21]. To date, FR13A has been recognized as the most submergence-tolerant cultivar and is therefore widely used in rice breeding programs for submergence tolerance [7,22–25]. Rice breeding was accelerated following the identification of QTL *Sub1* on chromosome 9 and understanding the mechanism of gene regulation [10,26,27]. QTL *Sub1* accounts for up to 70% of the submergence tolerance. The cluster genes, *Sub1A*, *Sub1B*, and *Sub1C*, are the different genes that were related to ethylene response factor (ERF)-like genes at the *Sub1* locus. Overexpression of a *Sub1A-1* full-length cDNA in a submergence-intolerant *japonica* rice conferred enhanced tolerance to the plants, which demonstrates that *Sub1A-1* is a primary determinant of submergence tolerance [10]. The 76 rice genotypes from the International Rice Germplasm Collection surveyed for *Sub1* specific markers, Sub1A and Sub1C, indicated that all accessions without the *Sub1A* gene were submergence intolerant. In the gene expression analyses, the results also demonstrate *Sub1C* expression did not associate with submergence tolerance [28]. In this study, the backcross population was constructed using the Swarna-*Sub1* near-isogenic line IR96321-315-240 as a submergence donor parent crossed with good grain quality and drought-tolerant line DT3.

Backcrossing is a conventional method of transferring one or more genes of interest from a donor parent into an elite variety. However, it cannot accurately select the target trait based on the phenotype that is controlled by a specific gene during each round of backcrossing. Modified backcrosses combined with a molecular-marker-assisted selection have already been demonstrated to improve the efficiency of plant breeding, leading to the development of genetic resources and the precise development of tolerance [7,29–32]. In this study, we successfully developed lines demonstrating submergence tolerance in the background selection of DT3 without altering the main features of the recipient variety. In addition, we accelerated the efficiency of the breeding program using gene-specific markers of *Sub1A* for foreground selection at BC_2F_1, BC_3F_1, and BC_3F_2 (Figure 1). MAB is considered more cost-effective and reduces the use of labor and resources compared to the classical backcross breeding approach. The modified MAB in our study used large population sizes (395 plants) for foreground selection in the BC_2F_1 generation. Foreground selection in BC_1F_1 was not validated, but 41 plants with the target gene were sufficient for the next-generation selection. The desired traits that were lacking in Taiwan's popular rice varieties can be improved in less time and with more precision using the results of our study.

In the BC_3F_2, a total of 230 plants were recovered from 800 plants after submergence treatment. Among them, 23 plants were shown to be without the Sub1A gene by marker analysis. The plants without *Sub1A* can survive, and we speculate that there are two possible reasons: (1) Although the *Sub1* QTL accounts for about 70% of the phenotype variation of the submergence tolerance [8], 30% of the phenotypic variation is controlled by other loci. Therefore, surviving plants cannot be ruled out as being affected by other loci. (2) Submergence is abiotic stress. Plants develop various physiological and biochemical mechanisms to adapt to stress, including the development of aerenchyma and adventitious roots for improved aeration, activation of internode and petiole elongation to outgrow submergence water, and conservation of energy until floodwater subsides [33–35].

High-density molecular markers used for background selection can greatly accelerate the recovery rate of the recurrent parental background. For DT3, an average of eight markers were used per chromosome, and the average distance between the two markers was 14.55 cm. In the BC_3F_2 generation, 800 plants were completely submerged for 14 days, and the surviving plants were subjected to foreground selection. A total of 207 BC_3F_2 plants that were *Sub1* homozygous or *Sub1/sub1* heterozygous were selected for background selection to evaluate the recovery rate. The best plant contained 99.6% of the recipient genome. The range of recovery rate was between 84.7 and 99.6%, with an average of 92.9% (Figure 3). Although we did not routinely use the background selection in each generation, the recovery rate was lower than that of the general MAB strategy [7]. The modified MAB in this study can therefore enhance the efficiency of breeding selection.

The development of rice varieties with submergence tolerance, high grain yield potential, and good grain and cooking quality can be immediately useful for flood-prone areas and help farmers increase their production and income [32]. Several lines exhibiting submergence-tolerance and jasmine-like cooking quality with low amylose have been developed in Thailand [30]. In this study, the agronomic performance evaluation of BC_3F_4 to BC_3F_5, a few elite lines were obtained and had important features, such as the genetic background of DT3. All the lines for most of the agronomical traits were, in general, like the recipient parent DT3. However, three candidate lines, SU40, 89, and 92, showed good grain yield and palatability (Tables 1 and 2), which was further confirmed by the multi-location evaluation. In addition, in previous studies, high taste quality was associated with a high peak viscosity and breakdown, and low setback viscosity, while white rice with low cold paste viscosity has a softer taste than cold rice [36,37]. In this study, SU92 had the highest peak and breakdown viscosity and the lowest cold paste viscosity. Therefore, SU92 may have better taste quality and cold rice quality than the parental line DT3 (Table 3).

Despite having gas spaces (aerenchyma) form as an adaptation to submergence, many lowland rice cultivars are still sensitive to complete submergence. If leaves of seedlings elongate their shoots rapidly to contact the aerial interface, they can escape the stress and successfully promote survival. However, their elongation growth can exhaust energy reserves and cause death during complete submergence [5,38,39]. In this study, donor parent IR96321-315-240 with the *Sub1A* gene caused it to enter into a "quiescent" state under submergence, and shoot elongation was no significant difference compared to normal condition (Figure 4). In addition, the donor parent and four lines with *Sub1A* introgression had a significantly higher survival rate compared to the recipient parent, DT3. In addition, we found that DT3 was markedly elongated compared to IR96321-315-240 and lodged severely after submergence. Although the survival rate of DT3 was lower than that of IR96321-315-240 and its backcross offspring after 14 days of complete submergence, its survival ability remained substantial, which coincided with previous findings of studies on upland rice [40].

Differences in the drought tolerance of plants can be evaluated by the visual scoring of leaf rolling, drying, and wiling symptoms, reflecting the dehydration status via simulated drought stress treatments of PEG [24,41]. In this study, the control, parent lines, and *Sub1A* lines were evaluated for their ability to tolerate dehydration in response to PEG treatment. IR64 and IR96321-315-240 exhibited a score of 7 after 5 days of 28% PEG treatment, suggesting two varieties are highly susceptible genotypes, while DT3 and *Sub1A* lines recorded a score of 4–5.5, and were moderately tolerant (Table 4 and Figure 5). In general, *Sub1A* lines had the successful recovery of good grain and eating quality from DT3 with improved tolerance against submergence and drought. This is a significant achievement that will provide farmers with new varieties for cultivation.

4. Materials and Methods

4.1. Plant Materials

The experiments were carried out to transfer the *Sub1* locus into DT3, an elite *japonica* line with drought tolerance, lodging tolerance, blast resistance, and good grain quality [19]. This line was derived from the cross TaiKen No. 9 (TK9, *Oryza sativa* ssp. *japonica*) and Huhan 15 (*Oryza sativa* ssp. *indica*). The yield potential of this line is 6.5 tons ha^{-1}. The submergence-tolerant rice genotype IR96321-315-240 was used as the *Sub1* donor parent. A cross was made between DT3 and IR96321-315-240, with F1 plants backcrossed thrice with DT3 to obtain BC_3F_1 plants, which were self-crossed to obtain the BC_3F_6 progeny. Selections based on foreground, background, and submergence tests were performed from BC_2F_1 to BC_3F_2 as a means of identifying lines similar to those of the recurrent parent.

4.2. Evaluation of Sub1A Lines for Submergence Tolerance

Before the submerged stress treatment, seeds were surface-disinfected with the recommended dose of the benomyl wettable powder (Fulon Chemical Industrial Co., Ltd.; Taoyuan City, Taiwan), which was diluted 1000 times for one day, and seeds were germinated in Petri plates. Germinated seedlings were transferred to seedling trays with soil and grown for up to 21 days. Seedlings were subjected to submergence stress by keeping the pots inside 1 m tanks filled with water. A randomized complete block design was used under the stress condition, with three reps and 25 plants for each replicate of each line. Plants were removed after two weeks of submergence treatment. The percentage of survival was assessed 14 days after de-submergence in the greenhouse.

4.3. Evaluation of Sub1A Lines for Osmotic Stress Tolerance

Seeds of DT3, IR64, and 4 BC_3F_5 lines were surface-disinfected with benomyl and germinated in plates. Germinated seeds were put in 96-well hydroponic trays and grown for up to 21 days under hydroponic conditions (Yoshida solution, Chiayi, Taiwan) [42]. Rice seedlings were stressed in a Yoshida solution containing 28% polyethylene glycol (PEG) 6000. Plants were scored for their responses to osmotic stress every day using a standard evaluation system [43].

4.4. Evaluation of Sub1A Lines for Blast Resistance

The rice blast nursery was conducted in an upland field in the second season of 2019. Natural *Magnaporthe grisea* infection was favored at the nursery by a high level of nitrogen fertilization and the planting of the susceptible rice cultivar 'Lomello', and resistant *japonica* rice cultivars 'Tainung No. 70' and *indica* rice 'Taichung Sen No. 10' in the field. In each replication, 5 g seeds for each rice variety (line) were planted in a row (50 cm row length with 10 cm spacing). Evaluation of leaf symptoms was performed on the leaves of every plant in a row using a standard visual scale (0–9) developed by the International Rice Research Institute [40]. The lines that had an average grade of 1–3, 3–5, 5–6, 6–7, and >7 were regarded as resistant (R), moderately resistant (MR), moderately susceptible (MS), susceptible (S), and highly susceptible (HS), respectively. The resistance reaction of each plant was scored from the two replications and averaged to represent the severity of leaf blast on each rice variety (line).

4.5. Evaluation of Agronomic Traits

During the first and second crop seasons of 2019, the thirty-day-old seedlings of the BC_3F_3 and BC_3F_4 lines and both parents were transplanted into three rows, with 24 plants per row, per entry, at 15 × 25 cm spacing, at the Chiayi Agricultural Experiment Station Farm of the Taiwan Agricultural Research Institute. The agronomic performance, grain quality, and palatability of the 28 BC_3F_3 and 11 BC_3F_4 lines were measured. In BC_3F_5, field planting followed a randomized complete block design with four replications (blocks). Four lines were selected, and two parents were transplanted into the plots. Each plot consisted of three rows with 20 plants per row at 30 × 15 cm. The following agronomic traits were recorded for each line: number of grains per panicle, panicle length (cm), panicle weight (g), spikelets/panicles, seed set (%), 1000-seed weight (g), grain yield (kg ha^{-1}), and yield index (%). In addition, the grain quality, including percentage of chalky rice (%), seed length (mm), seed width (mm), seed length-width ratio, seed thickness (mm), and palatability, were also investigated and analyzed. For palatability analyses, approximately 33 g of rice flour was hulled and ground into fine flour for the palatability evaluation, which was performed using a palatability analyzer system (Toyo Taste Meter, Model MA-30). The sample processing method was carried out in accordance with the manufacturer's operation manual (TRCM Co., Toyo Rice Polishing Machine Factory, Osaka, Japan), as previously described [44].

In addition, for the viscosity analysis of cooked rice grain, approximately 3 g rice flour was mixed with 25 mL water and used for RVA profile evaluation, which was performed using a Rapid Visco Analyzer (Model No. RVA-4, Newport Scientific, Sydney, Australia), according to the Standard Method AACC61-02 released by the American Association of Cereal Chemists. The sequential temperature curve for a 12.5 min test was as follows: (1) incubation at 50 °C for 1 min; (2) increased to 95 °C and held for 2.5 min; (3) cooling to 50 °C and held at 50 °C until the end. The RVA profiles were characterized by five parameters: peak viscosity (PKV), hot paste viscosity (HPV), breakdown viscosity (BDV = PKV − HPV), cool paste viscosity (CPV), and setback viscosity (SBV = CPV − PKV). Statistical analysis was performed with independent samples using least significant difference (LSD).

4.6. DNA Isolation and PCR Amplification

Rice genomic DNA extraction, with modifications, was adopted for mini preparation [44]. Approximately 5 cm of fresh leaf tissue from seedlings was homogenized with 300 µL extraction buffer (100 mm Tris-HCl, pH 9.0; 40 mm EDTA-2Na, pH 8.0; 1.67% SDS) at 30 strokes/s for 2 min using a TissueLyser (Qiagen Retsch GmbH & Co. KG, Haan, Germany). A total of 150 µL of benzyl chloride was added to the homogenized tissue and vortexed. After incubation in a 50 °C water bath for 15 min, 150 µL of 3 M sodium acetate (pH 5.2) was added. Supernatants were centrifuged at 15,000 rpm for 15 min at 4 °C, and 300 µL ice-cold isopropanol was added to precipitate DNA. After centrifugation at 15,000 rpm for 10 min, DNA pellets were saved and washed with 70% ethanol, air-dried, and dissolved in 50 µL TE buffer. A 10 µL PCR reaction containing 10 ng genomic DNA, 2.5 µM forward and reverse primers, and 5 µL Multiplex PCR Master Mix (QIAGEN, Inc., Redwood City, CA, USA) was performed using a thermocycler (GeneAmp PCR System 9700, Life Technologies Corp., Carlsbad, CA, USA) at 94 °C for 2 min for 1 cycle; 94 °C for 30 s, 55 °C for 20 s, 72 °C for 30 s for 35 cycles, and 72 °C for 2 min for 1 cycle. Following PCR, 1 µL of amplified DNA products was separated by 6% polyacrylamide gel in 0.5× TBE at 100 v (Dual Triple-Wide Mini-Vertical System, C. B. S. Scientific, San Diego, CA, USA) for 60 min.

4.7. Marker Analysis

Foreground selection was performed using functional and linkage markers, Sub1A1 and Sub1AB1 (Table S6) [15]. In addition, a total of 100 markers, including 66 SSRs, 8 STS, and 26 InDel, distributed evenly on the 12 chromosomes, were used for genotyping in BC_3F_2 with an average marker interval of 14.55 cm, and were used in a genome-wide survey to identify the chromosome segment substitution locations. These polymorphic markers were used for background selection to select plants with maximum recovery of the recurrent parent genome. The genotypes from polymorphic bands were recorded as A (IR96321-315-240), B (DT3), and H (DT3/IR96321-315-240). The Graphical Geno Types Version 2.0 [45] software program was used for the assessment of the recurrent parent genome (%RPG) in the selected recombinants, based on marker data.

5. Conclusions

Compared to conventional backcrossing, marker-assisted backcross breeding is an effective and reliable approach for transferring the *Sub1A* gene in rice. In this study, we constructed a low-cost, low-input, and high-efficiency molecular-marker-assisted selection platform for rice breeding programs. By using a few foreground selections, phenotyping of submergence tolerance, and background selection at the beginning of the breeding process, while combining the evaluation of agronomic traits in late generations, we successfully developed several submergence- and drought-tolerant lines, with high yield and good grain and eating quality. These *Sub1A* lines with a DT3 genetic background are expected to have a high impact on domestic rice production stability and reduce the risk of rice yield loss in flood-prone rain-fed areas caused by climate change.

Supplementary Materials: The following are available online at https://www.mdpi.com/article/10.3390/ijms222413365/s1.

Author Contributions: Y.-C.H. and Y.-P.W. conceived the study, designed the experiments, and helped draft the manuscript. Y.-C.C. and C.H. carried out the phenotyping and genotyping studies, prepared the figures and tables, and drafted the manuscript. Y.-C.H. and S.-M.W. analyzed and interpreted the data and contributed to the preparation of the manuscript. All authors have read and agreed to the published version of the manuscript.

Funding: This work was supported by the Ministry of Science and Technology (MOST 109-2313-B-415-007) of Taiwan.

Acknowledgments: We are grateful to the National Plant Resources Center for providing rice materials.

Conflicts of Interest: The authors declare that they have no conflict of interest.

References

1. Chantal, L.M.; Marie, D.G.; Olivier, M. *Land Use and Food Security in 2050: A Narrow Road. Éditions Quae*; Editions Quae: Paris, France, 2018.
2. Muthayya, S.; Sugimoto, J.D.; Montgomery, S.; Maberly, G.F. An overview of global rice production, supply, trade, and consumption. *Ann. N. Y. Acad. Sci.* **2014**, *1324*, 7–14. [CrossRef]
3. Schneider, P.; Asch, F. Rice production and food security in Asian Mega deltas—A review on characteristics, vulnerabilities and agricultural adaptation options to cope with climate change. *J. Agron. Crop Sci.* **2020**, *206*, 491–503. [CrossRef]
4. Loo, Y.Y.; Billa, L.; Singh, A. Effect of climate change on seasonal monsoon in Asia and its impact on the variability of monsoon rainfall in Southeast Asia. *Geosci. Front.* **2015**, *6*, 817–823. [CrossRef]
5. Bailey-Serres, J.; Fukao, T.; Ronald, P.; Ismail, A.; Heuer, S.; Mackill, D. Submergence tolerant rice: *SUB1*'s journey from landrace to modern cultivar. *Rice* **2010**, *3*, 138–147. [CrossRef]
6. Oladosu, Y.; Rafii, M.Y.; Arolu, F.; Chukwu, S.C.; Muhammad, I.; Kareem, I.; Salisu, M.A.; Arolu, I.W. Submergence tolerance in rice: Review of, mechanism, breeding and, future prospects. *Sustainability* **2020**, *12*, 1632. [CrossRef]
7. Neeraja, C.N.; Maghirang-Rodriguez, R.; Pamplona, A.; Heuer, S.; Collard, B.C.; Septiningsih, E.M.; Vergara, G.; Sanchez, D.; Xu, K.; Ismail, A.M.; et al. A marker-assisted backcross approach for developing submergence-tolerant rice cultivars. *Theor. Appl. Genet.* **2007**, *115*, 767–776. [CrossRef] [PubMed]
8. Xu, K.; Mackill, D.J. A major locus for submergence tolerance mapped on rice chromosome 9. *Mol. Breed.* **1996**, *2*, 219–224. [CrossRef]
9. Xu, K.; Xu, X.; Ronald, P.C.; Mackill, D.J. A high-resolution linkage map of the vicinity of the rice submergence tolerance locus Sub1. *Mol. Gen. Genet.* **2000**, *263*, 681–689. [CrossRef]
10. Xu, K.; Xu, X.; Fukao, T.; Canlas, P.; Maghirang-Rodriguez, R.; Heuer, S.; Ismail, A.M.; Bailey-Serres, J.; Ronald, P.C.; Mackill, D.J. *Sub1A* is an ethylene-response-factor-like gene that confers submergence tolerance to rice. *Nature* **2006**, *442*, 705–708. [CrossRef]
11. Fukao, T.; Xu, K.; Ronald, P.C.; Bailey-Serres, J. A variable cluster of ethylene response factor–like genes regulates metabolic and developmental acclimation responses to submergence in rice. *Plant Cell* **2006**, *18*, 2021–2034. [CrossRef]
12. Jia, W.; Ma, M.; Chen, J.; Wu, S. Plant morphological, physiological and anatomical adaption to flooding stress and the underlying molecular mechanisms. *Int. J. Mol. Sci.* **2021**, *22*, 1088. [CrossRef] [PubMed]
13. Iftekharuddaula, K.M.; Newaz, M.A.; Salam, M.A.; Ahmed, H.U.; Mahbub, M.A.A.; Septiningsih, E.M.; Collard, B.C.Y.; Sanchez, D.L.; Pamplona, A.M.; Mackill, D.J. Rapid and high-precision marker assisted backcrossing to introgress the *SUB1* QTL into BR11, the rainfed lowland rice mega variety of Bangladesh. *Euphytica* **2011**, *178*, 83–97. [CrossRef]
14. Septiningsih, E.M.; Hidayatun, N.; Sanchez, D.L.; Nugraha, Y.; Carandang, J.; Pamplona, A.M.; Collard, B.C.Y.; Ismail, A.M.; Mackill, D.J. Accelerating the development of new submergence tolerant rice varieties: The case of Ciherang-Sub1 and PSB Rc18-Sub1. *Euphytica* **2015**, *202*, 259–268. [CrossRef]
15. Septiningsih, E.M.; Pamplona, A.M.; Sanchez, D.L.; Neeraja, C.N.; Vergara, G.V.; Heuer, S.; Ismail, A.M.; Mackill, D.J. Development of submergence-tolerant rice cultivars: The *Sub1* locus and beyond. *Ann. Bot.* **2009**, *103*, 151–160. [CrossRef]
16. Dixit, S.; Singh, A.; Sandhu, N.; Bhandari, A.; Vikram, P.; Kumar, A. Combining drought and submergence tolerance in rice: Marker-assisted breeding and QTL combination effects. *Mol. Breed.* **2017**, *37*, 143. [CrossRef] [PubMed]
17. Iftekharuddaula, K.M.; Ahmed, H.U.; Ghosal, S.; Moni, Z.R.; Amin, A.; Ali, M.S. Development of new submergence tolerant rice variety for Bangladesh using marker-assisted backcrossing. *Rice Sci.* **2015**, *22*, 16–26. [CrossRef]
18. Chang, C.-P.; Yang, Y.-T.; Kuo, H.-C. Large increasing trend of tropical cyclone rainfall in Taiwan and the roles of terrain. *J. Clim.* **2013**, *26*, 4138–4147. [CrossRef]
19. Lin, K.H.; Kuo, C.W.; Hsu, Y.C.; Lin, Y.R.; Wu, Y.P. Increasing drought tolerance of rice cultivar 'TK9' using marker- assisted selection. *Crop Environ. Bioinform.* **2014**, *11*, 143–164.

20. Das, K.; Sarkar, R.; Ismail, A. Elongation ability and non-structural carbohydrate levels in relation to submergence tolerance in rice. *Plant Sci.* **2005**, *168*, 131–136. [CrossRef]
21. Singh, S.; Mackill, D.J.; Ismail, A.M. Responses of *SUB1* rice introgression lines to submergence in the field: Yield and grain quality. *Field Crops Res.* **2009**, *113*, 12–23. [CrossRef]
22. Mackill, D.J.; Amante, M.M.; Vergara, B.S.; Sarkarung, S. Improved semidwarf rice lines with tolerance to submergence of seedlings. *Crop Sci.* **1993**, *33*, 749–753. [CrossRef]
23. Sandhu, N.; Dixit, S.; Swamy, B.P.M.; Raman, A.; Kumar, S.; Singh, S.P.; Yadaw, R.B.; Singh, O.N.; Reddy, J.N.; Anandan, A.; et al. Marker assisted breeding to developmultiple stress tolerant varieties for flood and drought prone areas. *Rice* **2019**, *12*, 8. [CrossRef] [PubMed]
24. Muthu, V.; Abbai, R.; Nallathambi, J.; Rahman, H.; Ramasamy, S.; Kambale, R.; Thulasinathan, T.; Ayyenar, B.; Muthurajan, R. Pyramiding QTLs controlling tolerance against drought, salinity, and submergence in rice through marker assisted breeding. *PLoS ONE* **2020**, *15*, e0227421. [CrossRef] [PubMed]
25. Das, G.; Rao, G.J.N. Molecular marker assisted gene stacking for biotic and abiotic stress resistance genes in an elite rice cultivar. *Front. Plant Sci.* **2015**, *6*, 698. [CrossRef]
26. Fukao, T.; Bailey-Serres, J. Submergence tolerance conferred by *Sub1A* is mediated by SLR1 and SLRL1 restriction of gibberellin responses in rice. *Proc. Natl. Acad. Sci. USA* **2008**, *105*, 16814–16819. [CrossRef]
27. Fukao, T.; Yeung, E.; Bailey-Serres, J. The submergence tolerance regulator SUB1A mediates crosstalk between submergence and drought tolerance in rice. *Plant Cell* **2011**, *23*, 412–427. [CrossRef]
28. Singh, N.; Dang, T.T.; Vergara, G.V.; Pandey, D.M.; Sanchez, D.; Neeraja, C.N.; Septiningsih, E.M.; Mendioro, M.; Tecson-Mendoza, E.M.; Ismail, A.M.; et al. Molecular marker survey and expression analyses of the rice submergence-tolerance gene *SUB1A*. *Theor. Appl. Genet.* **2010**, *121*, 1441–1453. [CrossRef]
29. Luo, Y.; Ma, T.; Zhang, A.; Ong, K.H.; Luo, Z.; Li, Z.; Yang, J.; Yin, Z. Marker-assisted breeding of Chinese elite rice cultivar 9311 for disease resistance to rice blast and bacterial blight and tolerance to submergence. *Mol. Breed.* **2017**, *37*, 106. [CrossRef]
30. Jantaboon, J.; Siangliw, M.; Im-mark, S.; Jamboonsri, W.; Vanavichit, A.; Toojinda, T. Ideotype breeding for submergence tolerance and cooking quality by marker-assisted selection in rice. *Field Crops Res.* **2011**, *123*, 206–213. [CrossRef]
31. Mohapatra, S.; Panda, A.K.; Bastia, A.K.; Mukherjee, A.K.; Sanghamitra, P.; Meher, J.; Mohanty, S.P.; Pradhan, S.K. Development of submergence-tolerant, bacterial blight-resistant, and high-yielding near isogenic lines of popular variety, 'Swarna' through marker-assisted breeding approach. *Front. Plant Sci.* **2021**, *12*, 1441. [CrossRef] [PubMed]
32. Pandit, E.; Pawar, S.; Barik, S.R.; Mohanty, S.P.; Meher, J.; Pradhan, S.K. Marker-assisted backcross breeding for improvement of submergence tolerance and hrain yield in the popular rice variety 'Maudamani'. *Agronomy* **2021**, *11*, 1263. [CrossRef]
33. Bailey-Serres, J.; Voesenek, L.A.C.J. Flooding stress: Acclimations and genetic diversity. *Annu. Rev. Plant Biol.* **2008**, *59*, 313–339. [CrossRef] [PubMed]
34. Magneschi, L.; Perata, P. Rice germination and seedling growth in the absence of oxygen. *Ann. Bot.* **2009**, *103*, 181–196. [CrossRef] [PubMed]
35. Toojinda, T.; Siangliw, M.; Tragoonrung, S.; Vanavichit, A. Molecular genetics of submergence tolerance in rice: QTL analysis of key traits. *Ann. Bot.* **2003**, *91*, 243–253. [CrossRef] [PubMed]
36. Wei, H.Y.; Chen, Z.F.; Xing, Z.P.; Zhou, L.; Liu, Q.Y.; Zhang, Z.Z.; Jiang, Y.; Hu, Y.J.; Zhu, J.Y.; Cui, P.Y.; et al. Effects of slow or controlled release fertilizer types and fertilization modes on yield and quality of rice. *J. Integr. Agric.* **2018**, *17*, 2222–2234. [CrossRef]
37. Patindol, J.A.; Siebenmorgen, T.J.; Wang, Y.-J. Impact of environmental factors on rice starch structure: A review. *Starch-Stärke* **2015**, *67*, 42–54. [CrossRef]
38. Jackson, M.B.; Ram, P.C. Physiological and molecular basis of susceptibility and tolerance of rice plants to complete submergence. *Ann. Bot.* **2003**, *91*, 227–241. [CrossRef]
39. Nishiuchi, S.; Yamauchi, T.; Takahashi, H.; Kotula, L.; Nakazono, M. Mechanisms for coping with submergence and waterlogging in rice. *Rice* **2012**, *5*, 2. [CrossRef]
40. Yang, S.Y.; Wu, Y.S.; Chen, C.T.; Lai, M.H.; Yen, H.M.; Yang, C.Y. Physiological and molecular responses of seedlings of an upland rice ('Tung Lu 3') to total submergence compared to those of a submergence-tolerant lowland rice ('FR13A'). *Rice* **2017**, *10*, 42. [CrossRef]
41. Badr, A.; El-Shazly, H.H.; Tarawneh, R.A.; Börner, A. Screening for drought tolerance in maize (*Zea mays* L.) germplasm using germination and seedling traits under simulated drought conditions. *Plants* **2020**, *9*, 565. [CrossRef]
42. Yoshida, S.; Forno, D.A.; Cook, J.H.; Gomez, K.A. *Laboratory Manual for Physiological Studies of Rice*; IRRI: Los Banos, Philippines, 1976; pp. 61–65.
43. IRRI. *Standard Evaluation System for Rice*; International Rice Research Institute: Los Banos, Philippines, 2013.
44. Hsu, Y.C.; Tseng, M.C.; Wu, Y.P.; Lin, M.Y.; Wei, F.J.; Hwu, K.K.; Hsing, Y.I.; Lin, Y.R. Genetic factors responsible for eating and cooking qualities of rice grains in a recombinant inbred population of an inter-subspecific cross. *Mol. Breed.* **2014**, *34*, 655–673. [CrossRef] [PubMed]
45. Van Berloo, R. GGT: Software for the display of graphical genotypes. *J. Hered.* **1999**, *90*, 328–330. [CrossRef]

International Journal of
Molecular Sciences

MDPI

Article

Genetic Elucidation for Response of Flowering Time to Ambient Temperatures in Asian Rice Cultivars

Kiyosumi Hori [1,*,†], Daisuke Saisho [2,†], Kazufumi Nagata [1], Yasunori Nonoue [1], Yukiko Uehara-Yamaguchi [3], Asaka Kanatani [3], Koka Shu [1], Takashi Hirayama [2], Jun-ichi Yonemaru [1], Shuichi Fukuoka [1] and Keiichi Mochida [2,3,4,*]

1 National Agriculture and Food Research Organization, Institute of Crop Science, Tsukuba, Ibaraki 305-8518, Japan; kzfmngt@affrc.go.jp (K.N.); nonouey221@affrc.go.jp (Y.N.); zhengf@yahoo.co.jp (K.S.); yonemaru@affrc.go.jp (J.-i.Y.); fukusan@affrc.go.jp (S.F.)
2 Institute of Plant Science and Resources, Okayama University, Kurashiki 710-0046, Japan; saisho@rib.okayama-u.ac.jp (D.S.); hira-t@rib.okayama-u.ac.jp (T.H.)
3 RIKEN Center for Sustainable Resource Science, Yokohama 230-0045, Japan; yukiko.uehara@riken.jp (Y.U.-Y.); asaka.kanatani@riken.jp (A.K.)
4 Kihara Institute for Biological Research, Yokohama City University, Yokohama 244-0813, Japan
* Correspondence: horikiyo@affrc.go.jp (K.H.); keiichi.mochida@riken.jp (K.M.)
† These authors contributed to this work equally.

Abstract: Climate resilience of crops is critical for global food security. Understanding the genetic basis of plant responses to ambient environmental changes is key to developing resilient crops. To detect genetic factors that set flowering time according to seasonal temperature conditions, we evaluated differences of flowering time over years by using chromosome segment substitution lines (CSSLs) derived from *japonica* rice cultivars "Koshihikari" × "Khao Nam Jen", each with different robustness of flowering time to environmental fluctuations. The difference of flowering times in 9 years' field tests was large in "Khao Nam Jen" (36.7 days) but small in "Koshihikari" (9.9 days). Part of this difference was explained by two QTLs. A CSSL with a "Khao Nam Jen" segment on chromosome 11 showed 28.0 days' difference; this QTL would encode a novel flowering-time gene. Another CSSL with a segment from "Khao Nam Jen" in the region around *Hd16* on chromosome 3 showed 23.4 days" difference. A near-isogenic line (NIL) for *Hd16* showed 21.6 days' difference, suggesting *Hd16* as a candidate for this QTL. RNA-seq analysis showed differential expression of several flowering-time genes between early and late flowering seasons. Low-temperature treatment at panicle initiation stage significantly delayed flowering in the CSSL and NIL compared with "Koshihikari". Our results unravel the molecular control of flowering time under ambient temperature fluctuations.

Keywords: rice; flowering time; ambient temperature fluctuation; chromosome segment substitution line (CSSL): quantitative trait locus (QTL)

Citation: Hori, K.; Saisho, D.; Nagata, K.; Nonoue, Y.; Uehara-Yamaguchi, Y.; Kanatani, A.; Shu, K.; Hirayama, T.; Yonemaru, J.-i.; Fukuoka, S.; et al. Genetic Elucidation for Response of Flowering Time to Ambient Temperatures in Asian Rice Cultivars. *Int. J. Mol. Sci.* **2021**, *22*, 1024. https://doi.org/10.3390/ijms22031024

Received: 14 December 2020
Accepted: 17 January 2021
Published: 20 January 2021

1. Introduction

Global warming is likely to reach 1.5 °C between 2030 and 2052 if it continues at the current rate [1]. Cultivated rice (*Oryza sativa* L.) originated from a wild rice species (*Oryza rufipogon* Griff.), which grows mainly in the tropics [2,3]. Rice is inherently adaptable to hot environments. However, novel rice cultivars with resistance to very hot environments are required for regions where extreme warming is expected [1]. Climate change is causing large fluctuations in temperature, solar irradiation, precipitation and soil moisture, especially in the tropics [4]. At the high-latitude limits of rice cultivation; however, low temperatures at the seedling, panicle initiation and maturation stages severely decrease grain yield [5,6]. There is now an increasing demand for new cultivars that are adaptable to both effects of climate change. Therefore, to develop climate-change resilient crops, it is necessary to elucidate the genetic basis of plant response to ambient environmental changes.

Daylength is a primary environmental factor that determines maturation (harvest) time through induction or suppression of flowering and consequently adaptability to each growing area in plant species [7,8]. Rice is a short-day plant: flowering is promoted under short daylength. Most genes for flowering cloned so far are involved in response to daylength and photoperiod sensitivity in rice [9–13]. They are assigned to two major independent gene regulatory pathways, the *OsGI–Hd1–Hd3a* pathway and the *Ghd7–Ehd1–Hd3a/RFT1* pathway. Both pathways are regulated by light perception, daylength and the circadian clock. These previous molecular genetic studies show the importance of response to daylength in the determination of rice flowering time.

Temperature is another important ambient environmental factor that determines flowering time. Vernalization (low-temperature) treatment promotes flowering in several cereal and horticultural species such as wheat, barley and *Brassica* vegetables [14,15]. In Arabidopsis, higher temperature accelerates plant growth and development through signaling from *PhyB*, *PIFs*, *DELLA* and the evening complex, which includes *ELF3* [16–20]. Previous studies revealed that rice cultivars adapted to higher latitudes showed photoperiod insensitivity and high-temperature response [21–23]. Four flowering-time QTLs responsive to temperature were detected in a mapping population derived from crosses between *indica* and *japonica* cultivars [24]. In addition, other previous studies investigated relationships between rice flowering times under different temperature conditions and expression levels of previously isolated genes in rice mutant lines [25,26]. They showed that expression of *Ehd1*, *Hd3a* and *RFT1* was repressed under long daylength and low temperatures. The data suggest that temperature response is a decisive factor in determining flowering time in rice. However, despite their importance in relation to breeding programs, the molecular mechanisms involved in temperature response remain largely unknown. It is necessary to elucidate the genetic mechanisms underlying the determination of flowering time in response to ambient temperature in rice.

Here, we evaluated differences of flowering time over 9 years (2011 to 2019) in Asian rice cultivars and chromosome segment substitution lines (CSSLs) derived from crosses between *japonica* rice cultivars "Koshihikari" and "Khao Nam Jen". "Khao Nam Jen" had a large difference. Among the 40 CSSLs, two lines with a "Khao Nam Jen" segment on chromosomes (Chrs.) 11 and 3 (the latter including the *Hd16* region) had over 15 days' difference. The QTL on Chr. 11 would be a novel flowering-time gene, because no allelic differences have been found in genes previously isolated in this region. Low-temperature treatment at the panicle initiation stage delayed flowering time in both lines. RNA-seq analysis revealed that expression of the flowering-time genes *Ehd1*, *Hd3a* and *RFT1* was shifted later in the flowering season in the CSSL than in "Koshihikari". Through these analyses, we detected novel genetic factors involved in the control of flowering time under different ambient temperatures during the rice-growing season.

2. Results

2.1. Natural Variations in Flowering Time Among Years

We found wide variations in flowering time (days-to-heading, DTH) of 18 Asian rice cultivars during 2011 to 2019 (Figure 1A, Supplementary Table S1). For example, DTH varied from 87.6 $\pm$ 1.1 (extremely early) in "Qui Zhao Zong" to 201.4 $\pm$ 2.3 (extremely late) in "Khao Nam Jen" in 2011 and from 88.4 $\pm$ 2.3 (extremely early) in "Qui Zhao Zong" to 215.0 $\pm$ 0.5 (extremely late) in "Nona Bokra" in 2019. Among the 18 cultivars, "Khao Nam Jen" and "Bei Khe" had a large difference of flowering time: 36.7 and 23.2 days, respectively. We calculated coefficients of variations (CVs) to evaluate dispersions of flowering time in the 18 cultivars for nine years. "Khao Nam Jen", "Bei Khe", "Muha" and "Tupa 121-3" had large CVs: 5.6, 5.4, 5.7 and 5.2. On the other hand, "Bleiyo" had a small difference of flowering time: 3.2 days. "Bleiyo", "LAC23" and "Basilanon" had small CVs: 1.5, 2.7 and 2.8, respectively. Daylength is strictly stable, while ambient temperature fluctuates widely among years (Figure 1B,C). Flowering time in "Bleiyo", "LAC23′ and "Basilanon" was strongly controlled by daylength, while that in "Khao Nam Jen", "Bei Khe", "Muha" and "Tupa 121-3" was easily affected by ambient temperature during the cropping period. Therefore, we selected "Khao Nam Jen", with the largest flowering time difference, to genetically dissect the molecular basis for flowering time fluctuations among years through analysis of progeny derived through backcrossing to "Koshihikari", which had a moderate difference of flowering time (9.9 days; CV = 3.0).

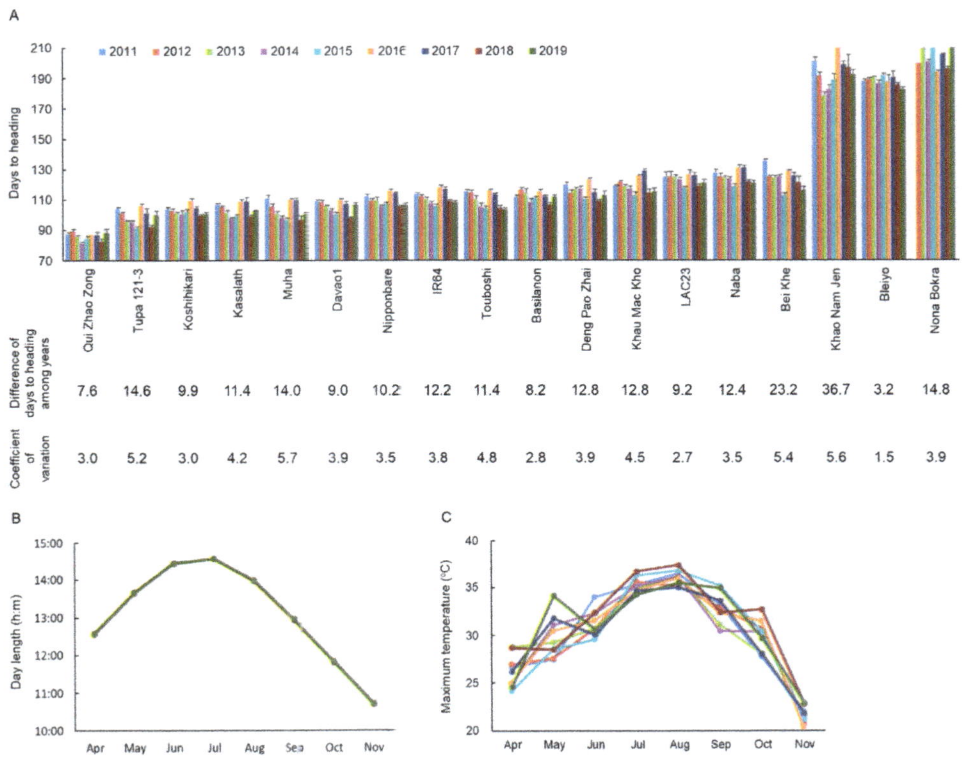

Figure 1. (**A**) Days-to-heading (DTH) of 18 Asian rice cultivars under natural field conditions from 2011 to 2019. Values are means ± SDs. Differences of DTH and coefficient of variation were based on the earliest and latest flowering times among years. (**B**) Daylength during rice-growing season from 2011 to 2019. (**C**) Maximum temperature during rice-growing season from 2011 to 2019.

2.2. Flowering Time Fluctuation in CSSLs of "Koshihikari" and "Khao Nam Jen"

To identify genetic factors related to the flowering time fluctuation, we evaluated flowering time in a set of CSSLs derived from crosses between "Koshihikari" and "Khao Nam Jen" from 2011 to 2019. The CSSLs, developed in a previous study, comprise 40 lines at the BC_4F_6 generation (SL2801–SL2840; Supplementary Figure S1; [27]). Each CSSL has a substituted segment of a target chromosomal region from "Khao Nam Jen" in the genetic background of "Koshihikari". Flowering times of most CSSLs were similar to those of "Koshihikari" in each year (Figure 2). However, several CSSLs had large differences in flowering time from "Koshihikari". Line SL2833, with a "Khao Nam Jen" segment on the short arm of Chr. 11, had a flowering time difference of 28.0 days and CV = 9.6 (Figures 2 and 3; Supplementary Table S1). SL2812, with a "Khao Nam Jen" segment on the long arm of Chr. 3, had a flowering time difference of 23.4 days and CV = 6.5. Most other CSSLs had a flowering time difference of ~15 days and CV ≤ 5, similar to those of "Koshihikari". Therefore, we assigned two major-effect QTLs for flowering time difference among years: one on the short arm of Chr. 11 and one on the long arm of Chr. 3. Both

"Khao Nam Jen" segments increased flowering time difference. QTLs for flowering time in the corresponding years were found in both regions (Supplementary Figure S2): on the long arm of Chr. 3 in all 9 years and on the short arm of Chr. 11 in 4 years.

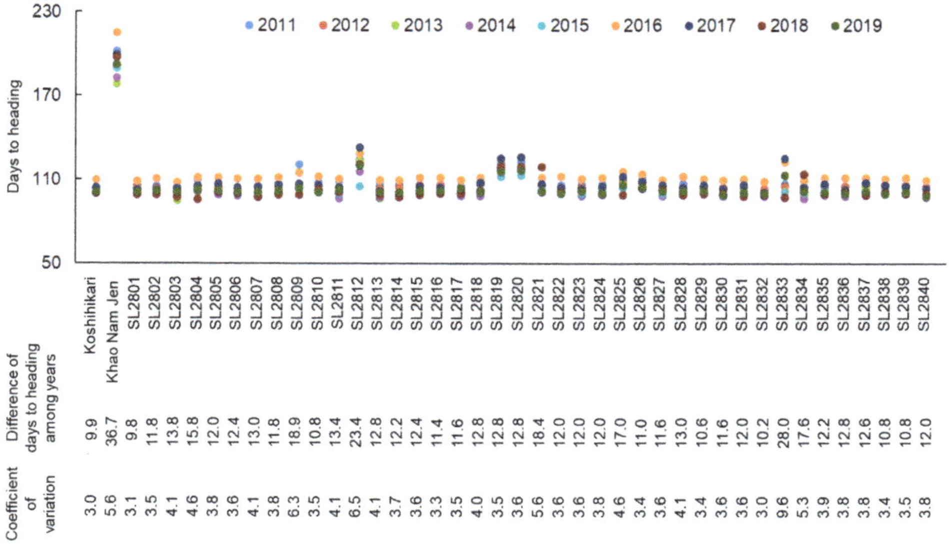

Figure 2. Days-to-heading (DTH) of a set of chromosome segment substitution lines (CSSLs) derived from "Koshihikari" × "Khao Nam Jen" under natural field conditions from 2011 to 2019. Differences of DTH among years were calculated from flowering times between earliest and latest years.

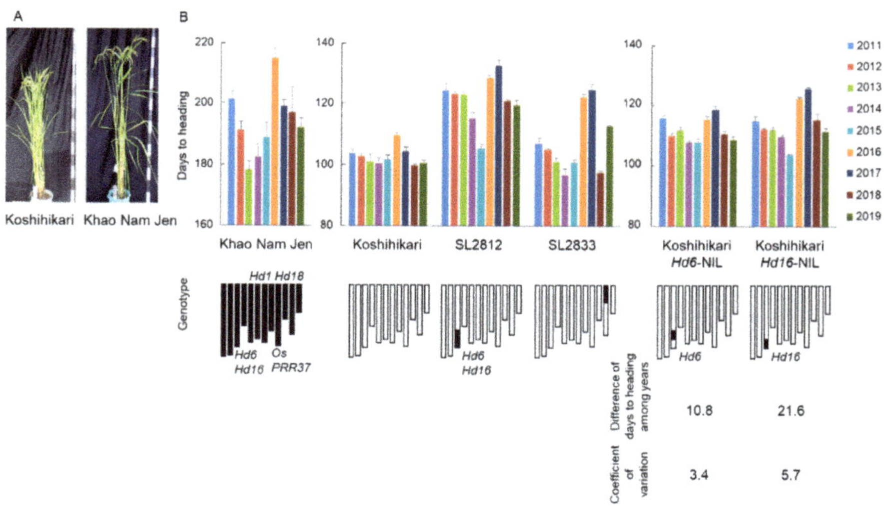

Figure 3. (**A**) "Koshihikari" and "Khao Nam Jen" grown under natural field conditions after flowering. (**B**) Days to heading under natural field conditions from 2011 to 2019 and its differences in "Khao Nam Jen", "Koshihikari", SL2812, SL2833, Koshihikari *Hd6*-NIL and Koshihikari *Hd16*-NIL.

No genes for flowering time showing DNA sequence variations between "Koshihikari" and "Khao Nam Jen" have been previously identified on the short arm of Chr. 11 (Supplementary Figure S2). Two flowering-time genes—*Hd6* and *Hd16*—have been identified on the long arm of Chr. 3 and their molecular functions in flowering time regulation have been elucidated [28,29]. "Koshihikari" has non-functional alleles at these two genes, while "Khao Nam Jen" has functional alleles [30,31]. Koshihikari *Hd16*-NIL, with a functional allele in the "Koshihikari" background, had a flowering time difference of 21.6 days and CV = 5.7 (Figure 3; Supplementary Table S1). Koshihikari *Hd6*-NIL, with a functional allele in the "Koshihikari" genetic background, had a flowering time difference of 10.8 days and CV = 3.4.

2.3. Transcriptome Analysis in Rice Plants from Juvenile to Mature Stages

To compare gene expression patterns between "Koshihikari" and SL2812 ("Khao Nam Jen" segment on long arm of Chr. 3), we carried out RNA-seq analysis in 2015 (early-flowering year) and 2016 (late-flowering year; Figure 3). "Koshihikari" showed a transcriptome phase transition from vegetative to reproductive stages at 10 weeks after transplanting in both 2015 and 2016 (Figure 4A). SL2812 showed it at 10 weeks in 2015 and at 12 weeks in 2016 (Figure 4B). Therefore, we focused on gene expression at 10 weeks. In 2015, 95 genes were up-regulated and 234 genes were down-regulated in "Koshihikari" compared with SL2812 (Figure 4C). In 2016, 207 were up-regulated and 436 were down-regulated in "Koshihikari" (Figure 4D). Among expressed genes without significant differences in 2015, 451 genes were differentially expressed (137 up- and 314 down-regulated) in 2016 (Figure 4E). Gene ontology analysis of these 451 genes showed that they included genes involved in responses to stress stimuli, flower development and embryo development (Figure 4F,G). Moreover, expression patterns of 88 genes for phytochromes and thermo-response signaling and 59 flowering-time regulation genes fluctuated much more after 10 weeks in 2016 than in 2015 (Figure 4H,I; Supplementary Table S2). Among these 147 genes, expression of 44 genes differed more than twofold between "Koshihikari" and SL2812 after 10 weeks in 2015 and expression of 75 genes differed more than twofold in 2016. Quantitative real-time PCR confirmed different expression peaks of *Ehd1*, *Hd3a* and *RFT1*, which function downstream of *Hd16* in the flowering-time gene network, at 8 to 12 weeks in 2016 between "Koshihikari" and SL2812 (Figure 4J).

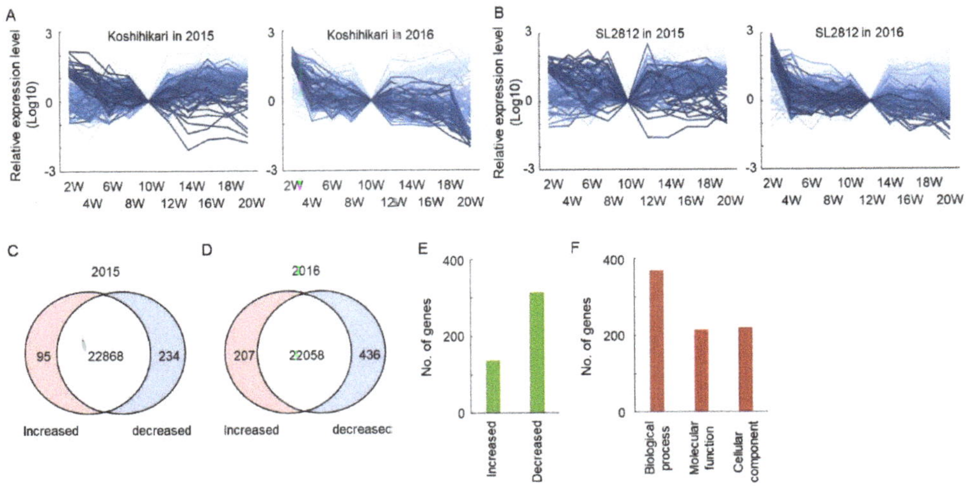

Figure 4. *Cont.*

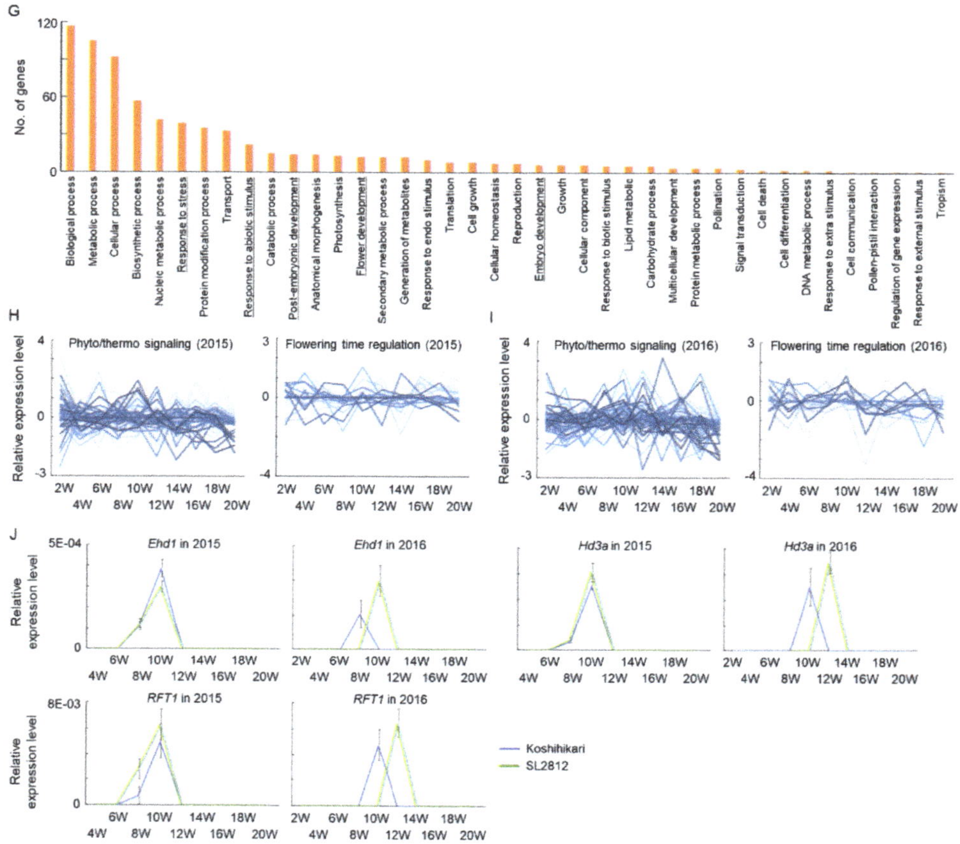

Figure 4. (**A**,**B**) Changes in transcriptome associated with the transition to reproductive stage in 2015 and 2016 in (**A**) "Koshihikari" and (**B**) SL2812. Relative expression patterns of 255 clusters based on 28 229 genes from 2 to 20 weeks after transplanting in the field. Dark blue lines, low expression at 20 weeks; light blue lines, high expression at 20 weeks. (**C**,**D**) Numbers of genes differentially expressed between "Koshihikari" and SL2812 at 10 weeks in (**C**) 2015 and (**D**) 2016. Increased and decreased genes were higher and lower expressions in "Koshihikari" as compared with SL2812. (**E**) Numbers of differentially expressed genes at 10 weeks in 2016 among genes showing ≤2-fold difference between "Koshihikari" and SL2812 in 2015. (**F**,**G**) Gene ontology (GO) analysis of differentially expressed genes in 2016 by (**F**) GO types and (**G**) GO categories. (**H**,**I**) Relative expression levels of phytochromes and thermo-response signaling genes and flowering time regulation genes of "Koshihikari"/SL2812 from 2 to 20 weeks in (**H**) 2015 and (**I**) 2016. (**J**) Confirmation of expression of three flowering-time genes—*Ehd1*, *Hd3a* and *RFT1*—in "Koshihikari" and SL2812 by quantitative real-time PCR.

2.4. Effect of Weather Factors on Flowering Time in Rice Plants

To reveal weather factors associated with the difference of flowering times, we tested correlations of flowering time with air temperature, relative humidity, precipitation, wind speed, wind direction, duration of sunshine, amount of insolation and soil temperature. Maximum temperatures in July were significantly negatively correlated with flowering time in "Koshihikari", SL2812, SL2833, Koshihikari *Hd6*-NIL and Koshihikari *Hd16*-NIL (Table 1; Supplementary Figure S3). Maximum temperatures in June, August and September were significantly negatively correlated with flowering time in "Khao Nam Jen" and Koshihikari *Hd16*-NIL.

Table 1. Correlations between monthly maximum temperature during rice-growing season and days-to-heading in "Koshihikari", "Khao Nam Jen", SL2812, SL2833, Koshihikari *Hd6*-NIL and Koshihikari *Hd16*-NIL. *** Significant correlation with days to heading at $p < 0.001$ by the Pearsons' test.

	Days to Heading					
	'Koshihikari'	'Khao Nam Jen'	SL2812	SL2833	Koshihikari *Hd6*-NIL	Koshihikari *Hd16*-NIL
Maximum temperature						
June	−0.07	0.34 ***	−0.08	−0.01	0.14	0.48 ***
July	−0.26 ***	−0.20	−0.35 ***	−0.45 ***	−0.34 ***	−0.44 ***
August	−0.16	−0.32 ***	−0.18	−0.24	−0.24	−0.42 ***
September	0.03	−0.43 ***	0.20	0.16	−0.02	−0.63 ***

Next, we evaluated the effects of high temperature (30 °C daytime / 24 °C nighttime) and low temperature (24/20 °C) on flowering in "Koshihikari", Koshihikari *Hd16*-NIL and SL2833. Under long daylength condition, low temperature delayed flowering in all three genotypes: the effect was 2.0 times that at normal temperature (27/24 °C) in "Koshihikari", 2.4 times in Koshihikari *Hd16*-NIL and 2.8 times in SL2833 (Figure 5). Low temperature delayed flowering significantly in Koshihikari *Hd16*-NIL and SL2833 relative to "Koshihikari". On the other hand, high temperature commonly led to early flowering: the effect was 0.7 times that at normal temperature in all three genotypes. Under short daylength condition, there was no significant difference in flowering time among lines, although low temperature somewhat delayed flowering time of all lines.

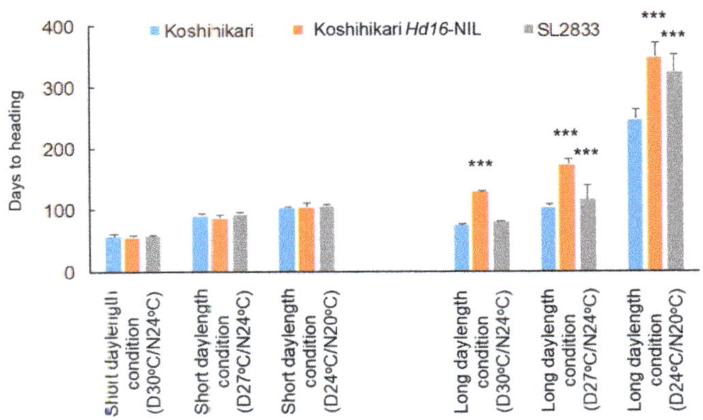

Figure 5. Days-to-heading of "Koshihikari", Koshihikari *Hd16*-NIL and SL2833 under high temperature (30 °C daytime / 24 °C nighttime), normal temperature (27/24 °C) and low temperature (24/20 °C), under short daylength (10 h light/14 h dark) and long daylength (14.5 h/9.5 h). Values are means ± SDs. *** Significant difference from "Koshihikari" at $p < 0.001$ by the *t*-test.

3. Discussion

3.1. Detection of QTLs for Response of Flowering Time to Ambient Temperature Fluctuations

Even though classical genetic studies suggest the importance of temperature response and photoperiod response in determination of flowering time in rice [21–23], few reports on genetic factors determine the response of flowering time to ambient temperatures. Here, we show that two major genetic factors—a novel QTL on Chr. 11 and photoperiod sensitivity gene *Hd16*—are associated with the response of flowering time to ambient temperatures during the rice-growing season.

The QTL for flowering time on the short arm of Chr. 11 was detected in only 4 out of the 9 years, with inconsistent directions of additive effects. The flowering time of SL2833, with the "Khao Nam Jen" allele at this QTL, was negatively correlated with maximum temperature in July in all 9 years; flowering was delayed in 2017 (low maximum temperature) and promoted in 2018 (high maximum temperature). Flowering time of SL2833 was significantly delayed under low temperature condition. Therefore, this QTL is associated with responsiveness to ambient temperature in regulating flowering time in rice. It was localized to the region containing *RCN1* and *RBS1* that encode phosphatidylethanolamine-binding protein and heterogeneous nuclear ribonucleoprotein R-type protein, respectively. However, the sequences of *RCN1* and *RBS1* were identical between "Khao Nam Jen" and "Koshihikari" and expression levels of both did not differ significantly in our RNA-seq dataset in 2015 and 2016. Therefore, this QTL may be a novel flowering-time gene in rice. Further genetic studies such as fine-mapping experiments are necessary to isolate the gene responsible for this QTL.

The QTL on the long arm of Chr. 3 corresponds to *Hd16*, which was isolated as a flowering time-gene involved in photoperiod sensitivity in rice [29,32,33]. SL2812 and Koshihikari-*Hd16* NIL had a wide difference of flowering time and a large CV among the 9 years. Additionally, low-temperature treatment delayed flowering time in Koshihikari-*Hd16* NIL. Our results indicate that *Hd16* is associated with both photoperiod sensitivity and response to ambient temperatures. *Hd16* encodes casein kinase-I; the "Koshihikari" allele has one non-synonymous substitution in the conserved kinase domain, which results in deficient function, causing decreased photoperiod sensitivity [29]. The functional *Hd16* allele from "Khao Nam Jen" promoted both strong photoperiod sensitivity and strong response to ambient temperatures in comparison with the non-functional "Koshihikari" *Hd16* allele. Biochemical characterization indicated that a functional Hd16 recombinant protein specifically phosphorylated the proteins encoded by rice flowering-time genes *Ghd7* and *OsPRR37* and the SLR1 protein in the gibberellin signaling pathway in rice [29,32,33]. The combined evidence suggests that *Hd16* alters the response of flowering time to ambient temperatures through both the photoperiod-sensitive flowering-time pathway and the gibberellin-dependent growth-regulatory pathway.

Several other CSSLs had moderately large flowering time fluctuations among the 9 years (Figure 2). SL2809, SL2821, SL2825 and SL2834 had more than 15 days' difference of flowering times or CV > 5 among years. Segments from "Khao Nam Jen" were introgressed on the long arms of Chr. 2 in SL2809, Chr. 6 in SL2821, Chr. 7 in SL2825 and Chr. 11 in SL2834. Therefore, multiple genetic factors play a role in controlling flowering time under ambient temperature fluctuations in rice.

3.2. Integration of Thermal Response into Photoperiod Flowering Pathway

In rice, most flowering-time genes isolated previously are associated with response to daylength and photoperiod sensitivity [9–13]. They function in two major independent pathways, the *OsGI–Hd1–Hd3a* pathway and the *Ghd7–Ehd1–Hd3a/RFT1* pathway. Both pathways promote flowering under short daylength but repress it under long daylength (Supplementary Figure S4). We found a novel QTL on Chr. 11 and *Hd16* on Chr. 3 that appear to control flowering time under different ambient temperatures. *Hd16* participates in the *Ghd7–Ehd1–Hd3a/RFT1* pathway [29,33]. Transcriptome analysis also showed fluctuations of *Ehd1*, *Hd3a* and *RFT1* expression between different ambient temperatures. Therefore, the response of flowering time to ambient temperatures would be controlled by members of the photoperiod sensitivity pathway of the *Ghd7–Ehd1–Hd3a/RFT1* pathway. The QTL on Chr. 11 also might function in the *Ghd7–Ehd1–Hd3a/RFT1* pathway, because the "Khao Nam Jen" allele did not alter the expression of *OsGI*, *Hd1*, or *Hd6* in the other pathway, or the difference of flowering time among years in the plants carrying the functional *Hd6* allele.

Extensive studies in model plant species have elucidated the details of the flowering-time genetic networks. Six major pathways have been reported in Arabidopsis: the photoperiod, vernalization, gibberellin, autonomous, age and ambient temperature pathways [34,35]. Efforts in wheat, barley, maize, sorghum, *Brassica* species and fruit trees have been also reviewed [36]. In Arabidopsis and temperate cereal crops such as wheat and barley, vernalization is a crucial factor that accelerates flowering time [15,37]. Arabidopsis has many vernalization-related genes, including *FLC*, *FRI*, *VRN1*, *VRN2*, *VIN2* and *VIN3*, but we did not detect rice homologs of these genes as QTLs in this study. Wheat and barley have three vernalization genes—*VRN1*, *VRN2* and *VRN3* (unrelated to the genes with the same names in Arabidopsis)—in each genome [38,39]. Wheat and barley *VRN1*, which encodes an AP1-like MADS-box transcription factor, is orthologous to *OsMADS14*, *OsMADS15* and *OsMADS18* in rice [40]. *VRN2*, with a zinc-finger motif and a CCT domain, seems to be homologous to rice *Ghd7* [41]. *VRN3* is homologous to *FT*-like genes such as *Hd3a* and *RFT1*. These studies suggest that homologs of rice photoperiod-sensitivity genes are associated with vernalization response in wheat and barley. Therefore, the genetic pathway that controls the response of flowering time to ambient temperatures might be conserved among cereal crop species.

High temperature accelerates plant growth and development in Arabidopsis through a signaling pathway that includes *PhyB*, *PIF4*, *YUC8*, *DELLA* and *ELF3* [16–20]. *PhyB* functions as a thermo-sensor in leaf epidermis and regulates *PIF4* activity and the auxin signaling pathway in response to ambient high temperature, as well as to light signals [17]. In our study, some rice orthologs of these genes showed similar expression patterns in 2015 (higher temperature year) between "Koshihikari" and SL2812, but large fluctuations in 2016 (lower temperature year) between the two. Therefore, these genes might also be associated with flowering time regulation in rice in response to ambient temperature fluctuations. Recently developed techniques in molecular biology can reveal the complex fundamental mechanisms involved in the control of the agronomically important traits [42]. To reveal the degree of conservation of the flowering-time pathway among rice, Arabidopsis and temperate cereal crops, it is necessary to identify more participant genes and to perform further molecular functional studies such as of phosphorylation, methylation and acetylation status at both gene and protein levels and to integrate those modifications into the gene pathway for flowering time regulation in rice.

3.3. *Application to Development of Climate-Resilient Crops*

Global warming is causing climate change with large temperature fluctuations, in addition to variations in solar irradiation, precipitation and soil moisture, in rice-growing regions [1,4]. Temperature fluctuations are strongly associated with crop yield losses. Novel rice cultivars resistant to both high and low temperatures are necessary for cultivation regions that are expected to experience large temperature fluctuations during the rice-growing season. In this study, we detected genetic factors for plant response to differences in ambient temperature. The two "Khao Nam Jen" alleles found change flowering time in response to ambient temperatures. If rice cultivars could promote or delay their growth and flowering time in response to ambient temperatures, we could reduce yield loss after exposure to temperature extremes at the panicle initiation, pollination and maturation stages. These "spontaneously flexible" alleles in "Khao Nam Jen" could be utilized for developing climate-resilient cultivars and thus achieving high productivity under climate change.

4. Conclusions

We found a wide difference of flowering time fluctuations among 9 years in Asian rice cultivars. We detected two QTLs in CSSLs with "Khao Nam Jen" chromosome segments in the "Koshihikari" genetic background. One QTL corresponds to *Hd16*, on the log arm of Chr. 3 and the other QTL, on the long arm of Chr. 11, might be a novel gene. Low temperature treatment at the panicle initiation stage delayed flowering in the two CSSLs with these QTLs. These results provide insights into thermo-sensitive flowering-time regulation in rice and will facilitate our understanding of the molecular basis of the control of flowering time variation under ambient temperature fluctuations and allow the development of climate-resilient crops that are adaptable to temperature fluctuations under climate change.

5. Materials and Methods

5.1. Plant Materials

We selected 18 rice cultivars: "Qui Zhao Zong", "Davao 1′, "IR64′, "Touboshi", "Deng Pao Zhai", "Naba", "Bei Khe", "Bleiyo", "Tupa 121-3′, "Kasalath", "Muha", "Koshihikari", "Nipponbare", "Basilanon", "Khao Mac Kho", "LAC 23′, "Khao Nam Jen" and "Nona Bokra". These cultivars (8 *indica*, 3 *aus*, 7 *japonica*), mainly from Asia, were selected on the basis of their geographical origin, cluster analysis of genome-wide genotype data and variation in DTH from a representative rice core collection [43]. Chromosome segment substitution lines (CSSLs) with "Khao Nam Jen" chromosome segments in the "Koshihikari" genetic background were developed in our previous study [27]. The CSSLs comprise a set of 40 lines at the BC_4F_6 generation, covering 95.9% of the "Khao Nam Jen" genome. Two near-isogenic lines (NILs) in the "Koshihikari" genetic background were developed in previous studies [29,44].

5.2. Evaluation of Flowering Time in Natural Field Conditions

All plants were grown in an experimental field at the Institute of Crop Science, National Agriculture and Food Research Organization (NARO), Tsukuba, Japan (36.03° N, 140.11° E), in 9 years (2011 to 2019). Month-old seedlings were transplanted in mid-May at one per hill in plots with a double row for each line, with 18 cm between plants and 36 cm between rows. The mean daylengths were 12.3 h in April, 13.4 h in May, 14.3 h in June, 14.4 h in July, 13.6 h in August and 12.6 h in September. Cultivation management followed the standard procedures used at the Institute of Crop Science. DTH of individual plants were scored as the number of days from sowing to the appearance of the first panicle in each of the 5 or 24 plants and mean values were calculated. The difference of flowering time of each line was calculated as the difference between the earliest and latest flowering times among the 9 years. The coefficient of variation (CV) was calculated as standard deviation ÷ average flowering time in each year.

5.3. QTL Detection in CSSLs and NILs

The difference of flowering time and CV among years in CSSLs and NILs were compared with those of "Koshihikari" by using Dunnett"s multiple comparison procedure in JMP v. 11.0.0 software. For QTL detection in the CSSLs, we used a total of 377 SNP and 112 SSR markers showing polymorphisms between "Koshihikari" and "Khao Nam Jen" reported by the previous studies [27,30]. QTLs were declared present when these scores were significantly different between a line and "Koshihikari". The scores were considered statistically significant if difference of flowering time >15.0 days and CV > 5.0.

5.4. Transcriptome at Panicle Initiation Stage

We extracted mRNA from the highest fully expanded leaf of "Koshihikari" and SL2812 ("Khao Nam Jen" segment on long arm of Chr. 3) at 2 to 20 weeks after transplanting in 2015 and 2016. Sequencing libraries for RNA-seq analysis were constructed by using the QuantSeq 3′ mRNA-seq Library Prep Kit (Lexogen, Greenland, NH, USA) with barcodes for Illumina sequencing according to the manufacturer's instructions. The libraries were sequenced on a HiSeq 4000 platform (Illumina, San Diego, CA, USA) using the TruSeq SBS Kit (Illumina, San Diego, CA, USA) and the paired-end sequencing method to obtain two 100 bp sequences. Individual reads were trimmed by cutting bases from the start and end of reads if quality ≤20 in Trimmomatic v. 0.36 software and by removing reads with final read length <50 in FASTQC v. 0.11.9 software. Trimmed reads were mapped on the IRGSP1.0 rice genome sequence and gene loci in RAP-DB [45]. Each gene locus read was counted by featureCounts v. 1.6.4 software. Reads per million (RPM) values of all genes were calculated from the read count data. Genes with RPM ≥ 1 in all three biological replicates were defined as expressed. Expressed genes were classified according to gene-ontology types and categories from the annotations in RAP-DB [45] and RiceNETDB [46]. The RNA-seq data are archived at the DNA Data Bank of Japan under accession number DRA011161. Quantitative real-time PCR were carried out by the methods of previous study [29]. Transcription levels of *Hd3a, RFT1, Ehd1* and *Rice ubiquitin2* (*UBQ*) were measured according to a SYBR Green-based method by using gene-specific primers.

5.5. Correlation between Flowering Time and Meteorological Conditions

Weather data at the paddy field from 2011 to 2019 were collected from the Weather Data Acquisition System of the Institute for Agro-Environmental Sciences, NARO ([47], http://www.naro.affrc.go.jp/org/niaes/aws/). We collected maximum, minimum and average values of air temperature, relative humidity, precipitation, wind speed, wind direction, duration of sunshine, amount of insolation and soil temperature each month from April to November. Pearson's correlation coefficient was used to assess relationships among DTH and the weather data in JMP.

5.6. Evaluation of Flowering Time under High and Low Temperatures

"Koshihikari", Koshihikari *Hd16*-NIL and SL2833 ("Khao Nam Jen" segment on Chr. 11) were grown in a controlled-environment cabinet under a short daylength (10 h light/14 h dark) or a long daylength (14.5/9.5 h). Relative humidity was maintained at 60% under a photosynthetic photon flux density of 500 µmol m^{-2} s^{-1} provided by metal halide lamps that covered the spectrum from 300 to 1000 nm. Ambient temperatures (light/dark) were 30/24 °C, 27/24 °C (usual rice cultivation condition), or 24/20 °C. DTH in 10 plants of each line was scored and mean values were calculated for each line.

Supplementary Materials: The following are available online at https://www.mdpi.com/1422-0067/22/3/1024/s1. Figure S1: Graphical representation of genotypes of CSSLs, Figure S2: Chromosome positions of QTLs for flowering time in each year in CSSLs, Figure S3: DTH and daily maximum temperatures at panicle initiation stage from 2011 to 2019, Figure S4: Genetic pathway controlling flowering time with ambient temperature fluctuations in rice, Table S1: DTH in rice cultivars, CSSLs and NILs from 2011 to 2019, Table S2: Loci of genes involved in flowering time regulation.

Author Contributions: K.H., D.S. and K.M. designed the experiments. K.H., D.S., K.N., Y.N., Y.U.-Y., A.K., K.S., T.H., J.-i.Y., S.F. and K.M. cultivated plant materials, evaluated flowering time, genotyped by DNA markers and performed RNA-seq analysis. K.H., D.S., S.F. and K.M. analyzed all of the data and wrote the manuscript. All authors have read and agreed to the published version of the manuscript.

Funding: This research was funded by grants from the Ministry of Agriculture, Forestry and Fisheries of Japan (Genomics for Agricultural Innovation GPN0001 and NVR0001 and Research Project for Genomics-based Technology for Agricultural Improvement RBS1003) and Joint Usage/Research Center, Institute of Plant Science and Resources, Okayama University, grant No. 2827.

Institutional Review Board Statement: Not applicable.

Informed Consent Statement: Not applicable.

Data Availability Statement: All datasets supporting the conclusions of this article are included in the article and supplementary files.

Acknowledgments: We are grateful to the technical staff of the Field Management Division at the National Agriculture and Food Research Organization for their assistance in field management and to Kaworu Ebana for distribution of seeds of Asian rice cultivars.

Conflicts of Interest: The authors declare no conflict of interest.

References

1. IPCC (Intergovernmental Panel on Climate Change). Global Warming of 1.5 °C. 2018. Available online: https://report.ipcc.ch/sr15/pdf/sr15_spm_final.pdf (accessed on 19 January 2021).
2. Londo, J.P.; Chiang, Y.C.; Hung, K.H.; Chiang, T.Y.; Schaal, B.A. Phylogeography of Asian wild rice, *Oryza rufipogon*, reveals multiple independent domestications of cultivated rice, *Oryza sativa*. *Proc. Natl. Acad. Sci. USA* **2006**, *103*, 9578. [CrossRef] [PubMed]
3. Huang, X.; Kurata, N.; Wei, X.; Wang, Z.X.; Wang, A.; Zhao, Q.; Zhao, Y.; Liu, K.; Lu, H.; Li, W.; et al. A map of rice genome variation reveals the origin of cultivated rice. *Nature* **2012**, *490*, 497–501. [CrossRef] [PubMed]
4. Bathiany, S.; Dakos, V.; Scheffer, M.; Lenton, T.M. Climate models predict increasing temperature variability in poor countries. *Sci. Adv.* **2018**, *4*, eaar5809. [CrossRef] [PubMed]
5. Takeuchi, Y.; Hayasaka, H.; Chiba, B.; Tanaka, I.; Shimano, T.; Yamagishi, M.; Nagano, K.; Sasaki, T.; Yano, M. Mapping quantitative trait loci controlling cool-temperature tolerance at booting stage in temperate *japonica* rice. *Breed. Sci.* **2001**, *51*, 191–197. [CrossRef]
6. Jiang, S.; Yang, C.; Xu, Q.; Wang, L.; Yang, X.; Song, X.; Wang, J.; Zhang, X.; Li, B.; Li, H.; et al. Genetic dissection of germinability under low temperature by building a resequencing linkage map in *japonica* rice. *Int. J. Mol. Sci.* **2020**, *21*, 1284. [CrossRef]
7. Thomas, B.; Vince-Prue, D. *Photoperiodism in Plants*, 2nd ed.; Academic Press: San Diego, CA, USA, 1997.
8. Izawa, T. Daylength measurements by rice plants in photoperiodic short-day flowering. In *International Review of Cytology*; Kwang, W.J., Ed.; Academic Press: Cambridge, MA, USA, 2007; Volume 256, pp. 191–222.
9. Shrestha, R.; Gomez-Ariza, J.; Brambilla, V.; Fornara, F. Molecular control of seasonal flowering in rice, Arabidopsis and temperate cereals. *Ann. Bot.* **2014**, *114*, 1445–1458. [CrossRef]
10. Hori, K.; Matsubara, K.; Yano, M. Genetic control of flowering time in rice: Integration of Mendelian genetics and genomics. *Theor. Appl. Genet.* **2016**, *129*, 2241–2252. [CrossRef]
11. Brambilla, V.; Martignago, D.; Goretti, D.; Cerise, M.; Somssich, M.; De Rosa, M.; Galbiati, F.; Simon, R.; Lazzaro, F.; Simon, R.; et al. Antagonistic transcription factor complexes modulate the floral transition in rice. *Plant Cell* **2017**, *29*, 2801. [CrossRef]
12. Matsubara, K.; Yano, M. Genetic and molecular dissection of flowering time control in rice. In *Rice Genomics, Genetics and Breeding*; Sasaki, T., Ashikari, M., Eds.; Springer: Singapore, 2018; pp. 177–190.
13. Liu, H.; Zhou, X.; Li, Q.; Wang, L.; Xing, Y. CCT domain-containing genes in cereal crops: Flowering time and beyond. *Theor. Appl. Genet.* **2020**, *133*, 1385–1396. [CrossRef]
14. Qüesta, J.I.; Song, J.; Geraldo, N.; An, H.; Dean, C. Arabidopsis transcriptional repressor VAL1 triggers Polycomb silencing at FLC during vernalization. *Science* **2016**, *353*, 485–488. [CrossRef]
15. Bloomer, R.H.; Dean, C. Fine-tuning timing: Natural variation informs the mechanistic basis of the switch to flowering in *Arabidopsis thaliana*. *J. Exp. Bot.* **2017**, *68*, 5439–5452. [CrossRef]
16. Jung, J.H.; Domijan, M.; Klose, C.; Biswas, S.; Ezer, D.; Gao, M.; Khattak, A.K.; Box, M.S.; Charoensawan, V.; Cortijo, S.; et al. Phytochromes function as thermosensors in *Arabidopsis*. *Science* **2016**, *354*, 886–889. [CrossRef] [PubMed]
17. Kim, S.; Hwang, G.; Kim, S.; Thi, T.N.; Kim, H.; Jeong, J.; Kim, J.; Kim, J.; Choi, G.; Oh, E. The epidermis coordinates thermoresponsive growth through the phyB-PIF4-auxin pathway. *Nat. Commun.* **2020**, *11*, 1053. [CrossRef]
18. Hahm, J.; Kim, K.; Qiu, Y.; Chen, M. Increasing ambient temperature progressively disassembles *Arabidopsis* phytochrome B from individual photobodies with distinct thermostabilities. *Nat. Commun.* **2020**, *11*, 1660. [CrossRef] [PubMed]
19. Blanco-Touriñán, N.; Legris, M.; Minguet, E.G.; Costigliolo-Rojas, C.; Nohales, M.A.; Iniesto, E.; García-León, M.; Pacín, M.; Heucken, N.; Blomeier, T.; et al. COP1 destabilizes DELLA proteins in *Arabidopsis*. *Proc. Natl. Acad. Sci. USA* **2020**, *117*, 13792–13799. [CrossRef] [PubMed]
20. Jung, J.H.; Barbosa, A.D.; Hutin, S.; Kumita, J.R.; Gao, M.; Derwort, D.; Silva, C.S.; Lai, X.; Pierre, E.; Geng, F.; et al. A prion-like domain in ELF3 functions as a thermosensor in *Arabidopsis*. *Nature* **2020**, *585*, 256–260. [CrossRef] [PubMed]
21. Hsu, C. Effect of seasonal changes of day-length and temperature to crop growth I. Studies on growth period of rice cultivars in Taiwan. *Agric. Hortic.* **1938**, *13*, 21–32.
22. Wada, E. Studies on the response of heading to day-length and temperature in rice plants. I. Response of varieties and the relation to their geographical distribution in Japan. *Jpn. J. Breed.* **1952**, *2*, 55–62. [CrossRef]
23. Oka, H. Varietal variation of the responses to day-length and temperature and the number of days to growth period. Phylogenic differentiation of the cultivated rice plant. III. Japan. *Jpn. J. Breed.* **1954**, *4*, 92–100. [CrossRef]
24. Nakagawa, H.; Yamagishi, J.; Miyamoto, N.; Motoyama, M.; Yano, M.; Nemoto, K. Flowering response of rice to photoperiod and temperature: A QTL analysis using a phenological model. *Theor. Appl. Genet.* **2005**, *110*, 778–786. [CrossRef]
25. Luan, W.; Chen, H.; Fu, Y.; Si, H.; Peng, W.; Song, S.; Liu, W.; Hu, G.; Sun, Z.; Xie, D.; et al. The effect of the crosstalk between photoperiod and temperature on the heading date in rice. *PLoS ONE* **2009**, *4*, e5891. [CrossRef]
26. Song, Y.; Gao, Z.; Luan, W. Interaction between temperature and photoperiod in regulation of flowering time in rice. *Sci. China Life Sci.* **2012**, *55*, 241–249. [CrossRef] [PubMed]
27. Nagata, K.; Nonoue, Y.; Mizobuchi, R.; Ono, N.; Shibaya, T.; Ebana, K.; Matsubara, K.; Ogiso, E. Development of twelve sets of chromosomal segment substitution lines that enhance allele mining in cultivated rice. *Breed. Res.* **2015**, *17*, 131.
28. Takahashi, Y.; Shomura, A.; Sasaki, T.; Yano, M. *Hd6*, a rice quantitative trait locus involved in photoperiod sensitivity, encodes the alpha subunit of protein kinase CK2. *Proc. Natl. Acad. Sci. USA* **2001**, *98*, 7922–7927. [CrossRef] [PubMed]

29. Hori, K.; Ogiso-Tanaka, E.; Matsubara, K.; Yamanouchi, U.; Ebana, K.; Yano, M. *Hd16*, a gene for casein kinase I, is involved in the control of rice flowering time by modulating the day-length response. *Plant J.* **2013**, *76*, 36–46. [CrossRef]
30. Hori, K.; Nonoue, Y.; Ono, N.; Shibaya, T.; Ebana, K.; Matsubara, K.; Ogiso-Tanaka, E.; Tanabata, T.; Sugimoto, K.; Taguchi-Shiobara, F.; et al. Genetic architecture of variation in heading date among Asian rice accessions. *BMC Plant Biol.* **2015**, *15*, 115. [CrossRef]
31. Itoh, H.; Wada, K.C.; Sakai, H.; Shibasaki, K.; Fukuoka, S.; Wu, J.; Yonemaru, J.-I.; Yano, M.; Izawa, T. Genomic adaptation of flowering-time genes during the expansion of rice cultivation area. *Plant J.* **2018**, *94*, 895–909. [CrossRef]
32. Dai, C.; Xue, H.W. Rice early flowering1, a CKI, phosphorylates DELLA protein SLR1 to negatively regulate gibberellin signalling. *EMBO J.* **2010**, *29*, 1916–1927. [CrossRef]
33. Kwon, C.T.; Yoo, S.C.; Koo, B.H.; Cho, S.H.; Park, J.W.; Zhang, Z.; Li, J.; Li, Z.; Paek, N.C. Natural variation in Early flowering1 contributes to early flowering in japonica rice under long days. *Plant Cell Environ.* **2014**, *37*, 101–112. [CrossRef]
34. Fornara, F.; de Montaigu, A.; Coupland, G. SnapShot: Control of flowering in Arabidopsis. *Cell* **2010**, *141*, 550–550.e2. [CrossRef]
35. Leijten, W.; Koes, R.; Roobeek, I.; Frugis, G. Translating flowering time from Arabidopsis thaliana to Brassicaceae and Asteraceae crop species. *Plants* **2018**, *7*, e111. [CrossRef] [PubMed]
36. Bentley, A.; Jensen, E.; Mackay, I.; Honicka, H.; Fladung, M.; Hori, K.; Yano, M.; Mullet, J. *Flowering Time: Genomics and Breeding for Climate-Resilient Crops*; Springer: Berlin, Germany, 2013; Volume 2, pp. 1–66.
37. Henderson, I.R.; Shindo, C.; Dean, C. The need for winter in the switch to flowering. *Annu. Rev. Genet.* **2003**, *37*, 371–392. [CrossRef] [PubMed]
38. Trevaskis, B.; Hemming, M.N.; Dennis, E.S.; Peacock, W.J. The molecular basis of vernalization-induced flowering in cereals. *Trends Plant Sci.* **2007**, *12*, 352–357. [CrossRef] [PubMed]
39. Xu, S.; Chong, K. Remembering winter through vernalisation. *Nat. Plants* **2018**, *4*, 997–1009. [CrossRef] [PubMed]
40. Li, C.; Lin, H.; Chen, A.; Lau, M.; Jernstedt, J.; Dubcovsky, J. Wheat VRN1, FUL2 and FUL3 play critical and redundant roles in spikelet development and spike determinacy. *Development* **2019**, *146*, dev175398. [CrossRef]
41. Woods, D.P.; McKeown, M.A.; Dong, Y.; Preston, J.C.; Amasino, R.M. Evolution of VRN2/Ghd7-like genes in vernalization-mediated repression of grass flowering. *Plant Physiol.* **2016**, *170*, 2124. [CrossRef]
42. Hori, K.; Shenton, M. Recent advances in molecular research in rice: Agronomically important traits. *Int. J. Mol. Sci.* **2020**, *21*, 5945. [CrossRef]
43. Fukuoka, S.; Nonoue, Y.; Yano, M. Germplasm enhancement by developing advanced plant materials from diverse rice accessions. *Breed. Sci.* **2010**, *60*, 509–517. [CrossRef]
44. Takeuchi, Y. Developing isogenic lines of Japanese rice cultivar "Koshihikari" with early and late heading. *JARQ* **2007**, *45*, 15–22. [CrossRef]
45. Sakai, H.; Lee, S.S.; Tanaka, T.; Numa, H.; Kim, J.; Kawahara, Y.; Wakimoto, H.; Yang, C.; Iwamoto, M.; Abe, T.; et al. Rice annotation project database (RAP-DB): An integrative and interactive database for rice genomics. *Plant Cell Physiol.* **2013**, *54*, e6. [CrossRef]
46. Liu, L.; Mei, Q.; Yu, Z.; Sun, T.; Zhang, Z.; Chen, M. An integrative bioinformatics framework for genome-scale multiple level network reconstruction of rice. *J. Integr. Bioinform.* **2013**, *10*, 94–102. [CrossRef]
47. Hayashi, Y.; Toritani, H.; Goto, S.; Yokozawa, M.; Seino, H. Climatic records of the National Institute of Agro-Environmental Sciences (1990–1996). *Misc. Publ. Natl. Inst. Agro-Environ. Sci.* **1998**, *23*, 1–168.

International Journal of
Molecular Sciences

MDPI

Article

New Insights into the Transcriptional Regulation of Genes Involved in the Nitrogen Use Efficiency under Potassium Chlorate in Rice (*Oryza sativa* L.)

Nkulu Rolly Kabange [1], So-Yeon Park [1], Ji-Yun Lee [1], Dongjin Shin [1], So-Myeong Lee [1], Youngho Kwon [1], Jin-Kyung Cha [1], Jun-Hyeon Cho [1], Dang Van Duyen [2], Jong-Min Ko [1] and Jong-Hee Lee [1,*]

1 Department of Southern Area Crop Science, National Institute of Crop Science, RDA, Miryang 50424, Korea; rollykabange@korea.kr (N.R.K.); f55261788@korea.kr (S.-Y.P.); minitia@korea.kr (J.-Y.L.); jacob1223@korea.kr (D.S.); olivetti90@korea.kr (S.-M.L.); kwon6344@korea.kr (Y.K.); jknzz5@korea.kr (J.-K.C.); hy4779@korea.kr (J.-H.C.); kojmin@korea.kr (J.-M.K.)

2 Molecular Biology Department, Agricultural Genetic Institute, Hanoi 11917, Vietnam; dangvanduyen79@gmail.com

* Correspondence: ccriljh@korea.kr; Tel.: +82-53-350-1168

Abstract: Potassium chlorate ($KClO_3$) has been widely used to evaluate the divergence in nitrogen use efficiency (NUE) between *indica* and *japonica* rice subspecies. This study investigated the transcriptional regulation of major genes involved in the NUE in rice treated with $KClO_3$, which acts as an inhibitor of the reducing activity of nitrate reductase (NR) in higher plants. A set of two $KClO_3$ sensitive nitrate reductase (NR) and two nitrate transporter (NRT) introgression rice lines (BC2F7), carrying the *indica* alleles of NR or NRT, derived from a cross between Saeilmi (*japonica*, P1) and Milyang23 (*indica*, P2), were exposed to $KClO_3$ at the seedling stage. The phenotypic responses were recorded 7 days after treatment, and samples for gene expression, physiological, and biochemical analyses were collected at 0 h (control) and 3 h after $KClO_3$ application. The results revealed that Saeilmi (P1, *japonica*) and Milyang23 (P2, *indica*) showed distinctive phenotypic responses. In addition, the expression of *OsNR2* was differentially regulated between the roots, stem, and leaf tissues, and between introgression lines. When expressed in the roots, *OsNR2* was downregulated in all introgression lines. However, in the stem and leaves, *OsNR2* was upregulated in the NR introgression lines, but downregulation in the NRT introgression lines. In the same way, the expression patterns of *OsNIA1* and *OsNIA2* in the roots, stem, and leaves indicated a differential transcriptional regulation by $KClO_3$, with *OsNIA2* prevailing over *OsNIA1* in the roots. Under the same conditions, the activity of NR was inhibited in the roots and differentially regulated in the stem and leaf tissues. Furthermore, the transcriptional divergence of *OsAMT1.3* and *OsAMT2.3*, *OsGLU1* and *OsGLU2*, between NR and NRT, coupled with the NR activity pattern in the roots, would indicate the prevalence of nitrate (NO_3^-) transport over ammonium (NH_4^+) transport. Moreover, the induction of catalase (CAT) and polyphenol oxidase (PPO) enzyme activities in Saeilmi (P1, $KClO_3$ resistant), and the decrease in Milyang23 (P2, $KClO_3$ sensitive), coupled with the malondialdehyde (MDA) content, indicated the extent of the oxidative stress, and the induction of the adaptive response mechanism, tending to maintain a balanced reduction–oxidation state in response to $KClO_3$. The changes in the chloroplast pigments and proline content propose these compounds as emerging biomarkers for assessing the overall plant health status. These results suggest that the inhibitory potential of $KClO_3$ on the reduction activity of the nitrate reductase (NR), as well as that of the genes encoding the nitrate and ammonium transporters, and glutamate synthase are tissue-specific, which may differentially affect the transport and assimilation of nitrate or ammonium in rice.

Keywords: nitrogen use efficiency; transcriptional regulation; nitrate reductase; nitrate transporter; glutamate synthase; potassium chlorate; rice

Citation: Kabange, N.R.; Park, S.-Y.; Lee, J.-Y.; Shin, D.; Lee, S.-M.; Kwon, Y.; Cha, J.-K.; Cho, J.-H.; Duyen, D.V.; Ko, J.-M.; et al. New Insights into the Transcriptional Regulation of Genes Involved in the Nitrogen Use Efficiency under Potassium Chlorate in Rice (*Oryza sativa* L.). *Int. J. Mol. Sci.* **2021**, *22*, 2192. https://doi.org/10.3390/ijms22042192

Academic Editor: Kiyosumi Hori

Received: 28 January 2021
Accepted: 17 February 2021
Published: 22 February 2021

1. Introduction

Nitrogen is an essential macronutrient that plays an important role in the growth and development of plants [1]. In soil, about 95–99% of the potentially available nitrogen exists in organic form (plant or animal residues), in the soil organic matter, and in living soil organisms [2]. Most of the nitrogen available to plants is in the inorganic form, such as ammonium (NH_4^+) and nitrate (NO_3^-) ions [3,4], also called mineral nitrogen [5], while a very small amount of organic compounds, such as urea, may be available to the plant. NH_4^+ ions have a high affinity for binding to the negatively charged soil cation exchange complex (CEC), and act much like other cations in the soil. In contrast, because soil carries negative charges, NO_3^- ions do not bind to soil particles, but are dissolved in the soil water, or precipitate in the form of soluble salts under dry conditions.

The nitrogen present in the soil that can be used by plants has two major sources: nitrogen-containing minerals, and the vast storehouse of nitrogen in the atmosphere. The biological conversion of the atmospheric nitrogen gas (N_2) to ammonia (NH_3^+) is exclusively performed by bacterial and archaeal species. It is said that biological nitrogen fixation is specific-dependent, and not limited to particular genera. Therefore, the detection of N_2 fixation is perceived as a complex task. For instance, in many legumes, such as legume symbionts, the fixation of nitrogen is strictly symbiotic [6].

The primary step of nitrogen acquisition by roots is the active transport across the plasma membrane of root epidermal and cortical cells. In rice, several members of nitrate transporter gene families (NPFs) have been identified, including 80 NPFs (NRT1/PRTs: NRT1, low-affinity nitrate transporter; PRT, di/tripeptide transporter), 5 NRT2s, 2 NAR2s members, and ammonium (NH_4^+) transporters (AMTs), and are reported to be involved in nitrogen use efficiency (NUE) [7]. Of this number, a few have been cloned and characterized [8–17]. Nitrogen assimilation is the formation of organic nitrogen compounds like amino acids from inorganic nitrogen compounds present in the environment. Organisms like fungi, certain bacteria, and the majority of plants that cannot fix nitrogen gas (N_2) depend on the ability to assimilate nitrate or ammonia for their needs. Other organisms, like animals, depend entirely on organic nitrogen from their food [18].

Plant growth, development, and productivity require the permanent availability of nutrients, and the plant nutrient needs increase with the growth stage. Nitrogen, a key macronutrient is required in the process, among others. However, external fluctuations in the supply of nitrogen to plants can affect its uptake, and lead to activation of various regulatory networks in order to optimize N uptake and utilization [19]. During these events, nitrate reductase (NR) and nitrite reductase (NiR) convert the exogenous nitrate to ammonium (NH_4^+). In the process, NH_4^+ is further assimilated by glutamine synthase and glutamate synthase into amino acids [18]. In addition, it was earlier reported that the activity of a large number of enzymes involved in nitrogen assimilation decreases under long-term abiotic stress conditions [20], while short-term stress increases others [21]. In the chloroplasts [22], glutamine synthase incorporates this ammonia as the amide group of glutamine, using glutamate as a substrate. Glutamate synthase (Fd-GOGAT and NADH-GOGAT) transfers the amide group onto a 2-oxoglutarate molecule producing two glutamates [23].

Nitrogen is also vital because it is the major component of chlorophyll molecules, the key compound by which plants use the energy from sunlight to yield sugars from water and carbon dioxide during photosynthesis.

In our recent study, a novel quantitative trait locus (QTL) for chlorate resistance was identified, and candidate genes were proposed to be associated with chlorate resistance in rice [24]. A study conducted by Gao and his colleagues [25] supported that the *indica* allele of nitrate reductase (*OsNR2*) is the major component for the nitrogen biological cycle, and confers a high nitrogen use efficiency (NUE) compared to the *japonica* allele [26]. In the same way, Duan et al. [27] indicated that the *indica* allele of NR or NRT confers a high NUE, and a mutation in the NRT1.1B resulted in an impaired NUE. One of the major challenges in crop production and plant biology research is how to maintain a balanced crop productivity and yield, while reducing the application of nitrogen fertilizers and utilization by plants.

Several reports have supported the use of potassium chlorate ($KClO_3$) as an effective and reliable strategy to investigate the nitrogen metabolism in higher plants, through the monitoring of nitrate uptake, transport, and assimilation, and knowing that nitrate is the main source of nitrogen. Chlorate (ClO_3^-) is a substrate for the nitrate reductase (NR) enzyme that reduces ClO_3^- to the toxic chlorite, leading to a quick degradation of NR [28–30]. In wheat [31], ClO_3^- resistance was used as a means of investigating the effect of NR antisense gene.

Potassium chlorate has been shown to induce stress in plants [24,32,33]. Generally, when plants are exposed to an environmental stimulus, they generate reactive oxygen species (ROS) [34,35] and reactive nitrogen species (RNS) [36]. Over accumulation of ROS or RNS may lead to oxidative damage [37] and lipid peroxidation [38], among other effects.

Therefore, this study aimed at investigating the transcriptional regulation of previously characterized genes reported to play important roles in nitrogen uptake, transport, assimilation, and remobilization; herein referred to as nitrogen use efficiency in rice. For this purpose, a set of nitrate reductase (NR) and nitrate transporter (NRT) introgression rice lines (BC2F7) and their related parental lines were exposed to potassium chlorate ($KClO_3$) at seedling stage, and the phenotypic responses were evaluated in different rice tissues. Additionally, changes in the physiological properties of the parental and NR or NRT introgression rice lines were assessed in response to $KClO_3$. Furthermore, the activity of key antioxidant enzymes and the change in the activity of the nitrate reductase (NR), as well as the extent of cell membrane degradation, were assessed under the same conditions.

2. Results

2.1. Distinctive Phenotypic Response between Parental Lines, and Identification of Introgression Lines

Initially, a population of 420 rice lines was screened through genotyping in order to identify NR or NRT introgression rice lines using insertion/deletion (InDel) markers (OsNR-IND2194 for nitrate reductase, and OsNRT-M10-22 for nitrate transporter). The results revealed that 26 lines and 59 lines carried the *indica* alleles of NR and NRT, respectively (Figure S2). These introgression lines amplified the expected band size of 200 bp for the *indica* allele of nitrate reductase, and 165 bp specific to the *indica* allele of nitrate transporter, respectively, on chromosomes 2 and 10, as reported earlier [24]. The *indica* alleles of the NR and NRT were previously suggested to explain the differences in the nitrogen use efficiency between *indica* and *japonica* rice varieties. Based on this evidence, the selected NR and NRT introgression lines (BC2F7) were exposed to 0.05% $KClO_3$ for 7 days. Of this number, the introgression rice lines that exhibited a high sensitivity soon after $KClO_3$ was applied (Figure S3), were used for physiological, biochemical, and molecular analyses.

Our data indicate that about 76.9% and 62.7% of NR and NRT introgression rice lines, showed a reduction in shoot growth, respectively, in response to $KClO_3$ (Figure 1A). Under the same conditions, about 46.2% and 61% of the NR and NRT introgression lines recorded a decrease in root growth, respectively (Figure 1B). In the same way, 96.2% of the NR introgression lines showed a highly sensitive phenotype under $KClO_3$, while for the NRT introgression lines, about 79.7% were $KClO_3$ sensitive. The parental lines, Saeilmi and Milyang23, showed distinctive $KClO_3$ responses. An increase in shoot growth was observed in Saeilmi (about 7.1%) and Milyang23 (about 0.1%) (Figures 1A and 2I,K). Saeilmi exhibited a significant increase in root growth (about 32%), but Milyang23 showed a significant decrease in root growth (17.6%) under the same conditions (Figures 1B and 2J,K).

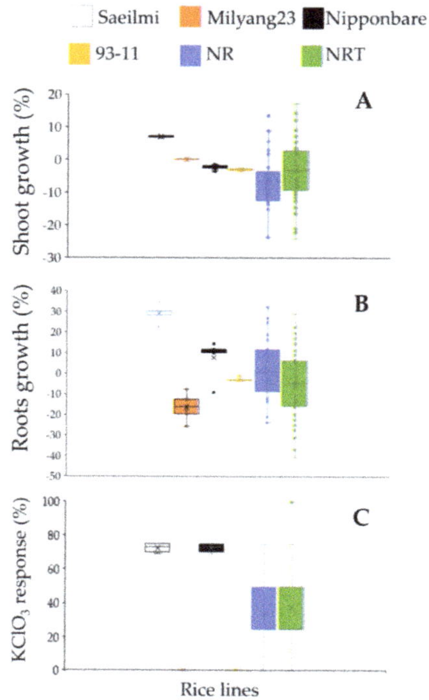

Figure 1. Genotype–phenotype correlation of the nitrate reductase and nitrate transporter introgression lines and parental line in response to potassium chlorate ($KClO_3$). (**A**) The box plots display the shoot growth patterns of the Saeilmi (P1, *japonica*), Milyang23 (P2, *indica*), Nipponbare (typical *japonica* cultivar), 93-11 (typical *indica* cultivar), nitrate reductase (NR, n = 26), and nitrate transporter (NRT, n = 59) introgression lines. (**B**) Box plots showing the root growth patterns, and (**C**) the $KClO_3$ response.

In addition, panels A and B of Figure 2 revealed a normal distribution of shoot growth for both NR and NRT introgression lines, while the root growth of the same rice lines showed a positive skewness (Figure 2C) and a normal distribution (Figure 2D). However, the recorded frequency distribution of the chlorate response indicated a positive skewness from both the NR and NRT introgression lines (Figure 2E,F). From another perspective, the data in panel G of Figure 2 groups Saeilmi (P1) and Nipponbare (typical *japonica*) in close clusters (right side of the figure) with regard to their similarity in phenotypic response towards $KClO_3$ treatment, while Milyang23 (P2) and 93-11 (typical *indica*) were assigned in the same cluster (left side of the figure). Moreover, the results of the principal component analysis (PCA) supported that root growth inhibition (RI) and chlorate were negatively correlated, whereas shoot growth inhibition positively correlated with the chlorate response of the NR and NRT introgression rice lines (Figure 2H). Principal component 1 (PC1) and 2 (PC2) explained 40.6% and 32.5% of the proportion of variance of the observed phenotypes, respectively, resulting in a cumulative proportion of 73.1%.

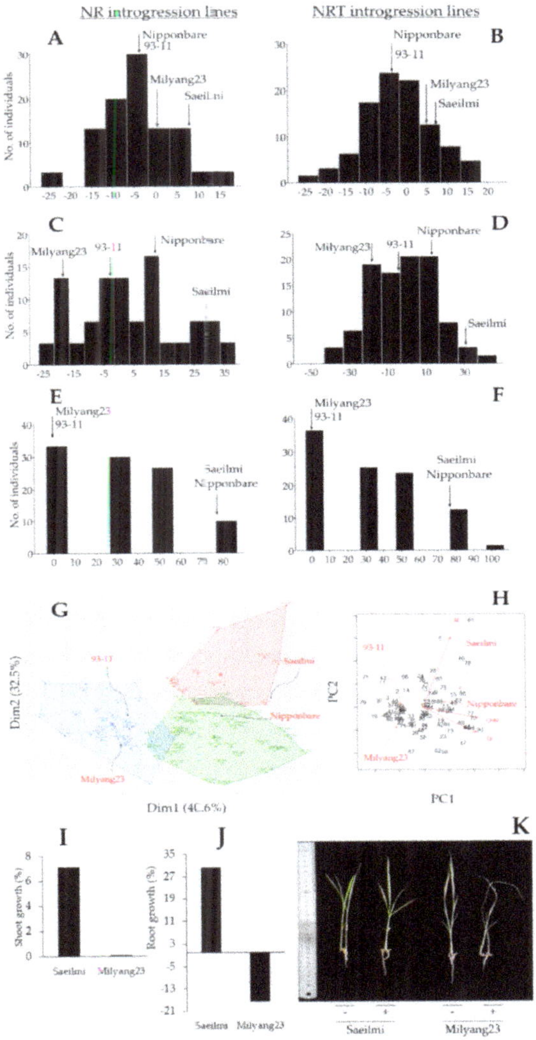

Figure 2. Frequency distribution, clustering, and principal component analysis results. Frequency distribution of the shoot growth of the nitrate reductase (NR, $n = 26$) (**A**), and nitrate transporter (NRT, $n = 59$) (**B**) introgression rice lines showing a normal distribution, (**C**) frequency distribution of the roots growth of NR, and (**D**) NRT introgression lines showing a positive skewness and a normal distribution, respectively. (**E**) frequency distribution of the $KClO_3$ response of NR, (**F**) NRT introgression rice lines showing a positive skewness, (**G**) Clusters showing the distinctive phenotypic response between Saeilmi (P1) and Milyang23 (P2), and that of the NR or NRT introgression lines, and (**H**) 2-D principal component analysis (PCA) indicating the correlation between traits, and (**I–K**) shoot and root growth pattern of Saeilmi (P1) and Milyang23 (P2) in response to $KClO_3$.

2.2. Exogenous Application of Potassium Chlorate Differentially Controlled the Activity of Nitrate Reductase in a Tissue-Specific Dependent Manner

Chlorates have been reported to have an inhibitory effect on the reducing activity of the nitrate reductase of nitrate (NO_3^-) to nitrite (NO_2^-). As expected, the activity of nitrate reductase (NR) was significantly suppressed by potassium chlorate ($KClO_3$) in the roots of Saeilmi (P1), Milyang23 (P2), and 93-11 (typical *indica*), as well as in both the NRT introgression lines (NRT39 and NRT383), while in NR24 and NR144, a significant increase and a non-significant change were observed, respectively (Figure 3A). In the stem, the NR activity was suppressed in Saeilmi and Milyang23 soon after $KClO_3$ was supplied, but significantly induced in 93-11 and in all the NR and NRT introgression lines (Figure 3B). Whereas, in the leaf tissues, Saeilmi, Milyang23, and 93-11 showed a similar increasing pattern of NR activity, but NR and NRT introgression lines differentially activated the NR enzyme under the same conditions (Figure 3C).

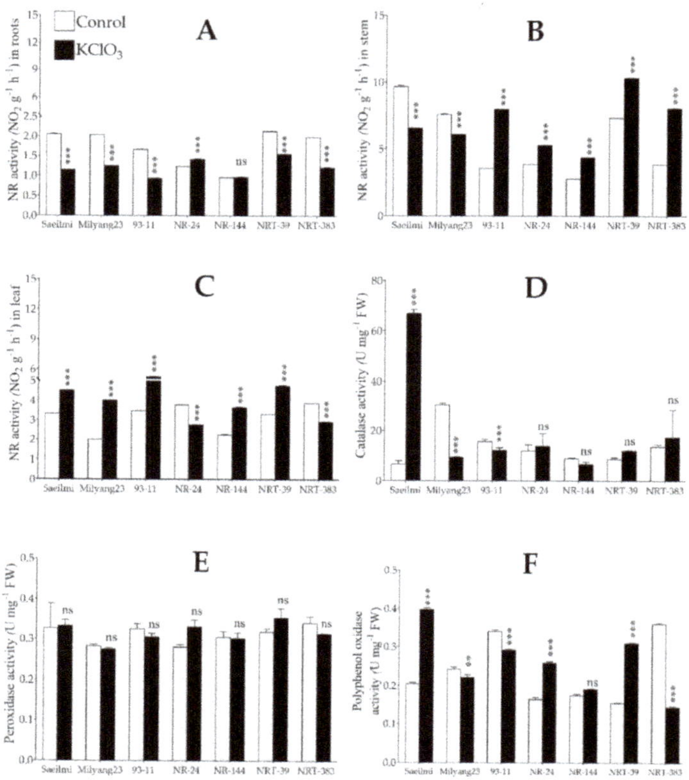

Figure 3. Changes in the activity of nitrate reductase (NR) and antioxidant enzymes. (**A–C**) Tissue-specific activity of the nitrate reductase (NR) enzyme in response to 0.5% potassium chlorate ($KClO_3$) 3 h after treatment, (**D**) catalase (CAT) activity, (**E**) peroxidase (POD) activity, and (**F**) polyphenol oxidase (PPO) activity in roots of rice seedlings exposed to $KClO_3$. White bars are controls, while black bars are $KClO_3$ treated seedlings. Bars are mean values of triplicates ± SE. *** $p < 0.001$, ** $p < 0.01$, *ns* non-significant.

Upon stress induction, plants initiate an intrinsic adaptive response mechanism to combat the stress. This includes the activation of various antioxidant systems, tending to lower the over accumulation of reactive oxygen species (ROS) and alleviate the oxidative stress. Our data show that catalase (CAT) activity significantly increased in Saeilmi (P1, $KClO_3$ resistant), but decreased in Milyang23 (P2, $KClO_3$ sensitive) and 93-11 (typical *indica* cultivar). However, no significant change in CAT activity was observed in all NR or NRT introgression rice lines (Figure 3D). Other enzymatic antioxidants that have been shown to be induced by environmental cues are peroxidase (POD) and polyphenol oxidase (PPO). Here, we observed that POD activity did not change significantly in all tested rice lines upon $KClO_3$ treatment (Figure 3E); however, PPO activity significantly increased in the

Saeilmi (P1) and decreased in Milyang23 (P2) and 93-11 lines, but increased in the NR and NRT introgression lines, except in NRT383 (Figure 3F).

2.3. Increased Chlorophyll and Carotenoids Content in Response to Potassium Chlorate

Under normal growth conditions, the chloroplast pigments, chlorophyll (Chl) and carotenoids have been shown to play a crucial role in the acquisition and supply of the energy required by plants to complete their life cycle [39]. Several studies have reported a change in the chlorophyll and carotenoids content in response to environmental stimuli [40]. Our findings show that the Chl *a* (Figure 4A), Chl *b* (Figure 4B), and total chlorophyll (Figure 5C), as well as the carotenoids relative to the chlorophyll (C*x* + *c*, Figure 4D), significantly increased in all the tested rice lines in response to $KClO_3$ application. Likewise, the pheophytin (Figure 4E–G), known as a derived molecule of chlorophyll lacking the magnesium at the reaction center, as well as the carotenoids relative to the pheophytin (Cph *x* + *c*, Figure 4H), showed a similar accumulation pattern to that observed for the Chl, but much more abundant compared to Chl.

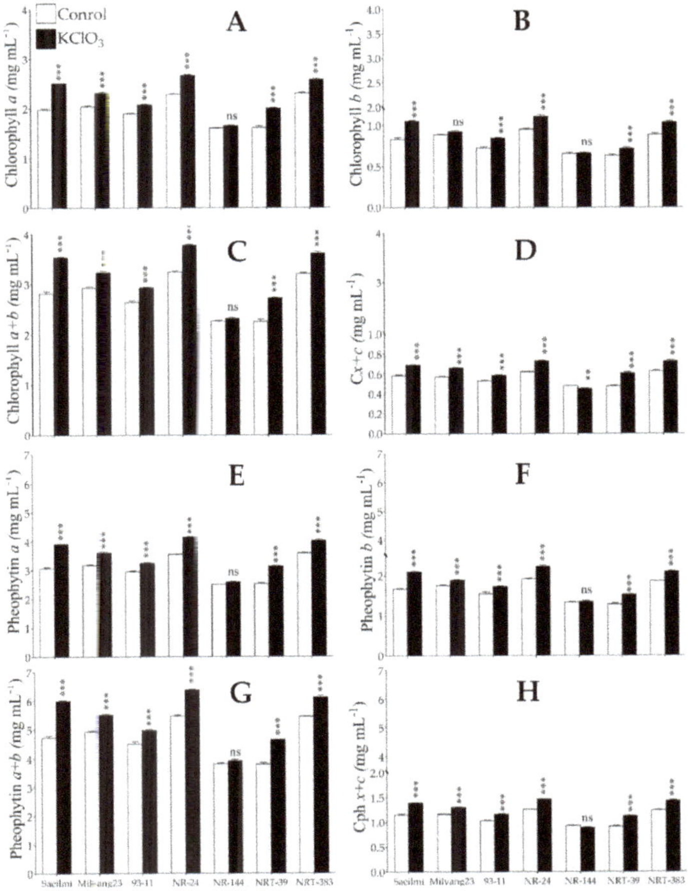

Figure 4. Pattern of chloroplast pigment content under potassium chlorate treatment. (**A–D**) Chlorophyll *a*, *b*, *a* + *b*, and carotenoids relative to chlorophyll (C*x* + *c*), respectively, (**E–H**) Pheophytin *a*, *b*, *a* + *b*, and carotenoid relative to pheophytin (C*ph x* + *c*), respectively. Bars are mean values of triplicates ± SE. *** $p < 0.001$, ** $p < 0.01$, *ns* non-significant.

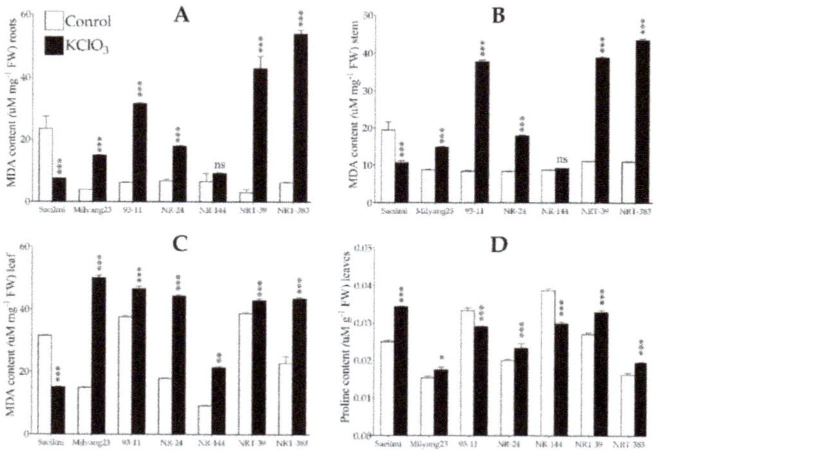

Figure 5. Lipid peroxidation and proline change in proline content. (**A**) Lipid peroxidation by malondialdehyde (MDA) content in the roots, (**B**) stem, and (**C**) leaf tissues in response to potassium chlorate ($KClO_3$) treatment, and (**D**) pattern of proline accumulation under $KClO_3$ treatment. Bars are mean values of triplicates ± SE. *** $p < 0.001$, ** $p < 0.01$, * $p < 0.05$, *ns* non-significant.

2.4. Potassium Chlorate Treatment Caused Lipid Peroxidation and Changes in Proline Content

Under stressful conditions, particularly caused by an environmental stimulus, various free radicals identified as reactive oxygen species (ROS) are generated. Over accumulation of ROS induces oxidative stress that may result in oxidative damage. In plants, malondialdehyde (MDA) is commonly used as a biomarker to estimate the lipid peroxidation due to damage to the cell membrane integrity. The results indicated that MDA content significantly decreased in the roots of Saeilmi (P1, chlorate resistant), while a significant increase was recorded in Milyang23 (P2, $KClO_3$ sensitive), 93-11 (typical *indica*), and all the NR and NRT introgression rice lines in response to $KClO_3$ (Figure 5A). A similar pattern of NR activity was observed in the stem (Figure 5B) and the leaf tissues (Figure 5C). Furthermore, proline has been reported to accumulate under abiotic stress in plants, as part of the adaptive response mechanism towards stress tolerance. Here, we observed a significant increase in the proline content in Saeilmi and NR24, as well as in NRT39 and NRT383, soon after $KClO_3$ was applied (Figure 6D). Under the same conditions, 93-11 and NR144 showed a significant reduction in proline content.

2.5. Potassium Chlorate Differentially Regulated the Expression of Genes Involved in Nitrogen Uptake, Transport, and Assimilation in Roots, Stem, and Leaf Tissues

Four $KClO_3$ sensitive NR (NR24, NR144) and NRT (NRT39 and NRT383) introgression rice lines and the parental lines, as well as 93-11 (the typical *indica* and $KClO_3$ sensitive), were used to investigate the transcriptional regulation of key genes involved in nitrogen uptake, transport, and assimilation, herein referred to as nitrogen use efficiency (NUE), in different plant organs (roots, stem, and leaf) under potassium chlorate ($KClO_3$) treatment. The relative expression levels of all analyzed genes were normalized to those of the housekeeping gene (*OsActin1*), and Milyang23 (P2, *indica*) was used to assess the statistical significance of the transcripts accumulation of the target genes in different rice lines. Our data showed that *OsNR2* was upregulated in the roots, stem, and leaf of Milyang23 (P2, about 1.5, 2.2, and 2.4-fold changes, respectively), while in Saeilmi (P1), a non-significant change was observed in the roots and leaf tissues (Figure 6A,C), but upregulated in the stem (2.2-fold change) (Figure 6B). When expressed in the NR introgression rice lines, the *OsNR2* was shown to be downregulated in the roots; while being upregulated in the stem and leaves. However, the transcripts accumulation of *OsNR2* was significantly decreased in all tested plant organs in both NRT39 and NRT383 rice lines, but was upregulated in 93-11 in the roots, and downregulated in the stem and leaves.

After being uptaken, nitrogen (N) is transported into the cell through the cell membrane in two major forms, nitrate (NO_3^-) and ammonium (NH_4^+). N transport in the form of nitrate, also known as the major form in which nitrogen is carried, is achieved through the combinational action of different enzymes, of which *OsNRT1.1B* has been suggested to play a preponderant role. Here, the expression of *OsNRT1.1B* was downregulated in the roots of Milyang23 (about 1-fold change), 93-11 (0.4-fold change), and NR24 (0.5-fold change), as well as in both NRT39 and NRT383 lines (0.4-fold change) (Figure 6D). A similar transcriptional pattern was observed when *OsNRT1.1B* was expressed in the stem, but in Saeilmi no significant change was recorded between the $KClO_3$ treated and control plants, with an exception being in the leaf tissues, where an increase in the transcripts accumulation was recorded in Saeilmi (0.7-fold change upregulated), Milyang23 (1.6-fold change upregulated), and 93-11 (0.7-fold change), as well as in the NR and NRT introgression lines (Figure 6E,F). However, *OsNRT1.1B* expression was downregulated in the NRT39 and NRT383 lines.

N transport in the form of NH_4^+ is facilitated by a group of well-identified NH_4^+ transporters, including *OsAMT1.3* and *OsAMT2.3* genes. Our data show that the expression of *OsAMT1.3* was downregulated in the roots of Saeilmi (P1, about 1-fold change), while being upregulated in Milyang23 (P2, about a 1.4-fold change) and 93-11 (1.2-fold change). However, *OsAMT1.3* was upregulated in the NR (1.0 and 1.1-fold change) and downregulated in the NRT (0.4 and 0.5-fold change) introgression lines (Figure 6G). Similarly, when expressed in the stem, the data in panel H of Figure 3 indicates that *OsAMT1.3* was downregulated in Saeilmi (0.4-fold change) and 93-11 (0.6-fold change), while showing an increased transcript accumulation pattern in Milyang23 (2.4-fold change) and NR (1.3 and 1.6-fold change) introgression lines, but being downregulated in the NRT (0.6 and 0.4-fold change) introgression lines (Figure 6H). Meanwhile, *OsAMT1.3* transcripts levels decreased in the leaf tissues of Saeilmi (about 1-fold change), while showing an increased pattern in 93-11 (about 1-fold change), as well as in all NR (4.3 and 2.3-fold change) and NRT (about 1.0 and 1.7-fold change) introgression lines, except NRT39 (Figure 6I), in response to $KClO_3$.

The other ammonium transporter of which the transcript accumulation was measured (*OsAMT2.3*) was shown to be induced by $KClO_3$ in the roots of the Saeilmi (1.1-fold change), Milyang23 (1.3-fold change), and 93-11 (1.7-fold change) cultivars. However, an opposite expression pattern was recorded in the NR (about 1-fold downregulated and 1.2-fold upregulated) and NRT (0.6 and 0.5-fold change, downregulated) introgression lines (Figure 6J). Under the same conditions, the transcripts accumulation of *OsAMT2.3* in the stem increased in Saeilmi (2.1-fold change), Milyang23 (1.3-fold change), and 93-11 (1.7-fold change), as well as in both NR introgression lines (1.7 and 1.4-fold change), but a decrease was observed in the NRT (0.7 and 0.8-fold change) introgression lines (Figure 6K). When expressed in the leaf tissues, *OsAMT2.3* exhibited a similar transcript accumulation pattern to that observed in the stem (Figure 6L).

The N that is uptaken by the roots, and transported through the vessels into to the leaf tissues via the stem is expected to be assimilated in the form of amino acids. In higher plants, glutamate synthase has been reported as being involved in the initial steps of the N assimilation. Therefore, in order to investigate the possible effect of $KClO_3$ on the N assimilation events, we monitored the transcript accumulation of glutamate synthase (*OsGLU1* and *OsGLU2*) genes. Our data indicate that *OsGLU1* was downregulated in the roots and stem of Saeilmi (P1) and Milyang23 (P2), as well as in NR24, NRT39, and NRT383 (Figure 6M,N). Whereas, a non-significant change in *OsGLU1* expression was observed (Figure 6O). In the same way, *OsGLU2* was downregulated in both the roots (Figure 6P) and the stem (Figure 6Q) of Saeilmi, but a slight increase was recorded in Milyang23 and 93-11, while a downregulation was recorded in the NR and NRT introgression lines. In the stem, *OsGLU2* transcript accumulation increased in 93-11 and decreased in Milyang23, as well as in all NR and NRT introgression lines (Figure 6P,Q). Meanwhile, when expressed in the leaves, the expression of *OsGLU2* was upregulated in Saeilmi and in the NR introgression lines, while being downregulated in Milyang23, 93-11, and the NRT introgression lines (Figure 6R).

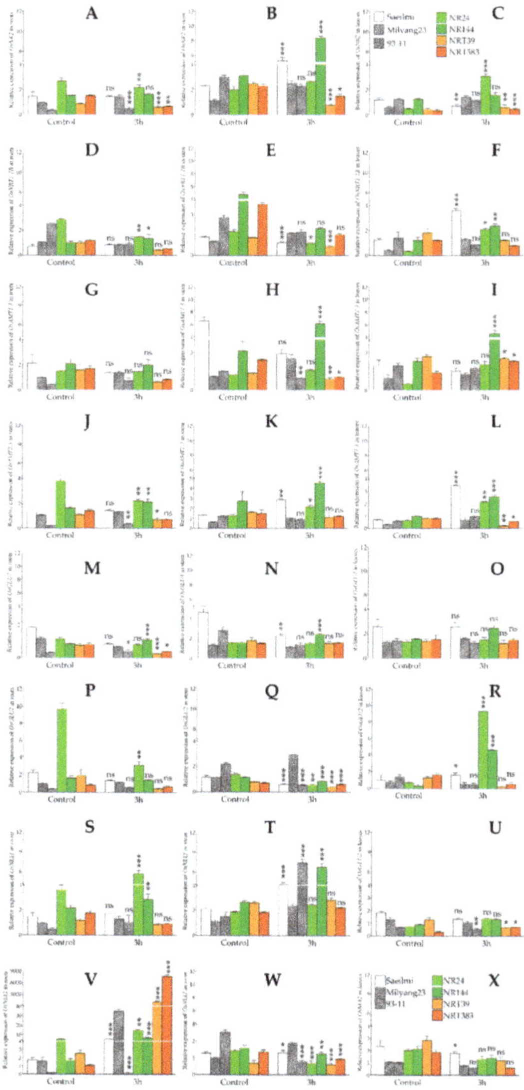

Figure 6. Transcriptional patterns of major nitrogen use efficiency (NUE) genes in different rice tissues under potassium chlorate ($KClO_3$) treatment. Transcriptional patterns of major NUE genes in different rice tissues under potassium chlorate treatment. (**A–C**) Transcriptional pattern of *OsNR2* in the roots, stem, and leaves of three week-old nitrate reductase (NR) and nitrate transporter (NRT) introgression rice lines in response to 0.05% potassium chlorate ($KClO_3$), soon after treatment. (**D–F**) Transcript accumulation of *OsNRT1.1B*, (**G–I**) *OsAMT1.3*, (**J–L**) *OsAMT2.3*, (**M–O**) *OsGLU1*, (**P–R**) *OsGLU2*, (**S–U**) *OsNIA1*, and (**V–X**) *OsNIA2* under the same conditions. Bars are mean values of triplicates ± SE. *** $p < 0.001$, ** $p < 0.01$, * $p < 0.05$, *ns* non-significant. Data were compared with the expression level of Milyang23 (P2, $KClO_3$ sensitive) for the statistical significance.

As part of the nitrogen (N) biological cycle, particularly during the denitrification process, where nitrate (NO_3^-) is reduced back to nitrous oxide (N_2O) and nitrogen (N_2), nitric oxide (NO) is released. For this reason, we investigated the changes in the transcripts accumulation of key NO biosynthetic genes, *OsNIA1* and *OsNIA2*, encoding a nitrate reductase enzyme, in response to $KClO_3$. Our findings revealed, on the one hand, that *OsNIA1* was upregulated in the roots of Saeilmi (1.2-fold change), Milyang23 (1.4-fold change), 93-11 (1.2-fold change), NR24, and NR144 (1.6 and 1.3-fold change, respectively), but downregulated by 0.7 and 0.5-fold change in NRT39 and NRT383, respectively (Figure 6S). Similarly, a significant upregulation of *OsNIA1* was recorded in the stems of both Saeilmi (P1) and Milyang23 (P2) (1.9 and 2.2-fold change, respectively), and the NR and NRT introgression lines (NR: 5.2 and 1.3-fold change; NRT: 1.1 and 1.2-fold change) (Figure 6T). However, *OsNIA1* transcript accumulation decreased in the leaf tissues of both P1 and P2 (0.4 and 0.3-fold change, respectively), and 93-11 (0.3-fold change), while being upregulated by about a 1-fold change in all the NR and NRT introgression lines (Figure 6U). On the other hand, *OsNIA2* transcript accumulation increased significantly in the roots of Saeilmi (3.2-fold change), and was much higher in Milyang23 (22.3-fold change) and in the NR (2.8 and 3.9-fold change) and NRT (424.8 and 6948.3-fold change) introgression lines (Figure 6V). The expression of *OsNIA2* was upregulated in the stem of Saeilmi (1.0-fold change), Milyang23 (1.9-fold change), but downregulated in 93-11 (0.3-fold change), as well as in NR (0.5 and 0.8-fold change) and NRT (0.8 and 0.7-fold change). Likewise, *OsNIA2* expression showed a similar downregulation pattern in the leaf tissues of all tested rice lines (Figure 6W,X).

3. Discussion

3.1. Potassium Chlorate Inhibits Shoot and Root Growth in a Cultivar Dependent Manner

Roots are important plant organs playing an essential role during plant nutrients and water uptake, as well as for anchoring the plant to the soil. Roots also serve as storage organs supporting the whole plants structure and fitness, among other things [41]. In the same way, the shoots (above ground part of the plant, including the stem or tillers and the leaves and grains) are important for the plant architecture, transport of nutrients and water, energy acquisition, and redistribution, and carry the target products of many food crops, such as rice. Nitrate (NO_3^-) and ammonium (NH_4^+) are widely known as the major forms of nitrogen (N) utilized by plants from the environment (soil) [42]. Under stressful conditions caused by environmental stimuli [43–45], roots are the first organs that sense the stress, and the availability of essential nutrients, including nitrogen (N), is affected, leading to impairment of root growth and/or shoot growth and productivity. Here, we recorded that about 46.2% and 61% of the NR or NRT introgression rice lines exhibited a reduction in their roots, and 76.9% and 62.7% showed a reduction in shoot growth, respectively, in response to $KClO_3$, suggesting a strong inhibitory effect of $KClO_3$ on the whole plant growth.

3.2. Potassium Chlorate Differentially Regulates Genes Involved in Nitrate Uptake, Transport, and Assimilation in a Tissue-Specific Dependent Manner

The *indica* allele of the nitrate reductase encoded by the *OsNR2* gene (chromosome 2) and the nitrate transport, *OsNRT1.1B* (chromosome 10) have been suggested to be key players in explaining the differences in the nitrate assimilation capacity and the nitrogen use efficiency (NUE) between *indica* and *japonica* rice subspecies [25]. According to the authors, this difference is conferred by allelic variation at the *OsNR2*, with the *indica* allele showing a high nitrate reductase (NR) activity compared to its *japonica* counterpart. Consequently, the *indica* allele was suggested to promote NO_3^- uptake, while interacting with the nitrate transporter encoding gene, *OsNRT1.1B*. However, NR activity has been reported to be inhibited by chlorates (ClO_3^-), which together with NO_3^- serve as substrates for the NR enzyme that reduces NO_3^- to the toxic chlorite [28–30,46]. Many studies have used ClO_3^- to isolate mutant plants that are defective in nitrate reduction [29], while others have suggested that when ClO_3^- is applied to plants, an increase in NR mRNA level is observed, but not the NR protein content [47]. Zhao and his colleagues [48] observed that a chlorate resistant mutant (lacking the NR activity) of rice exposed to ClO_3^- showed a similar level of NR activity to the wild type.

Here, the tissue-specific expression of the *OsNR2* gene in rice lines carrying the *indica* alleles of NR or NRT revealed a differential transcriptional regulation by $KClO_3$. When expressed in the roots, the expression of *OsNR2* was inhibited in all introgression lines, except in NRT144 line (Figure 6A), but its expression in the stem (Figure 6B) and the leaves (Figure 6C) was shown to be significantly upregulated in the NR introgression lines, while being downregulated in all the NRT introgression lines. Therefore, these results suggest that the previously reported inhibitory effect of ClO_3^- on the reducing activity of the nitrate reductase (NR) is tissue-specific, rather than systematic to all plant tissues and organs. This is also supported by the differential nitrate reductase activity, which was shown to vary between rice plants' tissues (roots, stem, and leaf tissues) (Figure 6A–C). In addition, the contrasting expression pattern of the *OsNR2* gene in the NRT introgression lines compared to that recorded in the NR introgression background suggests an antagonistic relationship, particularly in the stem and the leaf tissues. Furthermore, the differential transcriptional regulation of *OsNR2* in different plant tissues would imply that NR may also be involved in NO_3^- long-distance signaling and transport across the plant body.

Another set of genes known to code for nitrate reductase enzymes in plants are *NIA1* and *NIA2*. In rice, *OsNIA1* and *OsNIA2* have been shown to be induced by nitric oxide (NO), one of the compounds that is generated during the nitrogen metabolic process [49]. In *Arabidopsis*, mutations in the nitrate reductase structural genes *AtNIA1* and *AtNIA2* have been shown to result in an impairment in NO_3^- assimilation rate [50]. It could then be said that the recorded significant increase in the transcript accumulation of *OsNIA1* in the NR introgression rice lines background in all measured rice tissues, and the downregulation in the NRT introgression rice lines background in response to $KClO_3$ (Figure 6S–U), coupled with the upregulation or downregulation of *OsNIA2* expression in a tissue-dependent manner (Figure 6V–X), would suggest a possible role of NO in nitrogen use efficiency.

In the same way, *OsNRT1.1B*, known to function in NO_3^- uptake, transport, and signaling [51], exhibited a differential expression between the parental lines Saeilmi and Milyang23, as well as in different tissues of the NR or NRT introgression lines. Furthermore, the transcript accumulation of *OsNRT1.1B* was significantly lower in the NRT, compared to the NR, introgression line upon $KClO_3$ treatment, which would suggest that the NR enzyme prevails over the NRT in NO_3^- uptake as part of the NUE. From another perspective, *OsNRT1.1B* has been shown to be involved in the regulation of root microbiota to facilitate organic nitrogen mineralization in the soil, while mediating plant–microbe interactions [52]. *OsNRT1.1B* was also reported to transport N under both low and high NO_3^- levels, and had an essential single nucleotide polymorphism (SNP) between *indica* and *japonica* rice subspecies [53], where the *indica* variation was shown to contribute to the NUE divergence between rice *indica* and *japonica*, eventually by promoting NO_3^- uptake and translocation, and upregulation of NO_3^- response [17].

Likewise, nitrogen is also transported in the form of ammonium (NH_4^+). In the soil, NH_4^+ basically results from the mineralization of organic matter, and represents, besides NO_3^-, the quantitatively most important source of N for plant nutrition [54]. The authors supported that despite the low concentrations in soils, the uptake of NH_4^+ by the plant can be achieved at a very high rate, facilitated by the multiple transporters in the root plasma membrane. A recent report [55] suggested that, in addition to functioning in NH_4^+ uptake in the roots, *OsAMT1.3* may act as a signal sensor to regulate plant growth, carbon, and nitrogen use efficiency (NUE). Similarly, *OsAMT1.3* has been suggested as being involved in the adaptation ability of rice to low NH_4^+ supplies [56]. In the same way, *OsAMT2.3* was reported to be affected by N supply [57]. Therefore, the recorded upregulation of *OsAMT1.3* in the roots, stem, and leaves of NR introgression lines, and the downregulation in NRT introgression lines (Figure 6G–I) soon after $KClO_3$ was applied, coupled with the transcriptional patterns of *OsAMT2.3* under the same conditions (Figure 6J–L), would indicate an antagonistic relationship with NR, while suggesting a co-expression with NRT, in rice.

Glutamate synthase (GOGAT) is involved in the initial steps of nitrogen assimilation in plants. The GS/GOGAT cycle is widely known as the major route of NH_4^+ assimilation in higher plants [58]. Hence, the recorded transcriptional levels of Glutamate synthase 1 and 2 encoding genes (Figure 6M–O and Figure 2P–R) would imply that *OsGLU2* would prevail over *OsGLU1* in N assimilation.

3.3. Potassium Chlorate Differentially Regulates Antioxidant Enzymes Between Parental and NR or NRT Introgression Rice Lines

Generally, when plants are exposed to a changing environment, various ROS are generated [59,60]. To cope with the stress, plants activate an array of antioxidant systems as part of their adaptive response mechanism, which include catalase (CAT), peroxidase (POD), and polyphenol oxidase (PPO), in order to maintain a low level of ROS accumulation, and keep a balanced reduction–oxidation state within the cell [61,62]. However, over accumulation of ROS was shown to induce oxidative stress, which may result in oxidative damage, and may culminate in the induction of programmed cell death (PCD) [63,64]. Owing to the recorded increase in CAT and PPO activities in the *japonica* parental line (Saeilmi, chlorate resistant) and the decrease in the *indica* parental line (Milyang23, chlorate sensitive), coupled with the differential activity in the NR or NRT introgression lines, this study suggests that CAT and PPO play an important role in the initial adaptive response mechanism towards $KClO_3$ resistance.

During oxidative stress, plants accumulate ROS, which exacerbates oxidative damage [37]. In the process, malondialdehyde (MDA), commonly used as a marker to estimate lipid peroxidation in plant species, accumulates abundantly to the extent of the degradation of the cell membrane [38]. Therefore, the recorded MDA accumulation patterns in different plant organs (roots, stem, and leaf tissues) (Figure 5A–C) indicate the level of oxidative stress caused upon $KClO_3$ application, suggesting that the integrity of the cell membrane might have been affected as a result of oxidative damage.

3.4. Exogenous Application of Potassium Chlorate Triggers the Accumulation of Chloroplast Pigments

Chlorates (ClO_3^-) were earlier reported as potent inhibitors of the reducing activity of NR in plants, which may have a significant effect on N metabolism. As is well known, assimilation of the inorganic form of nitrogen (N) is as of fundamental importance to crop growth, as it is to crop productivity. It is said that N is the main plant mineral nutrient required for the production of chlorophyll (Chl), as well as other components of the plant's cells, such as proteins, nucleic acids, and amino acids [51]. As ClO_3^- is said to inhibit NR activity, which may impair NUE, and owing to the fact that N is essential for chlorophyll production, we were expecting to see a reduction in the chlorophyll content upon $KClO_3$ application. Rather, we recorded a significant increase in the chlorophyll *a* and *b* contents and their immediate precursors and degradation products, pheophytin [65] and carotenoids, in all tested rice lines (parental lines and BC2F7 NR or NRT introgression lines) (Figure 4A–H). In our recent studies, chlorophyll content was shown to be increased in response to abiotic stresses, such as drought stress [62] and salinity [66]. It is then thought that the observed Chl and pheophytin accumulation, as well as the carotenoids content, were a result of the oxidative stress caused by the application of $KClO_3$, which suggests these molecules as emerging biomarkers for assessing the overall plant health status under oxidative stress conditions, and in which proline would contribute as an osmoprotectant [67,68].

Therefore, from the perspective of providing a comprehensive view of the recorded morphological, physiological, and biochemical changes, as well as the molecular responses of the nitrate reductase (NR) or nitrate transporter (NRT) introgression rice lines (BC2F7) relative to their parental lines Saeilmi (P1, *japonica*) and Milyang23 (P2, *indica*), moving towards potassium chlorate treatment, in the roots, stem and the leaf tissues, a signaling model is proposed (Figure 7).

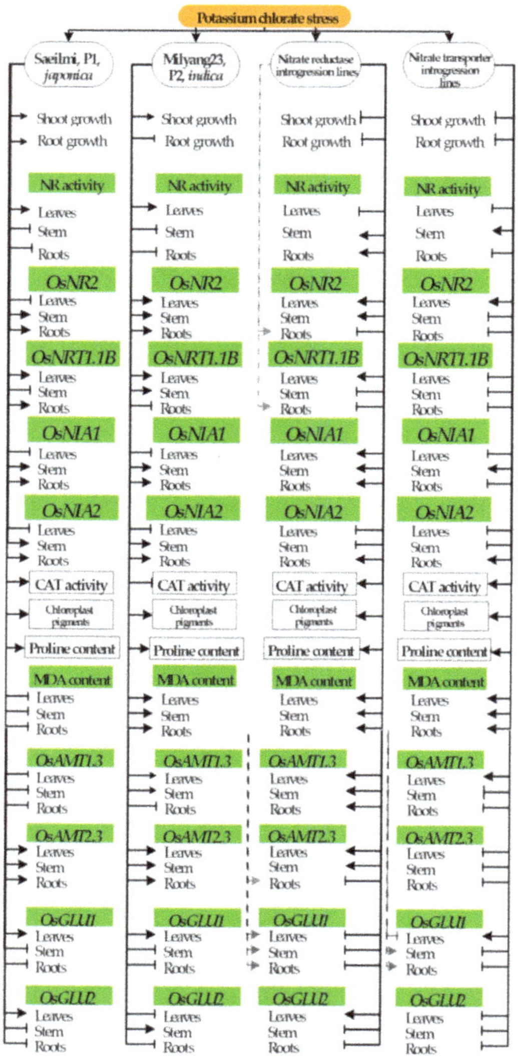

Figure 7. Signaling model summarizing the morphological, physio-biochemical, and molecular response of NR and NRT introgression rice lines towards potassium chlorate. The phenotypic responses, the physiological and biochemical changes, as well as the molecular responses of the parental lines Saeilmi (P1, *japonica*) and Milyang23 (P2, *indica*) and their derived introgression lines (nitrate reductase or nitrate transporter, BC2F7) are summarized in the above signaling model. Continuous lines with an arrow indicate upregulation (for gene expression) or induction/increase (for enzyme activity or accumulation of physiological components). Continuous lines with a perpendicular bar indicate downregulation or inhibition/decrease. Dotted lines with or without an arrow or a perpendicular bar indicate that one of the introgression lines differentially expressed a particular gene in the corresponding rice tissue. This signaling model was created using ConceptDraw Pro v.10.3.2.114 (CS Odessa Corp, San Jose, CA, USA).

4. Materials and Methods

4.1. Plant Materials and Potassium Chlorate Treatment

To perform the experiments, a total of seven rice lines, including two parental lines (Saeilmi: *japonica* and Milyang23, *indica*), the *indica* reference cultivar 93-11, and four introgression lines carrying the *indica* alleles of nitrate reductase (NR, two lines) or nitrate transporter (NRT, two lines) derived from a cross between Saeilmi and Milyang23, were used as genetic materials. Initially, cv. Nipponbare (typical *japonica* ssp.) and cv. 93-11 (typical *indica* ssp.) were used to identify an optimum potassium chlorate ($KClO_3$) concentration. For this purpose, the two cultivars were subjected to a gradient of concentrations of $KClO_3$ (0.01%, 0.025%, 0.05%, 0.07%, and 0.1%) for seven days at seedling stage. The control treatments were supplied with distilled water only. Based on the phenotypic responses, 0.05% $KClO_3$ was selected as the concentration to be used for downstream analyses. At 0.05% $KClO_3$, a significant difference of the shoot (Figure S1A) or root (Figure S1B) inhibition percentage between Nipponbare and 93-11 was obtained. Meanwhile, the gap in the biomass fresh weight (BFW) between Nipponbare and 93-11 was small at 0.05% $KClO_3$ level (Figure S1C). Therefore, 0.05% was used for downstream experiments.

Prior to germination, rice seeds were sterilized with 0.7% nitric acid (HNO_3) (CAS: 7697-37-2, Lot No. 2016B3902; Junsei Chemical Co. Ltd., Tokyo, Japan) overnight to break the dormancy [24], followed by incubation for 48 h at 27 °C to induce germination. Germinated seeds were grown in 50-well trays containing an enriched soil, and placed in a greenhouse until three-leaf stage. Seedlings with a uniform height (3-week old, six seedlings per treatment per rice line) were transferred into 50 mL falcon tubes prior to $KClO_3$ treatment. Then, the roots of seedlings were immerged into 5–10 mL of 0.05% potassium chlorate ($KClO_3$) (CAS: 3811-04-9, Lot No. BCBW5513; Sigma-Aldrich, St. Louis, MO, USA) solution (pH 5.6), and placed in a growth chamber under dark conditions for 7 days at ±25 °C [33]. The $KClO_3$ solution was replaced three days after the initial application by irrigation method. The control seedlings were supplemented with distilled water only. The phenotypic response was recorded 7 days after $KClO_3$ application. The chlorate resistance was calculated as the percentage of the ((total number of tested seedlings − number of dead seedlings in $KClO_3$)/total number of tested seedlings) × 100. However, shoot inhibition was estimated as the percentage of the ((shoot length under control (SLC)—shoot length under $KClO_3$ (SL_$KClO_3$)/SLC) × 100. The same formula was used to estimate the roots inhibition percentage. Samples (leaf, stem, and roots) for gene expression, biochemical, and physiological analyses were collected in triplicate at 0 h (untreated control) and 3 h after treatment, and immediately frozen in liquid nitrogen and kept in a −80 °C freezer for further processing.

4.2. Genomic DNA Extraction, Genotyping, and Molecular Marker Analysis

The genomic DNA was extracted from leaf samples using the previously described CTAB method with slight modifications [40]. Briefly, frozen leaf samples were crushed in 1.5 mL Eppendorf tubes (e-tubes). Then, 600 μL of 2× CTAB buffer (D2026, Lot D2618U12K; Biosesang, Seongnam-si, Korea) was added, and the mixture was vortexed and incubated for 30 min at 65 °C in a dry oven. A solution containing 500 μL of PCI (Phenol:Chloroform:Isoamylalcohol, 25:24:1, Batch No. 0888k0774; Sigma-Aldrich, St. Louis, MO, USA) was added, followed by gentle mixing by inversion. The tubes were centrifuged for 15 min at 13,000 rpm, and the supernatant was transferred to fresh e-tubes, followed by the addition of 500 μL of isopropanol (CAS: 67-63-0, Lot No. SHBC3600V; Sigma-Aldrich, St. Louis, MO, USA), mixing by inversion, incubation at −20 °C for 1 h, and centrifugation at 13,000 rpm for 7 min. The supernatant was removed and the pellets were washed with 70% ethanol (1 mL). Samples were centrifuged at 13,000 rpm for 2 min and ethanol was discarded, followed by drying at room temperature and re-suspension in 100 μL 1× TE buffer (Lot No. 0000278325; Promega, Madison, WI, USA).

A population consisting of 420 rice lines was genotyped in order to identify introgression lines carrying the *indica* alleles of the nitrate reductase (NR) or nitrate transporter

(NRT) genes through polymerase chain reaction (PCR) using OsNR-IND2194 and OsNRT-M10-22 insertion/deletion (InDel) markers. The reaction mixture (15 μL) consisted of 1.5 μL 10X reaction buffer, 0.8 μL 10 mM dNTP, 1 μL 10 pM primers (forward and reverse), 0.1 μL *Taq* polymerase, and adjusted to the final volume with nuclease-free water. A 3-step cycling reaction was performed including polymerase activation at 95 °C for 5 min, strand separation at 94 °C for 20 s, annealing at 56–59 °C for 30 s for 35 cycles, extension at 72 °C for 1 min/kb, and a final extension at 72 °C for 5 min. The amplicons were separated on 3% agarose gel electrophoresis, and the bands were visualized using a gel documentation system. The sequences of InDel primers can be found in Table S1.

4.3. Total RNA Extraction, cDNA Synthesis, and qPCR Analysis

Total RNA was extracted from leaf samples using the TaKaRa MiniBEST Universal Plant RNA Extraction Kit (TAKARA Bio Inc., Cat. No. 9769 v201309Da, Kusatsu, Japan) following the manufacturer's instructions. Briefly, frozen samples with liquid nitrogen were ground to fine powder, and 450 μL buffer RL (containing 50× dithiothreitol (DTT, 20 μL per 1 mL buffer RL) was added, and the mixture was pipetted up and down for few seconds until the lysate showed no precipitate, followed by centrifugation at 12,000 rpm for 5 min at 4 °C. The supernatant was transferred to fresh 1.5 mL Eppendorf tubes (e-tubes) (step 1). Then, 1 volume of ethanol (100%) was added to the mixture from step 1, followed by pipetting up and down to mix well, and 600 μL were transferred to a spin column with a 2 mL collection tube. The tubes were centrifuged at 12,000 rpm for 1 min, and the flow-through was discarded. Then, 500 μL buffer RWA was added, followed by centrifugation at 12,000 rpm for 30 s, and the flow-through was discarded. Next, 600 μL buffer RWB was added to the spin column, followed by centrifugation at 12,000 rpm for 30 sec (this step was repeated twice), and the flow-through was discarded. Then, empty spin columns with collection tubes were centrifuges for 2 min at 12,000 rpm, and the spin columns were placed on fresh 1.5 mL e-tubes, followed by the addition of 100 μL RNase free water. Samples were incubated at room temperature for 5 min, followed by centrifugation at 12,000 rpm for 2 min (the elution step was repeated twice).

For cDNA (complementary DNA) synthesis [69], 1 μg of RNA was used as a template and the ProtoScript® II First Strand cDNA Synthesis Kit (New England BioLabs Inc., NEB Labs, MA, USA) was employed, as described by the manufacturer. The cDNA was then used as a template for qPCR (quantitative polymerase chain reaction) to investigate the transcript accumulation of the selected genes (see Table S1).

For gene expression analysis, a reaction mixture comprising 10 μL Prime Q-Master Mix (with SYBR green I), 0.1× ROX (0.1 μL/50×) (GENETBIO Inc., Daejeon, Korea) along with 1 μL of template DNA and 10 pM of each forward and reverse primer in a final volume of 20 μL reaction. A no-template control (NTC) was used. A 3-step reaction including polymerase activation at 95 °C for 10 min, a denaturation at 95 °C for 20 s, annealing at 60 °C for 30 sec and extension at 72 °C for 30 sec was performed in a Real-time PCR machine (QuantStudio™ Design and Analysis Software v.1.3, Applied Biosystems, Thermo Fisher Scientific, Seoul, Korea). Total reaction cycles were 40 and the data were normalized with relative expression of rice Actin1. The list of the genes with their corresponding primers sequences used in the study is given in Table S1.

4.4. Proline Measurement Assay

Proline quantification was done following the colorimetric method described earlier [70]. Approximately 100 mg leaf samples were homogenized in 3% sulfosalicylic acid (5 μL × mg^{-1} FW) in Eppendorf tubes (e-tube). Then, the homogenate was centrifuged using a benchtop centrifuge at 13,000 rpm for 5 min. Then, 100 μL from the supernatant of the plant extract was added to the reaction mixture (100 μL of 3% sulfosalicylic acid, 200 μL glacial acetic acid, 200 μL acidic ninhydrin (1.25 g ninhydrin (1,2,3-indantrione monohydrate), 30 mL glacial acetic acid, 20 mL of 6 M orthophosphoric acid (H_3PO_4)), dissolved into double distilled water and stored at 4 °C). The mixture was incubated at

96 °C for 1 h and immediately cooled on ice to terminate the reaction. The samples were extracted with toluene by adding 1 mL toluene to the reaction mixture and vortexing for about 20 s, and then leave for 5 min on the bench in order to allow separation of the organic and water phases. Then the chromophore (upper phase colored light red) containing toluene was moved into the fresh cuvette and the absorbance was read at 520 nm using toluene as reference. The proline concentration was calculated on the fresh weight basis and expressed in μg g^{-1} FW.

4.5. Catalase, Polyphenol Oxidase, and Peroxidase Activity Assay

The activity of antioxidant enzymes was assayed using the spectrophotometric method as described by Elavarthi and Martin [71]. Briefly, 100 mg of leaves were ground to a fine powder using liquid nitrogen (N_2) and immediately homogenized in 50 mM phosphate buffer (pH = 7.5). The mixture was centrifuged for 10 min at 12,000 rpm at 4 °C after being kept on ice for about 10 min. The supernatant was then transferred to fresh 1.5 mL e-tubes. The homogenate was used as a crude enzyme source for catalase (CAT), peroxidase (POD), and polyphenol oxidase (PPO) activities.

Catalase activity was assayed as described earlier [49]. Briefly, a volume of 50 μL H_2O_2 (50 mM) was added to the crude enzyme extract, and the absorbance of the reaction mixture was measured at 240 nm wavelength after one minute. The activity of catalase was expressed in units per milligram of sample fresh weight [72].

The spectrophotometric method was also used to estimate the activity of peroxidase (POD) and polyphenol oxidase (PPO), as earlier described by Khan, et al. [73]. The supernatant obtained after centrifugation in the previous paragraph was used as crude enzyme source. For POD, the reaction contained 50 μL crude extract, 50 μL pyrogallol (50 μM), 25 μL H_2O_2 (50 mM), and 10 μL phosphate buffer (0.1 M), and incubated for 5 min in dark conditions. Then, 25 μL H_2SO_4 (5% v/v) was added to stop the reaction. The absorbance was measured at 420 nm wavelength [74]. For PPO activity analysis, the reaction mixture was composed of 50 μL pyrogallol (50 mM) and 100 μL phosphate buffer (0.1 M). The absorbance of the reaction was read at 420 nm wavelength, and calculations were done as previously described [75].

4.6. Chlorophyll, Pheophytin, and Carotenoids Content Measurements

The photosynthesis process has been shown to be crucial for plant growth and development, and productivity. Chlorophyll, pheophytin, and total carotenoids content were measured as described by Lichtenthaler [76]. About 200 mg leaf tissues were homogenized with acetone (80%), followed by centrifugation for 10 min at 3000 rpm. The supernatant was transferred to fresh falcon tubes, and the centrifugation was repeated until all chlorophyll was harvested in the solvent. The combined supernatant was made up to a known volume with 80% acetone. The absorbance of the extract was read at 645, 663, and 652 nm for chlorophyll *a*, *b*, and total chlorophyll; 480 and 510 nm for total carotenoids; and 665, 653, and 470 nm for pheophytin *a*, *b*, and total pheophytin. Acetone was used as a blank. The OD values at 645, 663, and 652 were used to calculate chlorophyll contents and total carotenoids as described earlier [77].

4.7. Lipid Peroxidation Assay

The lipid peroxidation level in the leaf tissue mentioned here was measured as the malondialdehyde (MDA) content, determined by thiobarbituric acid (TBARS) reaction [78]. About 200 mg of non-treated and treated samples were homogenized in 4 mL Trichloroacetic acid (TCA) (0.1%) with a porcelain mortar and pestle, followed by centrifugation for 15 min at 10,000 rpm. Then, 1 mL of the supernatant was harvested and 2 mL of TCA (20%) containing 0.5% TBA was added. The supernatant was incubated in a preheated water bath at 95 °C for 30 min and immediately cooled on ice. The reaction mixture was centrifuged for 10 min at 10,000 rpm, and the optical density (OD) was measured at 532 nm and 600 nm wavelength (OD_{600} as the non-specific absorbance that is subtracted from the

OD_{532} reading). MDA level was calculated using Lambert's equation (extinction coefficient of MDA 155 $nM^{-1} \times cm^{-1}$).

4.8. Nitrate Reductase Activity Assay

The nitrate reductase activity was assayed following an in vivo spectrophotometric method, as previously described [79], with slight modifications. NR enzyme activity was determined by measuring the amount of nitrite (NO_2^-) released from plant tissues. Briefly, about 2 g of samples were homogenized into 2 mL reaction buffer composed of 40 m*M* potassium nitrate (KNO_3), 0.08 M Sodium phosphate dibasic (Na_2HPO_4), 0.02 *M* Sodium phosphate monobasic (NaH_2PO_4), and 4% (*v*/*v*) n-propanol ($CH_3CH_2CH_2OH$), with a pH of 7.5. The mixture was centrifuged at 12,000 rpm for 10 min. The supernatant was incubated for 2 h in dark conditions at room temperature, and the reaction was stopped by adding of 200 µL of 1% sulfanilamide ($H_2NC_6H_4SO_2NH_2$) dissolved in 3N HCl and 200 µL of 0.05% *N*-(1-naphthyl) ethylenediamine hydrochloride ($C_{10}H_7NHCH_2CH_2NH_2 \cdot 2HCl$). We, therefore, determined NO_2^- concentration by reading the absorbance of the solution at A_{540} nm. In the case the absorbance was greater than 0.5, the reaction solution was diluted 10-fold with the reaction buffer and sulfanilamide.

4.9. Statistical Analysis

All the experiments were performed using a completely randomized design (CRD). The data were collected in triplicates and analyzed statistically with GraphPad Prism software (Version 7.00, 1992–2016 GraphPad, San Diego, CA, USA). Analysis of variance (ANOVA) for CRD was performed, and the Turkey's multiple comparison was employed at a significance level of 0.05.

5. Conclusions

Understanding nitrogen (N) metabolism in higher plants and all aspects involved in the process is essential to improving nitrogen use efficiency (NUE) and rationalizing N application to plants. Here, a set of potassium chlorate ($KClO_3$) sensitive rice lines (BC2F7) carrying the *indica* alleles of the nitrate reductase (NR) or the nitrate transporter (NRT), were exposed to $KClO_3$ at seedling stage. The parental lines recorded distinctive phenotypic responses 7 days after treatment. In addition, *OsNR2* expression was differentially regulated between the roots, stem, and leaf tissues, and between introgression lines. Similarly, the expression pattern of *OsNIA1* and *OsNIA2* in the roots, stem, and leaves indicated that they are differentially regulated by $KClO_3$. Furthermore, the transcriptional divergence of the ammonium transporters and that of the glutamate synthase encoding genes and associated with the pattern of NR activity in the roots, would indicate the prevalence of nitrate (NO_3^-) transport over ammonium (NH_4^+) transport. Moreover, the increase or decrease of catalase (CAT) and polyphenol oxidase (PPO) enzyme activities, coupled with the changes in the chloroplast pigments and proline contents, as well as the accumulation of malondialdehyde (MDA) revealed the extent of the oxidative damage by $KClO_3$, resulting in lipid peroxidation. All results suggest that the inhibitory effect of $KClO_3$ on the reducing activity of the nitrate reductase (NR), as well as that of the genes encoding the ammonium transporters and glutamate synthase are tissue-specific, rather than systematic to all plant tissues or organs in rice.

Supplementary Materials: The following are available online at https://www.mdpi.com/1422-0067/22/4/2192/s1, Figure S1: Optimization of potassium chlorate ($KClO_3$) concentration using Nipponbare and 93-11 rice cultivars, Figure S2: Identified nitrate reductase (NR) and nitrate transporter (NRT) introgression lines, Figure S3: Chord diagram showing the shoot and roots growth patterns of Saeilmi (P1) and Milyang23 (P2), and four BC2F7 introgression lines, Table S1: List of primer sequences for gene expression used in the study.

Author Contributions: Conceptualization, J.-H.L., J.-Y.L. and D.S.; methodology, J.-H.L. and N.R.K.; validation, J.-M.K., J.-H.L. and D.S.; formal analysis, N.R.K., S.-Y.P., S.-M.L., Y.K., J.-K.C. and D.V.D.;

investigation, N.R.K., J.-H.L. and S.-Y.P.; resources, J.-M.K., J.-H.L., J.-H.C. and J.-Y.L.; data curation, N.R.K.; writing—original draft preparation, N.R.K.; writing—review and editing, J.-H.L. and D.S.; visualization, and supervision, J.-H.L. and J.-M.K.; project administration, J.-H.L., J.-H.C. and J.-M.K.; funding acquisition, J.-M.K. and J.-H.L. All authors have read and agreed to the published version of the manuscript.

Funding: This work was supported by the Cooperative Research Program for Agriculture Science & Technology Development (Project No. PJ01506901).

Institutional Review Board Statement: Not applicable.

Informed Consent Statement: Not applicable.

Data Availability Statement: The data presented in this study are available in supplementary material.

Acknowledgments: This study was supported by 2021 the RDA Fellowship Program of NICS, Rural Development Administration, Korea.

Conflicts of Interest: The authors declare no conflict of interest.

References

1. Shi, W.M.; Xu, W.F.; Li, S.M.; Zhao, X.Q.; Dong, G.Q. Responses of two rice cultivars differing in seedling-stage nitrogen use efficiency to growth under low-nitrogen conditions. *Plant Soil* **2010**, *326*, 291. [CrossRef]
2. Mustapha, S.; Voncir, N.; Umar, S. Content and distribution of nitrogen forms in some black cotton soils in Akko LGA, Gombe State, Nigeria. *Int. J. Soil Sci.* **2011**, *6*, 275–281. [CrossRef]
3. Harmsen, G.; Van Schreven, D. Mineralization of organic nitrogen in soil. In *Advances in Agronomy*; Elsevier: Amsterdam, The Netherlands, 1955; Volume 7, pp. 299–398.
4. Cameron, R.; Posner, A. Mineralisable organic nitrogen in soil fractionated according to particle size. *J. Soil Sci.* **1979**, *30*, 565–577. [CrossRef]
5. Bock, E.; Wagner, M. Oxidation of inorganic nitrogen compounds as an energy source. *Prokaryotes* **2006**, *2*, 457–495.
6. Weidner, S.; Pühler, A.; Küster, H. Genomics insights into symbiotic nitrogen fixation. *Curr. Opin. Biotechnol.* **2003**, *14*, 200–205. [CrossRef]
7. Wang, J.; Lu, K.; Nie, H.; Zeng, Q.; Wu, B.; Qian, J.; Fang, Z. Rice nitrate transporter OsNPF7. 2 positively regulates tiller number and grain yield. *Rice* **2018**, *11*, 1–13. [CrossRef]
8. Xu, G.; Fan, X.; Miller, A.J. Plant nitrogen assimilation and use efficiency. *Annu. Rev. Plant Biol.* **2012**, *63*, 153–182. [CrossRef] [PubMed]
9. Tang, Z.; Fan, X.; Li, Q.; Feng, H.; Miller, A.J.; Shen, Q.; Xu, G. Knockdown of a rice stelar nitrate transporter alters long-distance translocation but not root influx. *Plant Physiol.* **2012**, *160*, 2052–2063. [CrossRef] [PubMed]
10. Yan, M.; Fan, X.; Feng, H.; Miller, A.J.; Shen, Q.; Xu, G. Rice OsNAR2.1 interacts with OsNRT2.1, OsNRT2.2 and OsNRT2.3a nitrate transporters to provide uptake over high and low concentration ranges. *Plant Cell Environ.* **2011**, *34*, 1360–1372. [CrossRef] [PubMed]
11. Fan, X.; Shen, Q.; Ma, Z.; Zhu, H.; Yin, X.; Miller, A.J. A comparison of nitrate transport in four different rice (Oryza sativa L.) cultivars. *Sci. China Ser. C Life Sci.* **2005**, *48*, 897–911.
12. Fan, X.; Feng, H.; Tan, Y.; Xu, Y.; Miao, Q.; Xu, G. A putative 6-transmembrane nitrate transporter OsNRT1. 1b plays a key role in rice under low nitrogen. *J. Integr. Plant Biol.* **2016**, *58*, 590–599. [CrossRef]
13. Chao, D.-Y.; Lin, H.-X. Nitrogen-use efficiency: Transport solution in rice variations. *Nat. Plants* **2015**, *1*, 1–2. [CrossRef] [PubMed]
14. Zhao, S.; Shi, W. Expression patterns of OsAMT (1.1–1.3), OsAMT3.1 and OsAMT4.1 in rice (Oryza sativa L.). *Soil* **2007**, *39*, 460–464.
15. Zhao, S.; Ye, X.; Shi, W. Expression of OsAMT1 (1.1–1.3) in rice varieties differing in nitrogen accumulation. *Russ. J. Plant Physiol.* **2014**, *61*, 707–713. [CrossRef]
16. Li, Y.; Ouyang, J.; Wang, Y.-Y.; Hu, R.; Xia, K.; Duan, J.; Wang, Y.; Tsay, Y.-F.; Zhang, M. Disruption of the rice nitrate transporter OsNPF2.2 hinders root-to-shoot nitrate transport and vascular development. *Sci. Rep.* **2015**, *5*, 1–10.
17. Hu, B.; Wang, W.; Ou, S.; Tang, J.; Li, H.; Che, R.; Zhang, Z.; Chai, X.; Wang, H.; Wang, Y. Variation in NRT1.1B contributes to nitrate-use divergence between rice subspecies. *Nat. Genet.* **2015**, *47*, 834. [CrossRef]
18. Wang, X.; Cai, X.; Xu, C.; Wang, Q.; Dai, S. Drought-responsive mechanisms in plant leaves revealed by proteomics. *Int. J. Mol. Sci.* **2016**, *17*, 1706. [CrossRef]
19. Xuan, W.; Beeckman, T.; Xu, G. Plant nitrogen nutrition: Sensing and signaling. *Curr. Opin. Plant Biol.* **2017**, *39*, 57–65. [CrossRef] [PubMed]
20. Ashoub, A.; Baeumlisberger, M.; Neupaertl, M.; Karas, M.; Brüggemann, W. Characterization of common and distinctive adjustments of wild barley leaf proteome under drought acclimation, heat stress and their combination. *Plant Mol. Biol.* **2015**, *87*, 459–471. [CrossRef]

21. Wendelboe-Nelson, C.; Morris, P.C. Proteins linked to drought tolerance revealed by DIGE analysis of drought resistant and susceptible barley varieties. *Proteomics* **2012**, *12*, 3374–3385. [CrossRef]
22. Tcherkez, G.; Hodges, M. How stable isotopes may help to elucidate primary nitrogen metabolism and its interaction with (photo) respiration in C3 leaves. *J. Exp. Bot.* **2007**, *59*, 1685–1693. [CrossRef]
23. Lea, P.J.; Miflin, B.J. Glutamate synthase and the synthesis of glutamate in plants. *Plant Physiol. Biochem.* **2003**, *41*, 555–564. [CrossRef]
24. Kabange, N.R.; Park, S.-Y.; Shin, D.; Lee, S.-M.; Jo, S.-M.; Kwon, Y.; Cha, J.-K.; Song, Y.-C.; Ko, J.-M.; Lee, J.-H.J.A. Identification of a Novel QTL for Chlorate Resistance in Rice (*Oryza sativa* L.). *Agriculture* **2020**, *10*, 360. [CrossRef]
25. Gao, Z.; Wang, Y.; Chen, G.; Zhang, A.; Yang, S.; Shang, L.; Wang, D.; Ruan, B.; Liu, C.; Jiang, H.; et al. The indica nitrate reductase gene OsNR2 allele enhances rice yield potential and nitrogen use efficiency. *Nat. Commun.* **2019**, *10*, 1–10. [CrossRef]
26. Zhang, Z.; Chu, C. Nitrogen-use divergence between indica and japonica rice: Variation at nitrate assimilation. *Mol. Plant* **2020**, *13*, 6–7. [CrossRef] [PubMed]
27. Duan, D.; Zhang, H. A single SNP in NRT1. 1B has a major impact on nitrogen use efficiency in rice. *Sci. China Life Sci.* **2015**, *58*, 827. [CrossRef]
28. Nakagawa, H.; Yamashita, N. Chlorate reducing activity of spinach nitrate reductase. *Agric. Biol. Chem.* **1986**, *50*, 1893–1894.
29. Solomonson, L.; Vennesland, B. Nitrate reductase and chlorate toxicity in Chlorella vulgaris Beijerinck. *Plant Physiol.* **1972**, *50*, 421–424. [CrossRef]
30. Roldan, M.; Reyes, F.; Moreno-Vivian, C.; Castillo, F. Chlorate and nitrate reduction in the phototrophic bacteriaRhodobacter capsulatus andRhodobacter sphaeroides. *Curr. Microbiol.* **1994**, *29*, 241–245. [CrossRef]
31. Mészáros, A.; Pauk, J. Chlorate resistance as a tool to study the effect of nitrate reductase antisense gene in wheat. *Cereal Res. Commun.* **2002**, *30*, 245–252. [CrossRef]
32. Kamal, N.M.; Gorafi, Y.S.A.; Mega, R.; Tsujimoto, H.J.A. Physiological response of wheat to chemical desiccants used to simulate post-anthesis drought stress. *Agronomy* **2018**, *8*, 44. [CrossRef]
33. Teng, S.; Tian, C.; Chen, M.; Zeng, D.; Guo, L.; Zhu, L.; Han, B.; Qian, Q.J.E. QTLs and candidate genes for chlorate resistance in rice (Oryzasativa L.). *Euphytica* **2006**, *152*, 141–148. [CrossRef]
34. Inzé, D.; Van Montagu, M. Oxidative stress in plants. *Curr. Opin. Biotechnol.* **1995**, *6*, 153–158. [CrossRef]
35. Bartosz, G. Oxidative stress in plants. *Acta Physiol.* **1997**, *19*, 47–64. [CrossRef]
36. Corpas, F.J.; Chaki, M.; Fernandez-Ocana, A.; Valderrama, R.; Palma, J.M.; Carreras, A.; Begara-Morales, J.C.; Airaki, M.; del Río, L.A.; Barroso, J.B.J.P.; et al. Metabolism of reactive nitrogen species in pea plants under abiotic stress conditions. *Plant Cell Physiol.* **2008**, *49*, 1711–1722. [CrossRef] [PubMed]
37. Bhattacharjee, S. Reactive oxygen species and oxidative burst: Roles in stress, senescence and signal transducation in plants. *Curr. Sci.* **2005**, *89*, 1113–1121.
38. Parankusam, S.; Adimulam, S.S.; Bhatnagar-Mathur, P.; Sharma, K.K. Nitric oxide (NO) in plant heat stress tolerance: Current knowledge and perspectives. *Front. Plant Sci.* **2017**, *8*, 1582. [CrossRef]
39. Pavlovic, D.; Nikolic, B.; Djurovic, S.; Waisi, H.; Andjelkovic, A.; Marisavljevic, D. Chlorophyll as a measure of plant health: Agroecological aspects. *Pestic. Fitomed.* **2014**, *29*, 21–34. [CrossRef]
40. Rebeiz, C.A. *Chlorophyll Biosynthesis and Technological Applications*; Springer: Berlin/Heidelberg, Germany, 2014.
41. Smith, S.; De Smet, I. Root system architecture: Insights from Arabidopsis and cereal crops. *R. Soc.* **2012**, *367*, 1595. [CrossRef] [PubMed]
42. Pandey, A.; Suter, H.; He, J.-Z.; Hu, H.-W.; Chen, D. Dissimilatory nitrate reduction to ammonium dominates nitrate reduction in long-term low nitrogen fertilized rice paddies. *Soil Biol. Biochem.* **2019**, *131*, 149–156. [CrossRef]
43. Kuiper, P.J.; Kuiper, D.; Schuit, J. Root functioning under stress conditions: An introduction. *Plant Soil* **1988**, 249–253. [CrossRef]
44. Hsiao, T.C.; Xu, L.K. Sensitivity of growth of roots versus leaves to water stress: Biophysical analysis and relation to water transport. *J. Exp. Bot.* **2000**, *51*, 1595–1616. [CrossRef]
45. Khan, M.; Gemenet, D.C.; Villordon, A. Root system architecture and abiotic stress tolerance: Current knowledge in root and tuber crops. *Front. Plant Sci.* **2016**, *7*, 1584. [CrossRef] [PubMed]
46. Borges, R.; Miguel, E.C.; Dias, J.M.; da Cunha, M.; Bressan-Smith, R.E.; de Oliveira, J.G.; de Souza Filho, G.A. Ultrastructural, physiological and biochemical analyses of chlorate toxicity on rice seedlings. *Plant Sci.* **2004**, *166*, 1057–1062. [CrossRef]
47. LaBrie, S.T.; Wilkinson, J.Q.; Crawford, N.M. Effect of chlorate treatment on nitrate reductase and nitrite reductase gene expression in Arabidopsis thaliana. *Plant Physiol.* **1991**, *97*, 873–879. [CrossRef] [PubMed]
48. Zhao, C.-m.; Hasegawa, H.; Ichii, M. A chlorate resistant mutant of rice (Oryza sativa L.) with normal nitrate uptake and nitrate reductase activity. *Breed. Sci.* **2000**, *50*, 9–15. [CrossRef]
49. Rolly, N.K.; Lee, S.-U.; Imran, Q.M.; Hussain, A.; Mun, B.-G.; Kim, K.-M.; Yun, B.-W. Nitrosative stress-mediated inhibition of OsDHODH1 gene expression suggests roots growth reduction in rice (*Oryza sativa* L.). *3 Biotech* **2019**, *9*, 273. [CrossRef]
50. Wilkinson, J.Q.; Crawford, N.M. Identification and characterization of a chlorate-resistant mutant of Arabidopsis thaliana with mutations in both nitrate reductase structural genes NIA1 and NIA2. *Mol. Gen. Genet.* **1993**, *239*, 289–297. [CrossRef]
51. Islam, M.S. Sensing and Uptake of Nitrogen in Rice Plant: A Molecular View. *Rice Sci.* **2019**, *26*, 343–355. [CrossRef]
52. Wang, W.; Hu, B.; Li, A.; Chu, C. NRT1.1s in plants: Functions beyond nitrate transport. *J. Exp. Bot.* **2020**, *71*, 4373–4379. [CrossRef]

53. Zhang, G.-B.; Meng, S.; Gong, J.-M. The expected and unexpected roles of nitrate transporters in plant abiotic stress resistance and their regulation. *Int. J. Mol. Sci.* **2018**, *19*, 3535. [CrossRef] [PubMed]
54. von Wirén, N.; Gojon, A.; Chaillou, S.; Raper, D. Mechanisms and regulation of ammonium uptake in higher plants. In *Plant Nitrogen*; Springer: Berlin/Heidelberg, Germany, 2001; pp. 61–77.
55. Bao, A.; Liang, Z.; Zhao, Z.; Cai, H. Overexpressing of OsAMT1-3, a high affinity ammonium transporter gene, modifies rice growth and carbon-nitrogen metabolic status. *Int. J. Mol. Sci.* **2015**, *16*, 9037–9063. [CrossRef] [PubMed]
56. Ferreira, L.M.; de Souza, V.M.; Tavares, O.C.H.; Zonta, E.; Santa-Catarina, C.; de Souza, S.R.; Fernandes, M.S.; Santos, L.A. OsAMT1.3 expression alters rice ammonium uptake kinetics and root morphology. *Plant Biotechnol. Rep.* **2015**, *9*, 221–229. [CrossRef]
57. Gaur, V.S.; Singh, U.; Gupta, A.K.; Kumar, A. Influence of different nitrogen inputs on the members of ammonium transporter and glutamine synthetase genes in two rice genotypes having differential responsiveness to nitrogen. *Mol. Biol. Rep.* **2012**, *39*, 8035–8044. [CrossRef]
58. Kamachi, K.; Yamaya, T.; Mae, T.; Ojima, K. A role for glutamine synthetase in the remobilization of leaf nitrogen during natural senescence in rice leaves. *Plant Physiol.* **1991**, *96*, 411–417. [CrossRef]
59. Gill, S.S.; Tuteja, N. Reactive oxygen species and antioxidant machinery in abiotic stress tolerance in crop plants. *Plant Physiol. Biochem.* **2010**, *48*, 909–930. [CrossRef]
60. Borsani, O.; Valpuesta, V.; Botella, M.A. Evidence for a role of salicylic acid in the oxidative damage generated by NaCl and osmotic stress in Arabidopsis seedlings. *Plant Physiol.* **2001**, *126*, 1024–1030. [CrossRef] [PubMed]
61. Corpas, F.J. What is the role of hydrogen peroxide in plant peroxisomes? *Plant Biol.* **2015**, *17*, 1099–1103. [CrossRef]
62. Rolly, N.K.; Imran, Q.M.; Shahid, M.; Imran, M.; Khan, M.; Lee, S.-U.; Hussain, A.; Lee, I.-J.; Yun, B.-W. Drought-induced AtbZIP62 transcription factor regulates drought stress response in Arabidopsis. *Plant Physiol. Biochem.* **2020**, *156*, 384–395. [CrossRef]
63. Sharma, P.; Jha, A.B.; Dubey, R.S.; Pessarakli, M. Reactive oxygen species, oxidative damage, and antioxidative defense mechanism in plants under stressful conditions. *J. Bot.* **2012**, *2012*. [CrossRef]
64. Mittler, R. Oxidative stress, antioxidants and stress tolerance. *Trends Plant Sci.* **2002**, *7*, 405–410. [CrossRef]
65. Fisher, H.; Orth, H. Die Chemie des pyrrols. *Acad. Verl. Leipz. Ger.* **1937**, *623*, 673.
66. Rolly, N.K.; Imran, Q.M.; Lee, I.-J.; Yun, B.-W. Salinity Stress-Mediated Suppression of Expression of Salt Overly Sensitive Signaling Pathway Genes Suggests Negative Regulation by AtbZIP62 Transcription Factor in Arabidopsis thaliana. *Int. J. Mol. Sci.* **2020**, *21*, 1726. [CrossRef] [PubMed]
67. Armengaud, P.; Thiery, L.; Buhot, N.; Grenier-de March, G.; Savouré, A. Transcriptional regulation of proline biosynthesis in Medicago truncatula reveals developmental and environmental specific features. *Physiol. Plant.* **2004**, *120*, 442–450. [CrossRef]
68. Newton, R.; Sen, S.; Puryear, J. Free proline changes in Pinus taeda L. callus in response to drought stress. *Tree Physiol.* **1986**, *1*, 325–332. [CrossRef]
69. Mun, B.-G.; Lee, S.-U.; Hussain, A.; Kim, H.-H.; Rolly, N.K.; Jung, K.-H.; Yun, B.-W. S-nitrosocysteine-responsive genes modulate diverse regulatory pathways in Oryza sativa: A transcriptome profiling study. *Funct. Plant Biol.* **2018**, *45*, 630–644. [CrossRef] [PubMed]
70. Ábrahám, E.; Hourton-Cabassa, C.; Erdei, L.; Szabados, L. Methods for determination of proline in plants. In *Plant Stress Tolerance*; Springer: Berlin/Heidelberg, Germany, 2010; pp. 317–331.
71. Elavarthi, S.; Martin, B. Spectrophotometric assays for antioxidant enzymes in plants. In *Plant Stress Tolerance*; Springer: Berlin/Heidelberg, Germany, 2010; pp. 273–280.
72. Aebi, H. Catalase in vitro. In *Methods in Enzymology*; Elsevier: Amsterdam, The Netherlands, 1984; Volume 105, pp. 121–126.
73. Khan, A.R.; Ullah, I.; Waqas, M.; Park, G.-S.; Khan, A.L.; Hong, S.-J.; Ullah, R.; Jung, B.K.; Park, C.E.; Ur-Rehman, S. Host plant growth promotion and cadmium detoxification in Solanum nigrum, mediated by endophytic fungi. *Ecotoxicol. Environ. Saf.* **2017**, *136*, 180–188. [CrossRef]
74. Shannon, L.M.; Kay, E.; Lew, J.Y. Peroxidase isozymes from horseradish roots I. Isolation and physical properties. *J. Biol. Chem.* **1966**, *241*, 2166–2172. [CrossRef]
75. Chance, B.; Maehly, A. Assay of catalases and peroxidases. *Methods Biochem. Anal.* **1954**, *1*, 357–424.
76. Lichtenthaler, H.K. Chlorophylls and carotenoids: Pigments of photosynthetic biomembranes. In *Methods in Enzymology*; Elsevier: Amsterdam, The Netherlands, 1987; Volume 148, pp. 350–382.
77. Sumanta, N.; Haque, C.I.; Nishika, J.; Suprakash, R. Spectrophotometric analysis of chlorophylls and carotenoids from commonly grown fern species by using various extracting solvents. *Res. J. Chem. Sci.* **2014**, *2231*, 606X.
78. Jambunathan, N. Determination and detection of reactive oxygen species (ROS), lipid peroxidation, and electrolyte leakage in plants. In *Plant Stress Tolerance*; Springer: Berlin/Heidelberg, Germany, 2010; pp. 291–297.
79. Márton, L.; Sidorov, V.; Biasini, G.; Maliga, P. Complementation in somatic hybrids indicates four types of nitrate reductase deficient lines in Nicotiana plumbaginifolia. *Mol. Gen. Genet. MGG* **1982**, *187*, 1–3. [CrossRef]

International Journal of
Molecular Sciences

MDPI

Article

Whole-Tissue Three-Dimensional Imaging of Rice at Single-Cell Resolution

Moeko Sato [1], Hiroko Akashi [1], Yuki Sakamoto [2,3], Sachihiro Matsunaga [2,4,5] and Hiroyuki Tsuji [1,*]

1 Kihara Institute for Biological Research, Yokohama City University, Maioka 641-12, Totsuka, Yokohama 244-0813, Kanagawa, Japan; n175255d@yokohama-cu.ac.jp (M.S.); hakashi@yokohama-cu.ac.jp (H.A.)
2 Imaging Frontier Center, Organization for Research Advancement, Tokyo University of Science, 2641 Yamazaki, Noda 278-8510, Chiba, Japan; yuki_sakamoto@bio.sci.osaka-u.ac.jp (Y.S.); sachi@edu.k.u-tokyo.ac.jp (S.M.)
3 Department of Biological Sciences, Graduate School of Science, Osaka University, Machikaneyama-cho 1-1, Toyonaka 560-0043, Osaka, Japan
4 Department of Applied Biological Science, Faculty of Science and Technology, Tokyo University of Science, 2641 Yamazaki, Noda 278-8510, Chiba, Japan
5 Department of Integrated Biosciences, Graduate School of Frontier Sciences, The University of Tokyo, 5-1-5 Kashiwanoha, Kashiwa 277-8562, Chiba, Japan
* Correspondence: tsujih@yokohama-cu.ac.jp

Abstract: The three-dimensional (3D) arrangement of cells in tissues provides an anatomical basis for analyzing physiological and biochemical aspects of plant and animal cellular development and function. In this study, we established a protocol for tissue clearing and 3D imaging in rice. Our protocol is based on three improvements: clearing with iTOMEI (clearing solution suitable for plants), developing microscopic conditions in which the Z step is optimized for 3D reconstruction, and optimizing cell-wall staining. Our protocol successfully 3D imaged rice shoot apical meristems, florets, and root apical meristems at cellular resolution throughout whole tissues. Using fluorescent reporters of auxin signaling in rice root tips, we also revealed the 3D distribution of auxin signaling events that are activated in the columella, quiescent center, and multiple rows of cells in the stele of the root apical meristem. Examination of cells with higher levels of auxin signaling revealed that only the central row of cells was connected to the quiescent center. Our method provides opportunities to observe the 3D arrangement of cells in rice tissues.

Keywords: three-dimensional imaging; shoot apical meristem; root tip; rice

Citation: Sato, M.; Akashi, H.; Sakamoto, Y.; Matsunaga, S.; Tsuji, H. Whole-Tissue Three-Dimensional Imaging of Rice at Single-Cell Resolution. *Int. J. Mol. Sci.* **2022**, *23*, 40. https://doi.org/10.3390/ijms23010040

Academic Editor: Luigi Cattivelli

Received: 28 May 2021
Accepted: 30 November 2021
Published: 21 December 2021

1. Introduction

The three-dimensional (3D) arrangement of cells in tissues provides an anatomical reference for monitoring the development and function of organisms [1]. For example, in the shoot apical meristem (SAM), the plant tissue that generates the above-ground organs, stem cells are located at the tip and differentiated cells are arranged around the peripheral zone that surrounds the stem cells [2]. In the root apical meristem (RAM), located at the root tip, stem cells are arranged around the quiescent center (QC) and differentiate into diverse cell types including the columella, stele, and endodermis [3]. The 3D arrangement of cells is important for understanding how the developmental processes of multicellular organisms function.

Topology and anisotropy are crucial characteristics to understand the development of organs with complex cellular patterns [4]. The relationship between the topology of cells and anisotropy of tissues in plants has been investigated [5]. Mechanical stress in the cell wall is caused by the 3D arrangement of cells, which is essential for proper organ differentiation [6]. These studies provided important insights about the topology and anisotropy of developing organs in plants; however, these studies have been limited to the analysis of two-dimensional sections. Since plant tissues are organized in three dimensions, a three-dimensional analysis is necessary to understand topology and anisotropy precisely.

Recent advances in tissue clearing methods have provided opportunities to observe the 3D arrangement of cells in tissues. For animal tissues, various clearing methods, such as ScaleS [7], CUBIC [8], and CLARITY [9], have enabled whole-tissue or whole-body imaging at the cellular resolution. For plant tissues, ClearSee [10], TOMEI [11], and PEA-CLARITY [12] have been proposed as protocols for tissue clearing and 3D imaging. These protocols have been used successfully for imaging pollen–pistil interaction [10], embryogenesis [13], and vascular development [14]. Recently, ClearSeeAlpha [15] and iTOMEI (see methods) were reported as improved methods for observing fluorescent proteins in fixed plant tissues and expanded the variety of plant species that can be cleared.

Rice is an important staple crop that provides 50% of the calories that the entire human population consumes and is a staple crop in more than 100 countries. Productivity improvements for rice are likely to come from advances in rice genetics, biotic and abiotic stress responses, and developmental biology [16,17]. In addition, rice is a model plant for monocotyledons with a growing number of rice genomic sequences, wider genetic diversity, and a long history of detailed study [18–20]. To improve our understanding of rice developmental biology, observation of cellular arrangements in 3D is a powerful tool; however, studies using tissue clearing and 3D imaging of rice have not been widely used due to the lack of established protocols. Three-dimensional imaging at single-cell resolution would allow quantitative analysis of cellular differentiation with modeling programs, e.g., MorphographX [21].

In this study, we optimized a protocol for tissue clearing and 3D imaging in rice. Our protocol is based on three improvements: clearing with iTOMEI, developing microscopic conditions in which the Z step is optimized for 3D reconstruction, and optimizing cell-wall staining. We observed SAMs, florets, and RAMs at cellular resolution throughout whole tissues. Using this protocol, we also revealed the 3D distribution of fluorescent reporters of auxin signaling in the root tip of rice.

2. Results

2.1. Development of a Tissue Clearing and 3D Imaging Protocol in Rice

To establish a protocol for tissue clearing and 3D imaging in rice, we improved methods for observation in four steps (Figure 1A). First, we optimized methods for sampling, fixation, and cell-wall staining of the dissected tissue. To sample SAMs, we carefully removed several leaves from the basal part of the seedlings by hand sectioning (Figure 1B). Then, the outermost P4 and P3 leaf primordia were removed with a scalpel while observing the samples with a stereoscopic microscope. The shoot apex, including leaf primordia P2 and P1 (plastochron number 2 and 1, which indicate the second youngest and youngest [20]) and the SAM, was exposed at this step and was excised by cutting at the middle of the stem. Excised shoot apices were placed into microtubes containing the fixative. This procedure enabled us to observe two leaf primordia and the SAM simultaneously. Occasionally, we sampled the shoot apex attached to the P3 leaf primordium. If necessary, P2 and P1 primordia were removed and only the SAM was sampled (Supplementary Materials Video S1). For fixatives, either 4% (w/v) paraformaldehyde or formaldehyde in phosphate-buffered saline can be used. For cell-wall staining, we used 0.1% (v/v) SCRI Renaissance 2200 (SR2200).

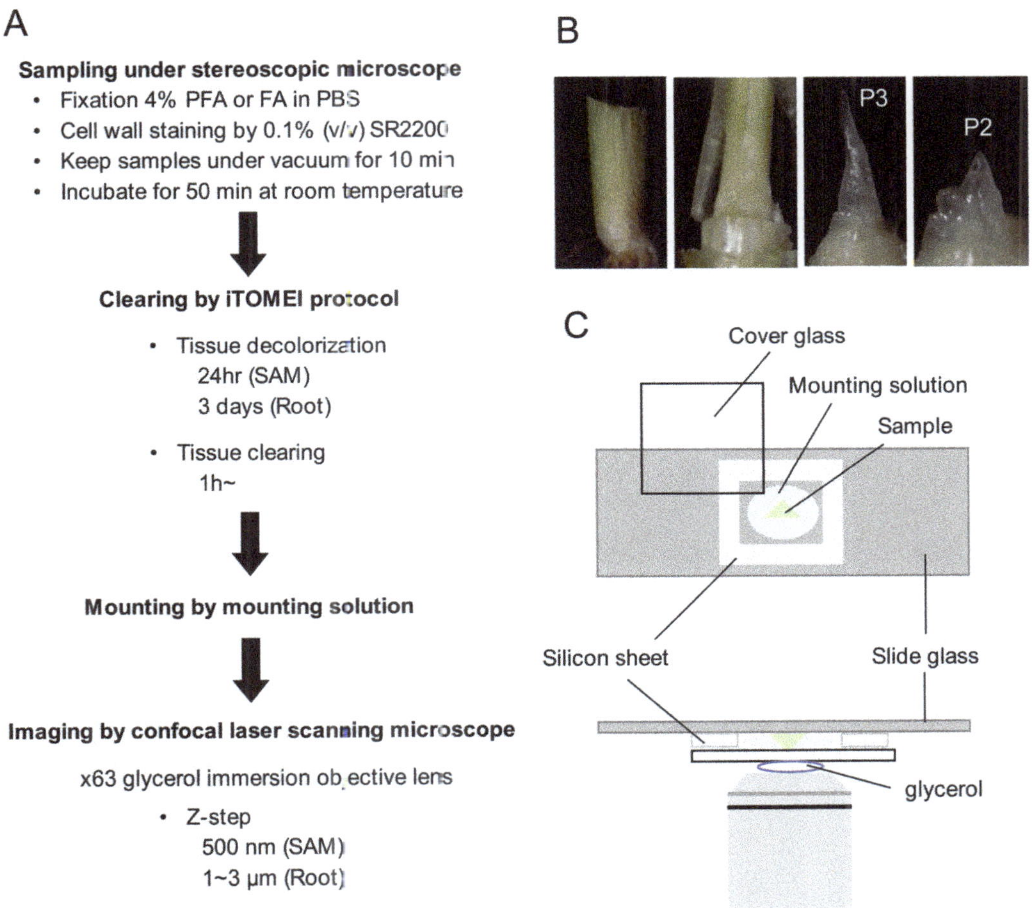

Figure 1. (**A**) Overview of the imaging protocol. (**B**) Sampling of a shoot apex. Far left: basal region of a seedling; middle left, after removing the outer three leaves; middle right, shoot apex with a P3 primordium; far right: shoot apex with a P2 primordium. (**C**) Mounting and preparation for imaging by confocal laser scanning microscopy. Upper: Samples were mounted on a silicon sheet with a square piece removed to maintain the 3D structure of tissues. Lower: a 63× glycerol-immersion objective lens was used. Bars; 100 μm.

Second, sampled tissues were cleared by iTOMEI (Figure 1A). Samples were placed in the decolorization solution of iTOMEI, followed by transfer to iTOMEI clearing solution. The time required for decolorization depended on the sample: 24 h for SAMs, 3 days for roots.

Third, the samples were mounted. The selection of a mounting solution depends on the refractive index of immersion for the lens to be used for observation. We used iTOMEI clearing solution as the mounting solution because it was optimal for our observation conditions. To maintain the 3D structure of tissues, samples were mounted on a 0.2 mm thickness silicon sheet with a square (8 mm × 8 mm) removed from the center (Figure 1C).

Fourth, the mounted samples were observed using a confocal laser scanning microscope equipped with a 63× glycerol-immersion objective lens. In order to collect a sufficient number and density of Z-stacked images for 3D reconstruction, we set the Z step at 500 nm for SAMs and 1–3 μm for root tips (Figure 1A).

2.2. Three-Dimensional Imaging of Rice SAMs with Single-Cell Resolution

We used our protocol to observe rice SAMs (Figure 2 and Supplementary Materials Video S2). We acquired longitudinal sections of SAMs at 500 nm Z steps; successfully obtained images of the entire tissue sample including the SAM, P1, P2, and part of P3; and reconstructed the 3D structure of the shoot apex (Figure 2A). We observed at cellular resolution that the SAM was wrapped around leaf primordia in a three-dimensional arrangement.

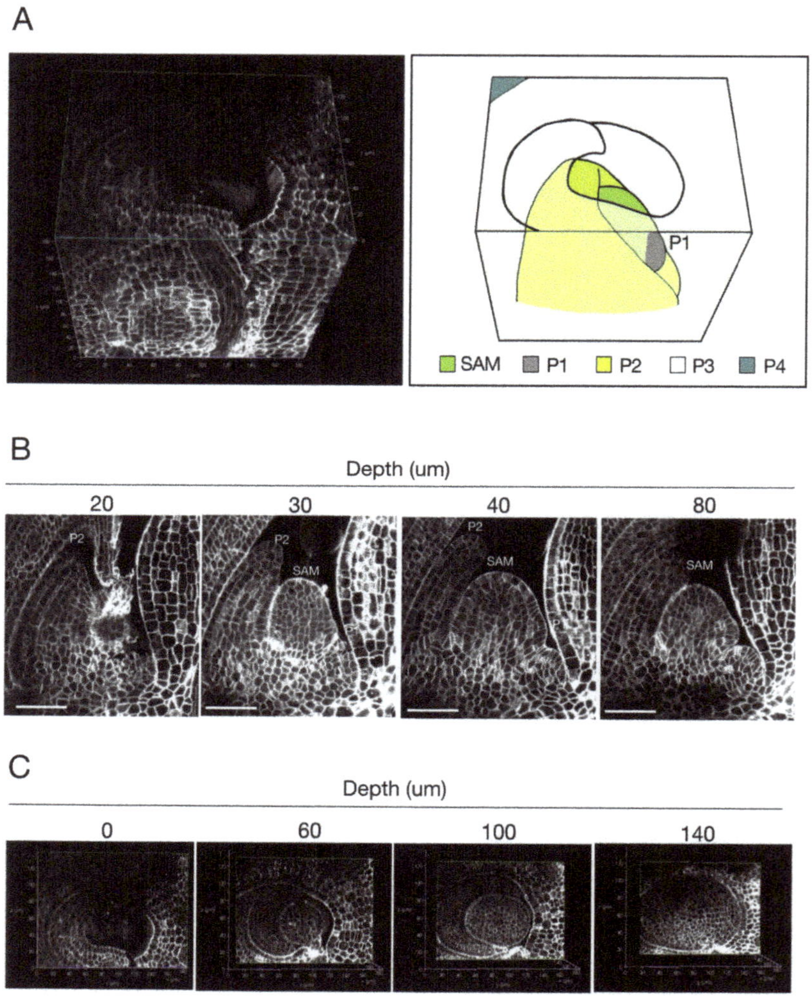

Figure 2. (**A**) Left: a 3D-reconstructed rice shoot apex. Right: a schematic presentation of the 3D-reconstructed shoot apex shown in the left panel. SAM; shoot apical meristem, P1; youngest leaf primordium (plastochron number 1), P2; second youngest leaf primordium (plastochron number 2), P3; third youngest leaf primordium (plastochron number 3), P4; fourth youngest leaf primordium (plastochron number 4). (**B**) Longitudinal sections of a rice SAM. Confocal optical sections were arranged according to the depth from the first section of serial imaging. The fluorescence intensity of cell-wall staining was adjusted for clear recognition of individual cells in the deeper sections. (**C**) Transverse sections of the 3D-reconstructed shoot apex. The transverse sections were arranged according to the depth from the top of the reconstructed image shown in (**A**). Bars; 50 μm.

The SAM in rice is composed of the tunica and the corpus, two cell populations that are derived from different cell lineages [2]. The tunica is the outermost single cell layer and is called the L1. The corpus is an inner cell population inside the L1. To examine whether this cellular arrangement can be observed using our protocol, we acquired longitudinal sections through the whole tissue (Figure 2B, Supplementary Materials Figure S1 and Video S2). Increasing depth resulted in a decreased fluorescence intensity in cell-wall staining (Supplementary Materials Figure S1). We adjusted the fluorescence intensity of cell-wall staining in all images in Figure 2B for well-defined recognition of individual cells in the deeper sections and clearly observed the cellular arrangement of the SAM. The single-cell layer of the L1 was distinguished from the inner cells throughout the SAM. We also found that the second layer of cells from the outermost subepidermal layer tended to be aligned.

We reconstructed the SAM in 3D from serial longitudinal sections and, using this reconstruction, we were able to obtain transverse sections of the same SAM (Figure 2C). We detected cell-wall staining throughout the transverse sections, enabling us to observe structures at cellular resolution. In summary, our protocol resulted in a series of images showing the precise cellular arrangement of the SAM and leaf primordia in 3D.

2.3. Three-Dimensional Imaging of Rice Florets with Single-Cell Resolution

Next, we used our protocol to observe rice florets (Figure 3 and Supplementary Materials Video S3). Rice florets are composed of the outermost structure, the lemma, and moving toward the interior, the palea, two lodicules, six stamens, and the innermost structure, the pistil, all generated by the floret meristem [20]. We observed the floret at the Sp5 stage when stamen primordia are developing. We acquired longitudinal sections of the floret and reconstructed a 3D image of the object (Figure 3A). Developing lemma and palea and four stamen primordia were visible from the front view, whereas the posterior side was covered by bract hair. To examine whether this cellular arrangement can be observed throughout the floret, we acquired longitudinal sections (Figure 3B and Supplementary Materials Video S3). Each cell was clearly distinguishable in the sections, and the internal cellular arrangement of the primordia for the lemma, palea, lodicule, and stamens was revealed. The same floret was observed from reconstructed images acquired from transverse sections (Figure 3C). We were able to observe the floret with cellular resolution from the apex to the base. The 3D arrangement of cells in the primordium for each organ was also found (Supplementary Materials Video S3).

2.4. Three-Dimensional Imaging of the Rice Anther with Single-Cell Resolution

We examined how large organs can be observed based on this protocol. We used a mature anther, which is composed of four pollen sacs [22]. The mature pollen sac is approximately 1 mm long and 0.2 mm wide. Because the object to be scanned is large, it is necessary to use a ×20 lens. Thus, the anther was cleared with ClearSee, which has a refractive index suitable for this lens [10]. In order to create a single image of the entire object, the images were concatenated and reconstructed (Figure 4A and Supplementary Materials Video S4). As a result, the entire image of the pollen sacs could be observed. In the central section of the pollen sacs, we observed aligned epidermal cells, degenerated second layer, and germination pore of the pollen grains. When the tip of the pollen sacs was magnified with a ×40 lens, the structure of the pollen sacs and pollen grain could be clearly observed (Figure 4B and Supplementary Materials Video S5). In particular, the morphology of the second layer of cells in the pollen sacs could be observed more clearly.

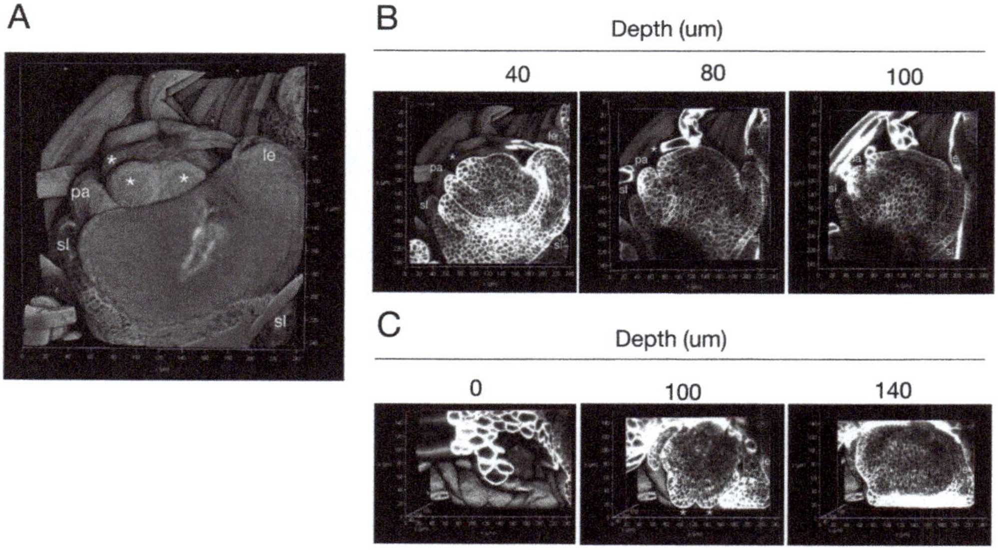

Figure 3. (**A**) A 3D-reconstructed rice floret. (**B**) Longitudinal sections of the floret. Confocal optical sections from the reconstructed image were arranged according to their depth from the first section of serial imaging. (**C**) Transverse sections of the 3D-reconstructed floret. The transverse sections were arranged according to the depth from the top of the reconstructed image shown in (**A**). sl; sterile lemma primordium, le; lemma primordium, pa; palea primordium, lo; lodicule primordium, *; stamen primordium.

2.5. Three-Dimensional Imaging of the Rice RAM with Single-Cell Resolution

Next, we examined whether our method could be used to image roots and whether the method was suitable for fluorescence observations. Rice roots consist of several tissues; from the interior to the exterior, the tissues include the central stele with vascular bundles, the endodermis, the cortex, the sclerenchyma, the exodermis, and the epidermis [3]. The columella and the lateral root cap are located in the root tip. We acquired longitudinal sections of the root tip and confirmed that these tissues were distinguishable in our observations (Figure 5A,B). We reconstructed a 3D image of a root tip from serial longitudinal sections and examined the transverse sections (Figure 5C, Supplementary Materials Video S6). The transverse sections showed that organ shape was maintained in this orientation, and the positional relationship of the central-lateral axis was preserved; however, we could not distinguish each cell in the reconstructed transverse sections of root tips because the Z-step spacing was not sufficiently dense.

To examine whether we could observe the 3D distribution of fluorescent proteins, we used the auxin response reporter DR5rev:NLS-3xVenus [23,24]. Without clearing and cell-wall staining, the DR5 signal was observed in the stele and the columella cells, although precise observation at cellular resolution could not be achieved (Supplementary Materials Video 6).

We acquired longitudinal sections of rice root tips in which fluorescent cell-wall staining and Venus expression simultaneously occurred in a DR5rev:NLS-3xVenus plant (Figure 5B, Supplementary Materials Video S6). Venus fluorescence was observed in columella cells, the QC, and several rows of cells in the stele. In the columella and the QC, three and two interconnected cells expressed Venus, respectively. In the stele, there was a row of Venus-expressing cells in the center and lateral rows of Venus-expressing cells one to two cells away from the center. There were five lateral rows of cells (Figure 5C) corresponding to the positions of metaxylem in the differentiation zone.

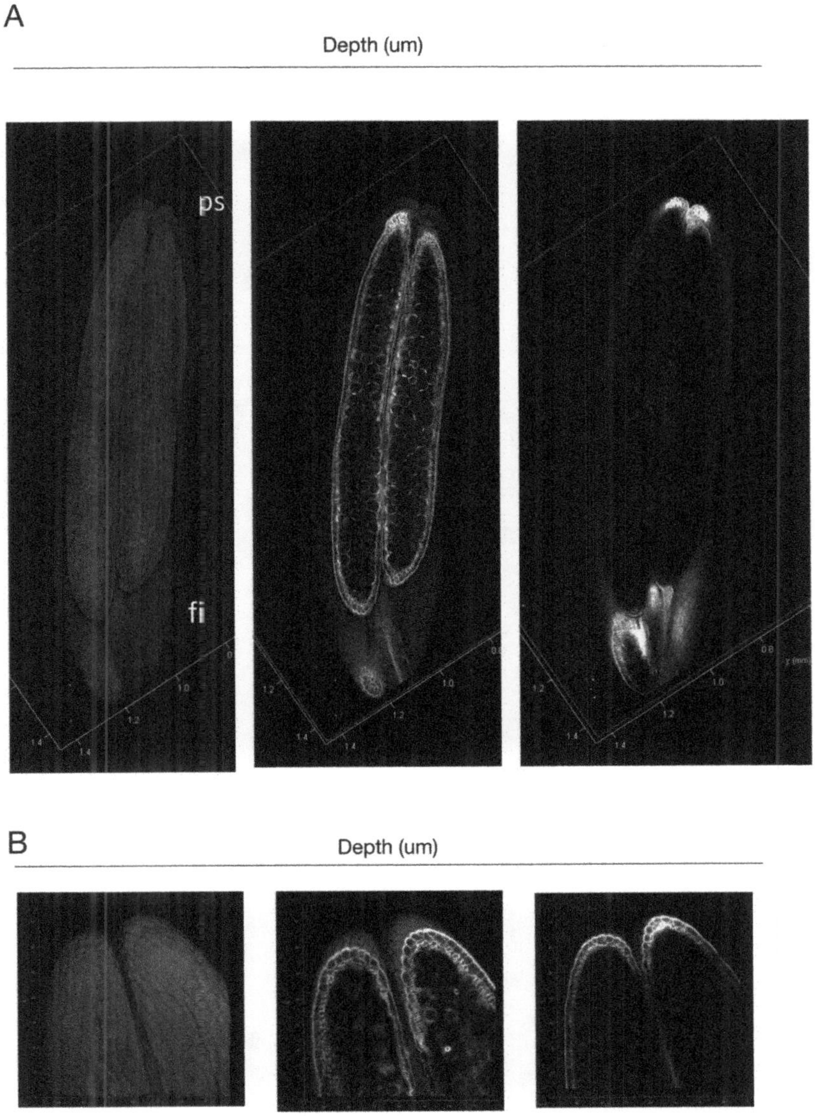

Figure 4. (**A**) Longitudinal sections of the 3D-reconstructed rice anther. (**B**) Longitudinal sections of the 3D-reconstructed rice tip of pollen sacs. ps; pollen sac, fl; filament, pg; pollen grain, gp; germination pore. Bars; 100 μm.

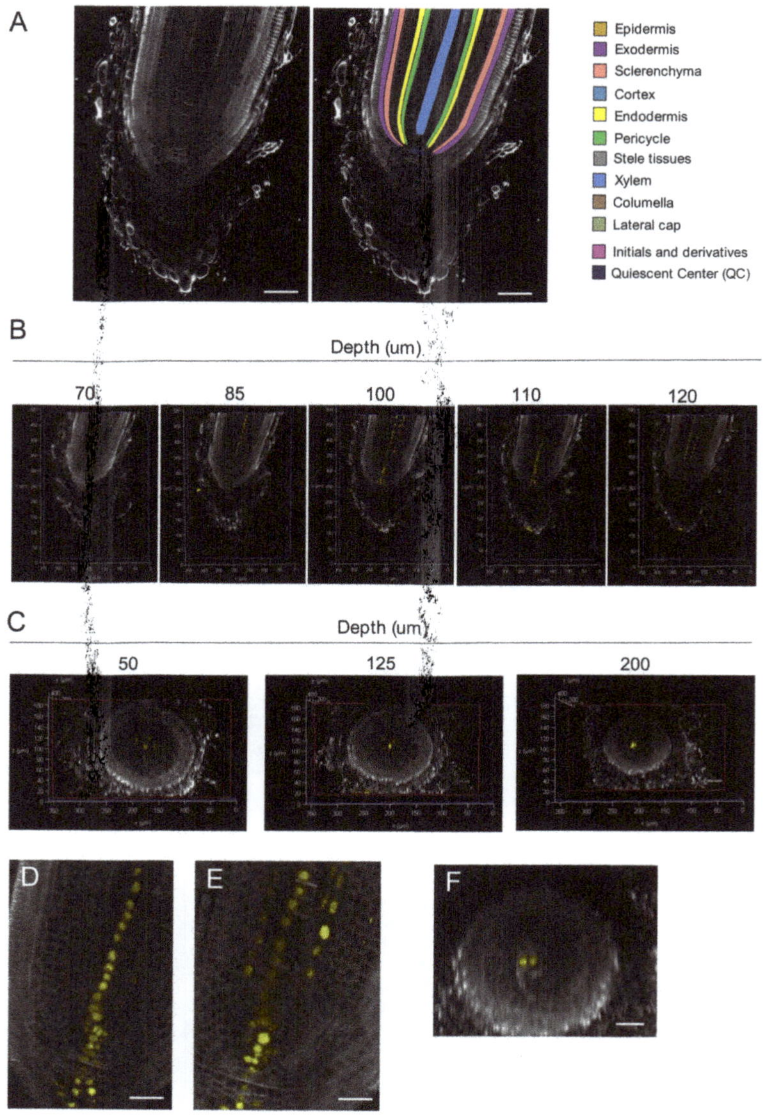

Figure 5. (**A**) Left: representative longitudinal section of a 3D-reconstructed rice root tip. Right: schematic presentation of the 3D-reconstructed root tip shown in the left panel. (**B**) Longitudinal sections of the root tip. Confocal optical sections from the reconstructed image were arranged according to the depth from the first section of serial imaging. (**C**) Transverse sections of the 3D-reconstructed root tip. The transverse sections were arranged according to the depth from the top of the reconstructed image. (**D**,**E**) Enlarged view of longitudinal sections with a central row of Venus-expressing cells connected to the QC (**D**) and with lateral rows of Venus-expressing cells that are not connected to the QC (**E**). (**F**) A transverse section across the QC. Bars; 50 µm in (**A**) and 25 µm in (**D**–**F**).

To determine whether the central and lateral rows of Venus-expressing cells were linked to the QC, we analyzed enlarged images of the sections (Figure 5D,E). The central row of Venus-expressing cells was linked to the QC. In contrast, the lateral rows of cells were not.

Close inspection of the QC region revealed that there were two Venus-expressing cells at the QC (Figure 5F). These results suggest different origins of the central and lateral metaxylem.

3. Discussion

3.1. Tissue Clearing and 3D Imaging Enabled the Precise Evaluation of the Cellular Arrangement of Rice Tissues

Resulting from this study, we propose a protocol to acquire images at cellular resolution of all layers of rice tissues. Using this protocol, we were able to obtain 3D reconstructed images. The advantages of this protocol can be summarized in five points.

First, we were able to visualize the 3D reconstruction of tissues and the internal arrangement of cells across all layers simultaneously. Conventionally, we observed 3D images of tissues by SEM but not the internal cell arrangement. Conventional cross-sectioning is useful for observing the interior, but it cannot be used to construct a 3D structure. For example, it is difficult to appreciate the complexity of the 3D arrangement and shape of lodicule primordia because the tissues are surrounded by the lemma primordium and stamen primordia (Figure 3A). By examining successive sections while referring to the 3D constructed image in this study, it became easier to appreciate the arrangement and morphology of the lodicule in three dimensions (Figure 3B,C and Supplementary Materials Video S3).

Second, our protocol allows longitudinal and transverse sections to be observed simultaneously from the same tissue. In conventional sectioning, transverse sections could not be obtained after longitudinal sections were created. In the method optimized by this study, transverse sections were obtained from the same tissue by serially acquiring longitudinal sections and 3D reconstructions. This advantage gave us a three-dimensional view of the number and arrangement of the central and lateral rows of DR5-expressing cells at the root tip (Figure 5, Supplementary Materials Video S6 and discussed below).

Third, 3D construction allows for a continuous view of the internal structure. This contiguity evoked the possibility of specialization for the second cell layer of the SAM. Our data suggest that the second layer tends to be aligned (Figure 2). The composition of cells in grass SAMs has been dichotomized into the outermost layer of tunica cells and the inner layer of corpus cells with no special identity proposed for the second layer from the outside [2]. Our data indicate that the second layer is well aligned and distinguishable, suggesting that it may be specialized, although its function is unknown at present.

Fourth, tissues with complex three-dimensional arrangements can be observed. It is difficult to observe accurate 3D arrangements of tissues that are wrapped together when using conventional sectioning methods. The method reported here enables us to acquire transverse sections continuously in the Z-axis direction, making it possible to observe the precise 3D arrangement of tissues. This advance made it possible to observe the arrangement of tissues, such as the leaf primordium in the SAM (Figure 2C, Supplementary Materials Video S2), the lemma, and the palea primordium (Figure 3C, Supplementary Materials Video S3).

Fifth, our 3D imaging method provides a unique opportunity to understand the mechanism for the maintenance of undifferentiated cells by supporting quantification of the velocity of cell replacement in the SAM and RAM. In the SAM, the organizing center is located at a certain distance from the stem cells. In the RAM, undifferentiated cells are maintained within a certain distance from the QC. Maintaining the proper distance between regions is essential to ensure the integrity of the SAM and RAM functions. Since the proper distance between regions is determined by the velocity of cell replacement, it is necessary to measure the velocity. So far, the velocity of cellular replacement has been measured on two-dimensional sections of the root tip of Arabidopsis [25]. The 3D imaging technique developed in this study will enable us to measure velocity comprehensively in three dimensions.

3.2. Spatial Distribution of DR5 Signals in the Root Tip

The fate of cells that differentiate into metaxylem is determined by the activation of auxin signaling at the root tip [26]. The location of these cells at the root tip was not well defined in rice. Our results showed that DR5 signals are predicted to be in five locations of the RAM stele before metaxylem differentiation (Figure 5 and Supplementary Materials Video S6).

Since diverse root cell types differentiate from stem cells in the vicinity of the QCs, rows of cells in which auxin signaling is activated and metaxylem differentiation occurs were thought to be connected to QCs. We found that in the root tip of a DR5rev:NLS-3xVenus plant, the central row of Venus-expressing cells was connected to the QC, but the other four rows of cells were not. The origin of the four unconnected DR5-expressing cells is not clear. DR5 expression may be repressed in the vicinity of the QC because DR5 fluorescence can be seen several cells away from the QC. Suppression of auxin signaling around the QC may be due to cytokinin localization because cytokinin is known to be distributed around the QC and is also known to inhibit auxin signaling [26–28].

4. Materials and Methods

4.1. Plant Materials and Growth Conditions

The Japonica rice cultivar Nipponbare was used for SAM and floret observations. The Japonica rice cultivar Norin 8 was used to generate DR5rev:NLS-3xVenus transgenic rice plants [23] and for observations of root tips. Plants were grown in climate-controlled chambers at 70% humidity under short-day conditions with daily cycles of 10 h of light at 28 °C and 14 h of dark at 25 °C. Light was provided by fluorescent white light tubes (400–700 nm, 100 μmol $m^{-2}{\cdot}s^{-1}$).

4.2. Sampling of Rice Shoot Apical Meristems and Florets

Plants of 25–35 days after germination were used to sample the SAMs. Slightly older plants of 35–45 days after germination were used to sample the florets. To sample SAMs, we carefully removed several leaves from the basal part of the seedlings by hand sectioning. The leaf primordia were removed carefully to expose the P2 leaf primordium as viewed with a stereoscopic microscope. The shoot apex was excised by cutting at a point 2–3 mm below the shoot apex, and the excised tissue was fixed in a microtube (Figure 1B, Supplementary Materials Video S1).

4.3. Fixation and Cell-Wall Staining

The samples were fixed in 4% (*v*/*v*) paraformaldehyde or formaldehyde in PBS supplemented with 0.1% (*v*/*v*) SR2200 (the solution from the supplier was considered to be 100%: Renaissance Chemicals, UK). Samples in the fixative were vacuumed infiltrated for 10 min on ice. After restoration to normal pressure, the samples were incubated for 50 min at room temperature.

4.4. Clearing by the iTOMEI Protocol

The samples were transferred to a decolorization solution (100 mM sodium phosphate buffer at pH 8.0 with 20% (*w*/*v*) caprylyl sulfobetaine (TCI, Tokyo, Japan)) and incubated for 24 h for SAMs and florets or 3 days for roots at room temperature to decolorize the tissues [29]. The samples were transferred to a clearing solution (56.2% (*w*/*w*) Histodenz (Merck, Darmstadt, Germany) in PBS buffer) and incubated for 1 h at room temperature for SAMs, florets, and roots [29]. For anther, fixed anther was transferred to ClearSee and incubated for 1 week at room temperature. After incubation, the anther was mounted and observed.

4.5. Mounting Samples and Imaging by a Confocal Laser Scanning Microscope

The samples were mounted on glass slides using the same product we used as a clearing solution. Small samples were supported by a silicon sheet with holes (8 mm × 8 mm square). The anther was mounted with water. The samples were visualized with a confocal laser-scanning microscope (TCS SP8; Leica Microsystems, Tokyo, Japan) equipped with a 405 nm and pulsed white-light laser (WLL) sources and a 63× glycerol-immersion objective lens (HC PL APO 63×/1.30 GLYC CORR CS2; Leica Microsystems), a 40× water-immersion objective lens (PL APO CS2 40×/1.10 W CORR HCX; Leica Microsystems), and a 20× objective lens (PL APO CS2 20×/0.75 IMM CORR HC; Leica Microsystems). For SR2200 fluorescence, images were captured at 410–480 nm after excitation at 405 nm with a solid-state laser. For Venus fluorescence, images were captured at 520–600 nm after excitation at 515 nm with WLL. We set the Z step at 500 nm for SAMs and 1–3 μm for root tips. After image acquisition, which took approximately 2 h for 150 μm in depth, the images were processed using LASX software (Leica Microsystems, Tokyo, Japan).

Supplementary Materials: The following are available online at https://www.mdpi.com/article/10.3390/ijms23010040/s1.

Author Contributions: M.S. and H.T. designed the research; M.S., H.A., Y.S. and S.M. performed the research; M.S. and H.T. analyzed the data; M.S. and H.T. wrote the paper. All authors have read and agreed to the published version of the manuscript.

Funding: This study was supported by a Grant-in-Aid for JSPS Research Fellow number 19J21998 to M.S., MEXT KAKENHI, Grants-in-Aid for Scientific Research on Innovative Areas (numbers 16H06464 and 16H06466 to H.T.), a Grant-in-Aid for Transformative Research Areas (A) (number 20H05911 to S.M.), Grants-in-Aid for Scientific Research (A) (number 16H02532 and 21H04728 to H.T.), Scientific Research (B) (number 19H03259 to S.M.), Scientific Research (C) (number 21K06247 to Y.S.) and by a Core Research for Evolutionary Science and Technology (CREST) grant from the Japan Science and Technology Agency (JST) JPMJCR16O4 to H.T.

Institutional Review Board Statement: Not applicable.

Informed Consent Statement: Not applicable.

Data Availability Statement: All datasets supporting the conclusions of this article are included in the article and supplementary files.

Conflicts of Interest: Yuki Sakamoto and Sachihiro Matsunaga are the inventors of a pending patent (No. 2020-026975) in Japan. They are unpaid advisory positions of plant tissue-clearing reagents in Tokyo Chemical Industry Cooperation (Tokyo, Japan).

References

1. Chan, C.J.; Heisenberg, C.-P.; Hiiragi, T. Coordination of Morphogenesis and Cell-Fate Specification in Development. *Curr. Biol.* **2017**, *27*, R1024–R1035. [CrossRef]
2. Hirano, H.-Y.; Tanaka, W. Stem Cell Maintenance in the Shoot Apical Meristems and during Axillary Meristem Development. *Cytologia* **2020**, *85*, 3–8. [CrossRef]
3. Coudert, Y.; Périn, C.; Courtois, B.; Khong, N.G.; Gantet, P. Genetic Control of Root Development in Rice, the Model Cereal. *Trends Plant Sci.* **2010**, *15*, 219–226. [CrossRef]
4. Lintilhac, P.M. The Problem of Morphogenesis: Unscripted Biophysical Control Systems in Plants. *Protoplasma* **2014**, *251*, 25–36. [CrossRef]
5. Piekarska-Stachowiak, A.; Szymanowska-Pułka, J.; Potocka, I.; Lipowczan, M. Topological Traits of a Cellular Pattern versus Growth Rate Anisotropy in Radish Roots. *Protoplasma* **2019**, *256*, 1037–1049. [CrossRef] [PubMed]
6. Lipowczan, M.; Borowska-Wykret, D.; Natonik-Bialon, S.; Kwiatkowska, D. Growing Cell Walls Show a Gradient of Elastic Strain across Their Layers. *J. Exp. Bot.* **2018**, *69*, 4349–4362. [CrossRef]
7. Hama, H.; Hioki, H.; Namiki, K.; Hoshida, T.; Kurokawa, H.; Ishidate, F.; Kaneko, T.; Akagi, T.; Saito, T.; Saido, T.; et al. ScaleS: An Optical Clearing Palette for Biological Imaging. *Nat. Neurosci.* **2015**, *18*, 1518–1529. [CrossRef] [PubMed]
8. Susaki, E.A.; Tainaka, K.; Perrin, D.; Kishino, F.; Tawara, T.; Watanabe, T.M.; Yokoyama, C.; Onoe, H.; Eguchi, M.; Yamaguchi, S.; et al. Whole-Brain Imaging with Single-Cell Resolution Using Chemical Cocktails and Computational Analysis. *Cell* **2014**, *157*, 726–739. [CrossRef] [PubMed]
9. Chung, K.; Wallace, J.; Kim, S.-Y.; Kalyanasundaram, S.; Andalman, A.S.; Davidson, T.J.; Mirzabekov, J.J.; Zalocusky, K.A.; Mattis, J.; Denisin, A.K.; et al. Structural and Molecular Interrogation of Intact Biological Systems. *Nature* **2013**, *497*, 332–337. [CrossRef]
10. Kurihara, D.; Mizuta, Y.; Sato, Y.; Higashiyama, T. ClearSee: A Rapid Optical Clearing Reagent for Whole-Plant Fluorescence Imaging. *Development* **2015**, *142*, 4168–4179. [CrossRef] [PubMed]

11. Hasegawa, J.; Sakamoto, Y.; Nakagami, S.; Aida, M.; Sawa, S.; Matsunaga, S. Three-Dimensional Imaging of Plant Organs Using a Simple and Rapid Transparency Technique. *Plant Cell Physiol.* **2016**, *57*, 462–472. [CrossRef]
12. Palmer, W.M.; Martin, A.P.; Flynn, J.R.; Reed, S.L.; White, R.G.; Furbank, R.T.; Grof, C.P.L. PEA-CLARITY: 3D Molecular Imaging of Whole Plant Organs. *Sci. Rep.* **2015**, *5*, 13492. [CrossRef] [PubMed]
13. Imoto, A.; Yamada, M.; Sakamoto, T.; Okuyama, A.; Ishida, T.; Sawa, S.; Aida, M. A ClearSee-Based Clearing Protocol for 3D Visualization of Arabidopsis Thaliana Embryos. *Plants* **2021**, *10*, 190. [CrossRef]
14. Hirai, R.; Higaki, T.; Takenaka, Y.; Sakamoto, Y.; Hasegawa, J.; Matsunaga, S.; Demura, T.; Ohtani, M. The Progression of Xylem Vessel Cell Differentiation Is Dependent on the Activity Level of VND7 in Arabidopsis Thaliana. *Plants* **2019**, *9*, 39. [CrossRef] [PubMed]
15. Kurihara, D.; Mizuta, Y.; Nagahara, S.; Higashiyama, T. ClearSeeAlpha: Advanced Optical Clearing for Whole-Plant Imaging. *Plant Cell Physiol.* **2021**, *62*, 1302–1310. [CrossRef]
16. Xing, Y.; Zhang, Q. Genetic and Molecular Bases of Rice Yield. *Annu. Rev. Plant Biol.* **2010**, *61*, 421–442. [CrossRef]
17. Bailey-Serres, J.; Parker, J.E.; Ainsworth, E.A.; Oldroyd, G.E.D.; Schroeder, J.I. Genetic Strategies for Improving Crop Yields. *Nature* **2019**, *575*, 109–118. [CrossRef]
18. Hori, K.; Shenton, M. Recent Advances in Molecular Research in Rice: Agronomically Important Traits. *Int. J. Mol. Sci.* **2020**, *21*, 5945. [CrossRef]
19. Wang, W.; Mauleon, R.; Hu, Z.; Chebotarov, D.; Tai, S.; Wu, Z.; Li, M.; Zheng, T.; Fuentes, R.R.; Zhang, F.; et al. Genomic Variation in 3010 Diverse Accessions of Asian Cultivated Rice. *Nature* **2018**, *557*, 43–49. [CrossRef] [PubMed]
20. Itoh, J.-I.; Nonomura, K.-I.; Ikeda, K.; Yamaki, S.; Inukai, Y.; Yamagishi, H.; Kitano, H.; Nagato, Y. Rice Plant Development: From Zygote to Spikelet. *Plant Cell Physiol.* **2005**, *46*, 23–47. [CrossRef]
21. Barbier de Reuille, P.; Routier-Kierzkowska, A.-L.; Kierzkowski, D.; Bassel, G.W.; Schüpbach, T.; Tauriello, G.; Bajpai, N.; Strauss, S.; Weber, A.; Kiss, A.; et al. MorphoGraphX: A Platform for Quantifying Morphogenesis in 4D. *Elife* **2015**, *4*, 05864. [CrossRef] [PubMed]
22. Toriba, T.; Suzaki, T.; Yamaguchi, T.; Ohmori, Y.; Tsukaya, H.; Hirano, H.-Y. Distinct Regulation of Adaxial-Abaxial Polarity in Anther Patterning in Rice. *Plant Cell* **2010**, *22*, 1452–1462. [CrossRef] [PubMed]
23. Lucob-Agustin, N.; Kawai, T.; Takahashi-Nosaka, M.; Kano-Nakata, M.; Wainaina, C.M.; Hasegawa, T.; Inari-Ikeda, M.; Sato, M.; Tsuji, H.; Yamauchi, A.; et al. WEG1, Which Encodes a Cell Wall Hydroxyproline-Rich Glycoprotein, Is Essential for Parental Root Elongation Controlling Lateral Root Formation in Rice. *Physiol. Plant.* **2020**, *169*, 214–227. [CrossRef] [PubMed]
24. Ulmasov, T.; Murfett, J.; Hagen, G.; Guilfoyle', T.J. Aux/lAA Proteins Repress Expression of Reporter Genes Containing Natural and Highly Active Synthetic Auxin Response Elements. Available online: http://www.plantcell.org/content/plantcell/9/11/1963.full.pdf (accessed on 16 April 2021).
25. Nakielski, J.; Lipowczan, M. A Method to Determine the Displacement Velocity Field in the Apical Region of the Arabidopsis Root. *Planta* **2012**, *236*, 1547–1557. [CrossRef] [PubMed]
26. Bishopp, A.; Help, H.; El-Showk, S.; Weijers, D.; Scheres, B.; Friml, J.; Benková, E.; Mähönen, A.P.; Helariutta, Y. A Mutually Inhibitory Interaction between Auxin and Cytokinin Specifies Vascular Pattern in Roots. *Curr. Biol.* **2011**, *21*, 917–926. [CrossRef] [PubMed]
27. Zürcher, E.; Tavor-Deslex, D.; Lituiev, D.; Enkerli, K.; Tarr, P.T.; Müller, B. A Robust and Sensitive Synthetic Sensor to Monitor the Transcriptional Output of the Cytokinin Signaling Network in Planta. *Plant Physiol.* **2013**, *161*, 1066–1075. [CrossRef]
28. Antoniadi, I.; Plačková, L.; Simonovik, B.; Doležal, K.; Turnbull, C.; Ljung, K.; Novák, O. Cell-Type-Specific Cytokinin Distribution within the Arabidopsis Primary Root Apex. *Plant Cell* **2015**, *27*, 1955–1967. [CrossRef]
29. Sakamoto, Y.; Ishimoto, A.; Sakai, Y.; Sato, M.; Nishihama, R.; Abe, K.; Sano, Y.; Furuichi, T.; Tsuji, H.; Kohchi, T.; et al. Improved clearing method contributes to deep imaging of plant organs. *Commun. Biol.* **2021**, *in press*. [CrossRef]

MDPI
St. Alban-Anlage 66
4052 Basel
Switzerland
Tel. +41 61 683 77 34
Fax +41 61 302 89 18
www.mdpi.com

International Journal of Molecular Sciences Editorial Office
E-mail: ijms@mdpi.com
www.mdpi.com/journal/ijms

www.ingramcontent.com/pod-product-compliance
Lightning Source LLC
Chambersburg PA
CBHW061607210926
51850CB00009B/2247
9783036549439